ONE PLANET
MANY PEOPLE

Atlas of Our Changing Environment

United Nations Environment Programme
PO Box 30552, Nairobi, Kenya
Tel: +254 20 621234
Fax: +254 20 623943/44
http://www.unep.org
http://www.unep.net

United Nations Environment Programme
Division of Early Warning and Assessment-North America
47914 252nd Street, USGS National Center for Earth Resources
Observation and Science (EROS)
Sioux Falls, SD 57198-0001 USA
Tel: 1-605-594-6117
Fax: 1-605-594-6119
info@na.unep.net
www.na.unep.net

For bibliographic and reference purposes this publication should be referred to as:

UNEP (2005), "One Planet Many People: Atlas of Our Changing Environment."
Division of Early Warning and Assessment (DEWA)
United Nations Environment Programme (UNEP)
P.O. Box 30552
Nairobi, Kenya

This book is available from Earthprint.com, http://www.earthprint.com.

Reprinted by Progress Press Company Limited, Malta.
Distribution by SMI (Distribution Services) Ltd. UK.

This publication is printed on chlorine free, acid free paper made of wood pulp from sustainable managed forests.

ONE PLANET
MANY PEOPLE

Atlas of Our Changing Environment

United Nations Environment Programme
2005

Editorial and Production Team

UNEP

Ashbindu Singh, *Team Coordinator*

USGS

Thomas R. Loveland, *Writer*

SAIC, TSSC to the USGS

Mark Ernste, *Remote Sensing/GIS Scientist*

Kimberly A. Giese, *Design and Layout*

Rebecca L. Johnson, *Editor*

Jane S. Smith, *Editorial Assistant/Support*

John Hutchinson, *Cartographer*

Eugene Fosnight, *Writer*

Consultant

H. Gyde Lund, *Lead Writer*

Tejaswi Giri, *Project Manager*

Jane Barr, *Writer*

Eugene Apindi Ochieng, *Remote Sensing/GIS Analyst*

UNEP–Nairobi

Audrey Ringler, *Cover Design*

To obtain a copy of this publication, please contact:
Earthprint Limited
P.O. Box 119
Stevenage, Hertfordshire, SGI 4TP
ENGLAND

Phone: +44 1438 748111
Fax: +44 1438 748844
E-mail: orders@earthprint.com
Order online at: www.earthprint.com

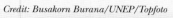

Credit: Busakorn Burana/UNEP/Topfoto

Table of Contents

Foreword

People affect the environment as they interact with it, using it for food, shelter, and recreation and making changes to better suit their needs, purposes, and inclinations. Through our ability to adapt natural resources to our use, we have altered the environment in ways that can now be objectively measured. Our presence on the Earth can be seen through changes on the landscape, as viewed from space. This publication presents images from space that portray the nature and extent of our impact on the planet.

Change is inevitable and an integral part of our planet, our environment, and even us. Our ability to adapt to diverse surroundings has allowed us to overcome many environmental constraints and tailor the planet to our benefit. We harnessed fire, cultivated plants and domesticated animals. We built homes, then villages, and then cities. We became "hewers of wood and drawers of water." We built tools and discovered how to quarry rocks and later metals. Each advance allowed us to further adapt to and affect the environment that shaped us.

Our ability to act positively to safeguard our heritage and natural wealth may be affected by the consequences of our success, however. As our numbers have increased, we have also evolved socially and culturally, applying different beliefs and practices to living in and using the environment. What we do affects those far away from us, even those separated from us by mountains, deserts, and oceans. Our activities change the planet in ways that affect our health as well as the health of the plants and animals upon which we depend. We harvest the seas, consume water and energy resources, and convert forests into pasture and cropland. We must be ever conscious of the potential to overuse the land and stress it in ways that it cannot bear.

Our growing populations and settlements make life easier in some ways, but also make us more vulnerable to massive earthquakes, volcanic eruptions, and other disasters. Imagine what would happen in Italy today if Vesuvius erupted on the same scale it did when it destroyed Pompeii. We have gravitated to the shores, making ourselves more vulnerable to storms and hurricanes. We have settled along rivers, making ourselves more vulnerable to floods. We have spread into marginal climates, making ourselves more vulnerable to drought. Wildfires threaten some of our cities and settlements, just as they do our forests and croplands. Each of these events can affect hundreds of thousands of people, and the cost of protecting ourselves and reducing the risk of disaster continues to increase. Our own activities can also lead to disasters such as oil spills and nuclear and industrial accidents that can devastate as much as any natural event.

Our dilemma is to avoid the most problematic consequences without constraining our need and ability to provide the world's inhabitants with the environment and resources that will enable every person to pursue an equitable life with all that such a life entails.

The images presented here show both the positive and negative impacts of human life on Earth. We hope also they will provide food for thought, as we seek ways to balance our use of the Earth's resources with the need to sustain the environments that produce them and support the living systems that we value so highly.

Klaus Töpfer, Ph.D.
Executive Director
United Nations Environment Programme

Ghassem Asrar, Ph.D.
Science Deputy
Associate Administrator
National Aeronautics
and Space Administration

Charles G. Groat, Ph.D.
Director
United States Geological Survey

John Townshend, Ph.D.
University of Maryland
Chair, Advisory Committee
UNEP/DEWA–North America

Preface

Our population is growing, yet our land base is currently fixed. With each new inhabitant comes a need to make more modifications to the Earth's environment. The impacts of these modifications may be both detrimental and beneficial. For example, we estimate that the Earth is losing 15 million hectares of tropical forest land per year, a loss that has a negative effect on biodiversity. At the same time, much of this deforested land is being converted to agricultural land to feed our growing population; this is a positive effect.

In the past 30 years—since the United Nations Conference on the Human Environment in Stockholm in 1972—we have made a concerted effort to understand the limits of the Earth's bountiful resources and have taken actions to preserve and sustain them. This publication illustrates some of the changes we have made to the environment in the recent past. It serves both as an early warning for things that may come and as a basis for developing policy decisions that can help sustain the Earth's and our own well-being.

The first chapter of this atlas provides a short environmental history of the world, one that illustrates how we got to where we are. Chapter 2 looks at the people and the planet today, covering status and trends over the last 30 years. Chapter 3 examines common issues regarding the Earth's land cover and provides examples that illustrate environmental status, trends, causes and consequences of change in the atmosphere, oceans and coastal zones, fresh-water ecosystems, forests, cropland, grasslands, urban areas, and tundra including polar regions. Chapter 4 illustrates changes that are the result of extreme events, both natural and human-induced.

These examples raise many questions. What is our likely environmental future? Are we better or less prepared for environmental change? What can people do to create a better future? The answers depend on the actions we choose to take.

Steve Lonergan, Ph.D.
Director
United Nations Environment Programme
Division of Early Warning and Assessment

Photo Credits (left to right):
Kern Khianchuen/UNEP/Topfoto
Dirk Heinrich/UNEP/Topfoto
Aimen Al-Sayya/UNEP/Topfoto

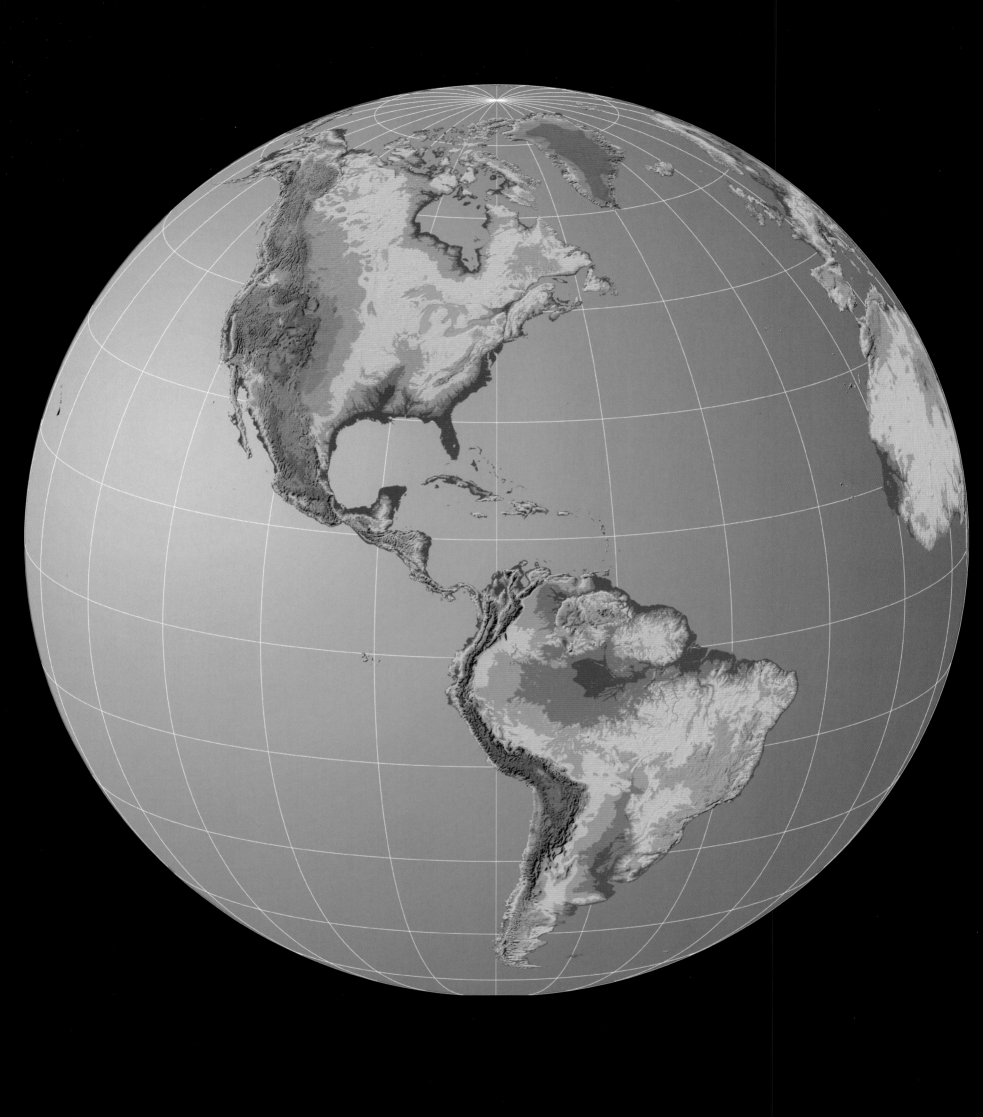

Topographic Map of the World
Credit: UNEP/NASA–GTOPO30

Introducing the Planet
A Story of Change

In our solar system, a single planet—the Earth—supports human life. World population is increasing. Yet for the moment, the Earth remains the only home for the human species. The way in which we care for this planet will affect our future and the future of our children for generations to come.

Seen from space, the Earth is largely a blue planet around which swirls of white clouds constantly move. The Earth's blue areas are its oceans. Oceans account for approximately 70 per cent of the Earth's total surface area; the remaining 30 per cent is land. The total size of the terrestrial surface is approximately 149 million km² (59.6 million square miles) (McNeill 2000; Grace n.d.).

The Earth's land surface is rich in its variety. The highest point on the Earth's land surface is Mount Everest, a breathtaking 8 850 m (29 035 ft) above sea level. The lowest point is the Dead Sea, which is, on average, about 400 m (1 312 ft) below sea level. Terrestrial surfaces gain and lose heat much more quickly than oceans and a region's distance from the equator dramatically affects its climate. Lands nearest the equator tend to be the warmest. Those that lie in the middle latitudes typically have cooler climates, but are not as cold as lands near the poles. Some 20 per cent of the Earth's terrestrial surface is covered by snow. Another 20 per cent is mountainous. Just 30 per cent of the Earth's land surface is suitable for farming.

Most people are accustomed to seeing the world around them as a relatively stable place, a generally nurturing environment that has allowed the human race to expand and develop in countless ways. In fact, the Earth is constantly changing, as is our understanding of it (Figure 1.1). Some changes to the Earth's surface occur on microscopic levels. Other changes take place on a scale so large as to be almost

> ## "The only thing permanent is change."
>
> — The Buddha (Siddartha Gautama)

inconceivable. Some types of change are instantaneous, while other types occur slowly, unfolding over centuries, millennia, and even eons. Some changes are caused by the actions of people. Many others are part of natural, inexorable cycles that can only be perceived when cataclysmic events occur or through painstaking research.

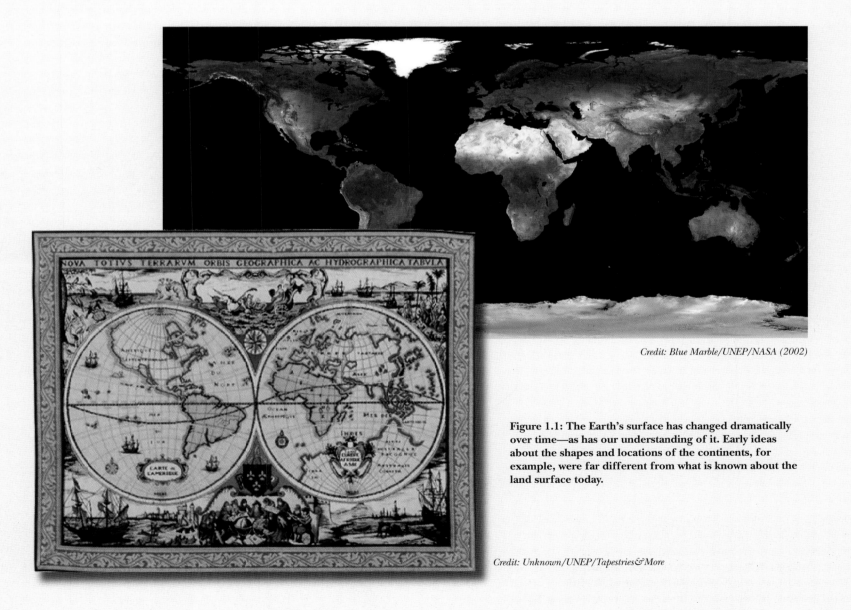

Credit: Blue Marble/UNEP/NASA (2002)

Figure 1.1: The Earth's surface has changed dramatically over time—as has our understanding of it. Early ideas about the shapes and locations of the continents, for example, were far different from what is known about the land surface today.

Credit: Unknown/UNEP/Tapestries&More

Table 1.1 – Approximate change of the Earth's global vegetative cover in relation to human population (Adapted from McNeill 2000).

| Year | Per cent of the Earth's Vegetated Land Area | | | | Human Population (Billions) |
	Forest and Woodland	Grassland	Pasture	Cropland	
8000 B.C.	51	49	0	0	0.005
1700 A.D.	47	47	4	2	0.6
1900	43	40	10	6	1.6
1920	43	38	12	7	1.9
1940	41	35	16	8	2.3
1960	40	31	20	9	3.0
1980	38	26	25	11	4.4
1990	36	27	26	11	5.3

Credit: Paul Fusco/UNEP/NRCS

Credit: Chatree Wanasan/UNEP/Topfoto

Agents of Environmental Change

From the Earth's earliest beginnings, forces such as climate, wind, water, fire, earthquakes, volcanic eruptions, and the impacts of meteors and comets have shaped the Earth's terrestrial environments. These same forces are at work today and will continue far into the future. In addition, every living thing influences its environment and is influenced by it. One species may lessen the chances for survival of the organisms it consumes for food. That same species, in turn, is affected by the actions of other organisms.

In order to survive, every organism must either adapt to its environment or modify the environment to make it more hospitable. Humans are particularly adept at modifying their environments. By their actions and interactions with the landscape, for example, people can increase the range of certain plant species, either by modifying existing environments or by dispersing seeds into new ones. Environmental modifications made by people may be beneficial or detrimental to a few or many other species. Large-scale environmental changes may not benefit or be to the liking of people themselves (Nott 1996). As world

population has increased and the scope and nature of technology has changed, people have brought about environmental changes that may seriously impact their future well-being and even survival.

Humans began modifying their environment a long time ago (Table 1.1). Evidence of the existence of our first humanoid ancestors dates to the Pliocene Epoch, which extended from roughly 5 million to 1.8-1.6 million years ago (Wikipedia n.d.). These protohumans sought protection from the elements and from predators in natural shelters such as caves and rock overhangs. Over time—and possibly

Five Major Events in the History of the Earth

Throughout the Earth's history, events have occurred that dramatically impacted life on our planet. Five of those events stand out as having resulted in widespread extinctions, in some cases destroying more than 90 per cent of all living things (Eldredge 2001):

- Around 440 million years ago, a relatively severe and sudden global cooling caused a mass extinction of marine life (little terrestrial life existed at that time). An estimated 25 per cent of the existing taxonomic families were lost. (A family may consist of a few to thousands of species.)

- Near the end of the Devonian Period, some 370 million years ago, a second major extinction occurred. Roughly 19 per cent of the existing taxonomic families were wiped out. It is uncertain whether climate change was a driving factor.

- About 245 million years ago, a third major extinction took place. Scientists estimate that more than half (54 per cent) of all taxonomic families were lost. Climate change may have played a role, and that change may have been caused by a comet or meteor impacting the Earth.

- At the end of the Triassic Period, around 210 million years ago, roughly 23 per cent of existing taxonomic families suddenly became extinct. This event occurred shortly after the appearance of the first dinosaurs and mammals. Its causes are not yet fully understood.

- The fifth major extinction is the most well-known. It occurred about 65 million years ago at the end of the Cretaceous Period. The event led to the extinction of all terrestrial dinosaurs and marine ammonites, along with many other species occupying many different habitats. All told, approximately 17 per cent of all taxonomic families vanished in a very short time. Currently, the most widely accepted hypothesis to explain this mass extinction is that a comet or other large extra-terrestrial object struck the Earth. Another view proposes that a great volcanic event, or series of events, disrupted ecosystems so severely worldwide that many terrestrial and marine species rapidly succumbed to extinction.

Credit: Unknown/UNEP/Bigfoto

Fire—A Tool for Humankind

For thousands of years humans have used fire for:

Hunting

By setting fire to parts of the landscape, people were able to drive game animals into smaller, more confined areas that made hunting easier. Fire was also used to drive animals into impoundments, chutes, river or lakes, or over cliffs. Fires also helped maintain open prairies and meadows by killing bushes and trees and encouraging rapid growth of grasses.

Improving plant growth and yields

Setting fires was a way to improve grass for grazing animals, both wild and domestic, and to promote the growth of certain desirable plant species.

Protection

Fire was used to protect human habitations.

Collecting insects

Some tribes used "fire surrounds" to collect and roast crickets, grasshoppers, and moths. People also used fire smoke to quiet bees while collecting honey.

Managing pests

Fire was a handy tool for reducing or driving away insect pests such as flies and mosquitoes as well as rodents. Fire was also effective for eliminating undesirable plants.

Warfare and signaling

Fire was both an effective defensive and offensive weapon. Offensively, it was used to deprive enemies of hiding places in tall grasses or underbrush. Used defensively, fire could provide cover during an escape. Smoke signals helped alert tribes to the presence of possible enemies or to gather forces to combat a foe. Large fires were set to signal a tribal gathering.

Clearing areas for travel

Fires were sometimes started to clear trails through dense vegetation. Burning helped to improve visibility in forests or grasslands for hunting and warfare.

Felling trees

Singed or charred trees were easier to fell and to work with.

Clearing riparian areas

Fire was used to clear vegetation from the edges of lakes and rivers.

Managing crops

Burning was later used to harvest crops and collect grass seeds. Fire also helped prevent abandoned fields from becoming overgrown and was employed to clear areas for planting.

Credit: Jeff Vanuga/UNEP/NRCS

influenced by the onset of colder weather during the Ice Ages—they created dwelling places for themselves in locations that had no natural shelter.

The oldest surviving traces of such a human-made habitation date to about 2 million years ago from Olduvai Gorge in central Africa. There, a small circle of stones was found stacked in such a way as to apparently have held branches in position. This early example of modification of the environment was the work of *Homo habilis*, a tool–making human ancestor (Kowalski n.d.).

The Pleistocene Epoch, including the Paleolithic and Mesolithic Periods (Wikipedia n.d.), is usually dated from the end of the Pliocene to 10 000 years ago. The Paleolithic Period, or Old Stone Age, is a term coined in the 19th century to define

the oldest period in the history of humankind. It lasted for some 2.5 million years, from the time human ancestors created and used the first stone tools to the end of the last glacial period some 10 000 years ago. *Homo erectus*, thought by many to be the direct ancestor of modern humans, lived from approximately 2 million to around 400 000 years ago. As a species, *Homo erectus* was very successful in developing tools that helped in adapting to new environments. They were pioneers in developing human culture, ultimately moving out of Africa to populate tropical and sub-tropical environmental zones in the Old World, possibly as early as 1.8 million years ago.

Homo erectus may also have mastered the use of fire around 1.6 million years ago (Mcrone 2000). Fire is an exception-

ally powerful tool. Since most animals, including large predators, are afraid of fire, early humans quickly discovered that campfires offered protection from attack during the night. Control of fire allowed them to move into colder regions as it provided warmth as well as security. Fire also changed the way food was prepared. Food that is cooked is less likely to carry disease organisms and its softer texture makes it easier to eat, enhancing the survival of young children and old members of a population.

The use of fire almost certainly increased during the Paleolithic Period. At that time, humans were primarily hunter-gatherers. The role of fire in modern hunter-gatherer cultures gives us some idea of its importance during the Paleolithic and how people then most likely used it (Williams 2001). Fire was an important tool in everyday life. It was also a tool with great potential to modify and change the environment.

The transition from *Homo erectus* to *Homo sapiens*—our modern human species—occurred approximately 300 000 to 400 000 years ago. Throughout the Paleolithic Period, humans survived by exploiting resources in their environment through subsistence activities such as fishing, hunting, and plant gathering. From ethnographic studies of modern hunter-gatherers, we can infer that the basic social unit of Paleolithic times was the band: a loosely bound, relatively small group (25 people on average) formed by the voluntary aggregation of a few families. Bands were mobile, regularly changing residence inside a delimited territory according to fluctuations in the abundance of different food sources. Inter-regional migrations, such as those within the Western Hemisphere, probably took place during this time.

This mobile life was punctuated by episodes of reunion, when several allied bands would meet and perform religious ceremonies. Such occasions would have provided a time and place for the transmission of techniques and artistic fashions, promoting their spread across vast expanses of territory.

Some 10 000 years ago, a new geological time period, the Holocene, began. World

sea levels rose about 35 m (116 ft) in the early Holocene due to melting glaciers. As the ice melted and glaciers receded, many land areas that had been depressed by glacial weight slowly rose as much as 180 m (594 ft) above their late-Pleistocene and early-Holocene levels. Both sea-level rise and depressions in the landscape allowed temporary ocean-water incursions into regions that are today far from any sea. Climatic shifts also were very large during this period. Habitable zones expanded northwards. Large, mid-latitude areas such as the Sahara that were previously productive became deserts. At the start of the Holocene, large lakes covered many areas that are now quite arid.

Animals and plants did not undergo major evolutionary changes during the Holocene, but there were significant shifts in their distribution. Several types of large mammals including mammoths, mastodons, saber-toothed cats, and giant sloths went extinct in the late Pleistocene and early Holocene. Ecological "islands" of isolated species were created throughout the world, including high-altitude remnants of cooler, previously regional climate ecosystems.

The period from 10 000 to 5 000 or 4 000 years ago is recognized as the Neolithic Period (New Stone Age). It was preceded by the Mesolithic Period (Middle Stone Age), which roughly corresponds with the beginning of the Holocene. The

Mesolithic Period formed a transition between the Paleolithic and Neolithic Periods, that is, from the end of the Pleistocene to the introduction of agriculture in any given geographical region (Wikipedia n.d.).

The size of the world's human population 10 000 years ago is estimated to have been around 5 million (IPC 2003a). This period saw the beginning of agriculture and the domestication of animals. People learned to cultivate crops rather than to simply gather what nature provided in the wild and to tame and raise animals such as sheep, goats, cattle, horses, and dogs, rather than hunt them. These activities helped ensure better food supplies and resulted in wide-ranging cultural consequences. Permanent communities were now established, since people were no longer dependent on following wild animals or moving with the seasons. Day-to-day existence changed from a life of nomadic foraging to one of permanence. This allowed some individuals to explore tool production to refine rude stone tools and improve implements such as stone-blade knives, bow drills with flints for starting fires, fish hooks, axes, and plows. Other people found time to improve agricultural and pastoral techniques, enabling communities to grow more food and tend livestock more effectively.

The cultivation of plants and the domestication of animals dramatically

impacted human lifestyles during the Neolithic Period. People left their temporary rock and wooden shelters and began to build more permanent homes in close proximity to their farms and gardens, where they started producing cereal grains which became an important part of their diet (Wadley and Martin 1993).

The Neolithic Period marked the beginning of true civilization, laying the foundations for major developments in social evolution such as permanent settlements, village life, formalized religion, art, architecture, farming, and the production of advanced tools and weapons.

Agriculture

The first cultivation of wild grains some 12 000 to 10 000 years ago turned hunter-gatherers into farmers. The transition gave people a more abundant and dependable source of food and changed the world forever (Wilford 1997). The practice of agriculture first developed in the Fertile Crescent of Mesopotamia (part of present-day Iraq, Turkey, Syria, and Jordan). This region, which was much wetter then than it is today, was home to a great diversity of annual plants and 32 of the 56 largest seed-producing grasses (Primal Seeds n.d.).

Around 11 000 years ago, much of the Earth experienced long dry seasons, probably as a consequence of the major climate change that took place at the end of the last Ice Age. These conditions favored annual plants that die off in the long dry season, leaving a dormant seed or tuber. Such plants put more energy into producing seeds than into woody growth. An abundance of readily storable wild grains

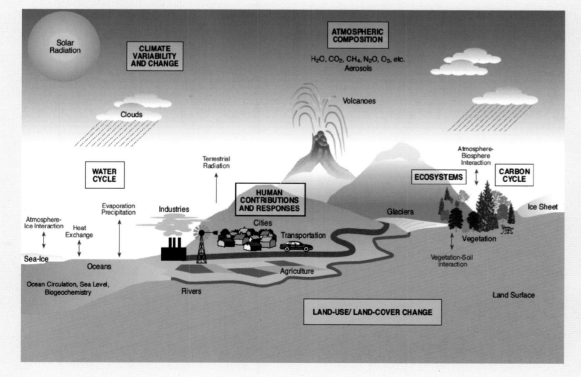

Figure 1.2: The Earth's climate system involves complex interactions among many elements and processes. *Source: http://www.usgcrp.gov/usgcrp/Library/ocp2004-5/ocp2004-5.pdf*

and other edible seeds enabled hunter-gatherers in some areas to form permanently settled villages at this time (Primal Seeds n.d.).

Theories vary as to how agriculture came into being. Some scientists argue that rising global temperatures created favorable conditions for agriculture. Others propose that an increase in seasonality after the last Ice Age encouraged people to domesticate plants. Still other researchers maintain that ecological changes, social development, or a growing human population intensified the exploitation of specific plant species (Baldia 2000).

Another suggestion is that an increase in carbon dioxide (CO_2) on a global scale may have played a critical role in bringing about the synchrony of agricultural origins around the globe (Sage 1995). Studies have shown that a rise in atmospheric CO_2 levels would have increased productivity of many plants by as much as 50 per cent. Furthermore, the water efficiency of cultivated plants increased, giving these plants a competitive advantage over wild species.

A few scientists have proposed that climatic changes at the end of the last glacial period led to an increase in the size and concentration of patches of wild cereals in certain areas (Wadley and Martin 1993). Increased availability of cereal grains provided people with an incentive to make a meal of them. Those who ate sizable amounts of cereal grains inadvertently discovered the rewards of consuming the various chemical compounds that cereal grains contain. As processing methods

Credit: Ed Simpson/UNEP/PhotoSpin

such as grinding and cooking made cereal grains more palatable, greater quantities were consumed.

At first these patches of wild cereals were protected and harvested. People began to settle around these food sources. They gradually abandoned their nomadic lifestyle and began working together more cooperatively. Later, land was cleared, seeds were planted, and seedlings tended to increase the quantity and reliability of cereal grain supply.

The rise of more permanent settlements intensified the domestication of

Credits: Michael Van Woert/UNEP/NOAA

animals. The first candidate for domestication, around 11 000 years ago, was probably the dog. The cow was domesticated around 10 000 years ago. Goats, sheep, and pigs were added to the growing list of domesticated animals around 8 000 years ago in western Asia. The horse was first domesticated in northern Russia around 4 000 years ago. Local equivalents and smaller species were increasingly domesticated from 2 500 years ago (Wikipedia n.d.).

Farming and herding facilitated the growth of larger settled human populations and led to increased competition for productive lands, laying the foundation for organized warfare. Food surpluses freed people to specialize in various crafts, such as weaving, and, in larger communities, supported the emergence of a privileged elite class. Archaeologists and historians agree that the rise of agriculture, including the domestication of animals for food and labor, produced the most important transformation in the interaction between the environment and human culture since the last Ice Age—perhaps the most significant development in human history since the control of fire (Wilford 1997).

Other milestones in human history that benefited people and changed the environment include:

The Bronze and Iron Ages (roughly 3300 B.C. to 0 A.D.)

The world population approximately 5 000 years ago is estimated to have been about 7 million (IPC 2003a). This period saw the introduction of metallurgy and mining, the invention of the wheel, and the domestication of the horse.

Classical Greece and Rome (0 to about 500 A.D.)

The world population at the beginning of this period was roughly 200 million (IPC 2003a). During this period, glass was invented and map-making developed.

Middle Ages to the Renaissance (500 to about 1700)

By this point, world population had grown to about 250 million (IPC 2003a). The clock, compass, telescope, thermometer, and barometer were developed, enabling people to expand their knowledge of the Earth and the Universe.

The Industrial Revolution (1700 – present)

By 1700, world population had risen to about 600 million (IPC 2003a). This period witnessed the development of mechanization and the beginning of serious air pollution. Industrial changes also led to an agricultural revolution.

The Agricultural Revolution (1750 – 1900)

By 1750, world population had risen to 790 million (IPC 2003a). In many countries the way in which farmers produced food began to change. New crops were exploited using new technologies such as the seed drill and the iron plow. These methods of production produced greater quantities of more nutritious foods, thereby improving peoples' diets and health. Better, more efficient farming methods also meant that fewer people were needed to farm. As a result, unemployed farmers formed a large new labor force.

The Green Revolution (1944 – present)

In 1944, world population reached 2 350 million (Anon n.d.). A breakthrough in wheat and rice production in Asia in the mid-1960s, which came to be known as the Green Revolution, symbolized the progress of agricultural science as it developed modern techniques for use in developing countries. The Green Revolution had its origin in Mexico, where a "quiet" wheat revolution began in the 1940s (Borlaug 2000).

The goal of the Green Revolution is to enhance the efficiency of agricultural processes in order to increase the productivity of crops, thereby helping developing countries to meet the needs of their growing populations. The Revolution consisted of three primary elements: continuing expansion of farming areas, double-cropping existing farmlands, and using genetically improved seeds. Thanks to the Green Revolution, we are able to grow more crops on less land.

However, the Green Revolution has impacted biodiversity and in some areas water quality and coastal ecosystems. The new techniques encouraged large-scale industrial agriculture at the expense of small farmers who were unable to compete with high-efficiency Green Revolution crops (Wikipedia n.d.). Nevertheless, the Green Revolution is a success. We are able to feed more people now, than ever before.

The Present Day

World population now stands at 6 billion people (IPC 2003b). While global resources were sufficient to support the Earth's human population as a whole prior to the Industrial Revolution, individual groups or even entire civilizations sometimes reached environmental limits for a particular resource; a number collapsed as a result of unsustainable hunting, fishing, logging, or land use practices. The ever-increasing cultural globalization of the 20th and 21st centuries has brought with it globalization of resource degradation, making current environmental problems an issue for the entire world rather than for individual, isolated groups. Although perceived environmental limits can sometimes be overcome, neither science nor technology has yet made possible unlimited supplies of natural resources or depositories for waste (Casagrande and Zaidman 1999).

Moderate projections put world population at around 8 300 million by 2025 (Figure 1.3), with the hope that it will stabilize at roughly 10 000 to 11 000 million by the end of the century. It took approximately 10 000 years to expand global food production to the current level of about 5 000 million metric tonnes per year. By 2025, production must be nearly doubled. In order to feed the world's people through 2025, an additional 1 000 million metric tonnes of grain must be produced annually. Most of this increase will have to be supplied by improving crop yields on land already in production.

This will not be possible unless farmers worldwide have access to existing

Credit: Lee Tsunhua/UNEP/Topfoto

high-yield crop production methods as well as biotechnological breakthroughs that increase the yield, dependability, and nutritional quality of our basic food crops (Borlaug 2000).

Credit: Paulus Suwito/UNEP/Topfoto

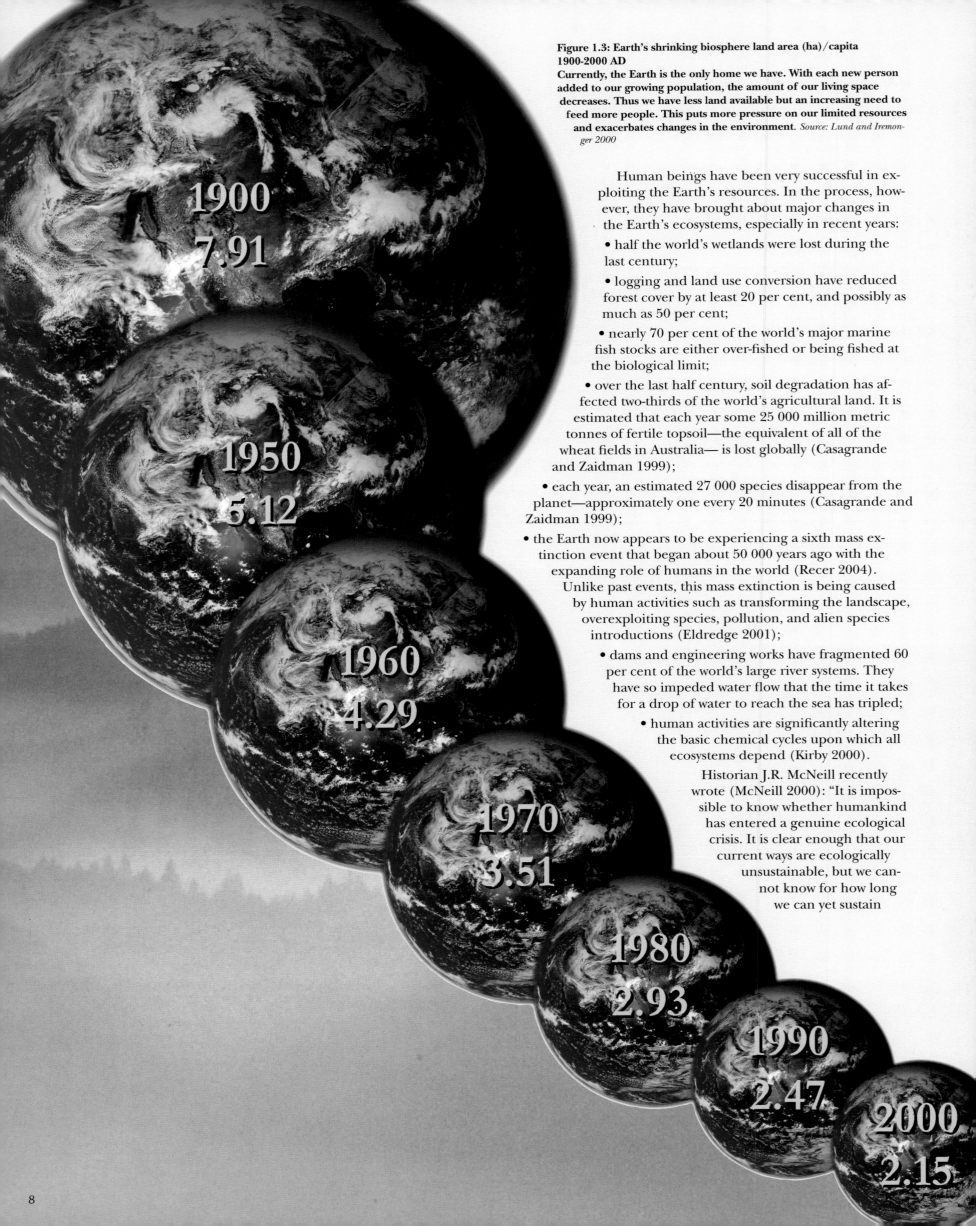

Figure 1.3: Earth's shrinking biosphere land area (ha)/capita 1900-2000 AD
Currently, the Earth is the only home we have. With each new person added to our growing population, the amount of our living space decreases. Thus we have less land available but an increasing need to feed more people. This puts more pressure on our limited resources and exacerbates changes in the environment. *Source: Lund and Iremonger 2000*

1900 7.91

1950 5.12

1960 4.29

1970 3.51

1980 2.93

1990 2.47

2000 2.15

Human beings have been very successful in exploiting the Earth's resources. In the process, however, they have brought about major changes in the Earth's ecosystems, especially in recent years:

• half the world's wetlands were lost during the last century;

• logging and land use conversion have reduced forest cover by at least 20 per cent, and possibly as much as 50 per cent;

• nearly 70 per cent of the world's major marine fish stocks are either over-fished or being fished at the biological limit;

• over the last half century, soil degradation has affected two-thirds of the world's agricultural land. It is estimated that each year some 25 000 million metric tonnes of fertile topsoil—the equivalent of all of the wheat fields in Australia— is lost globally (Casagrande and Zaidman 1999);

• each year, an estimated 27 000 species disappear from the planet—approximately one every 20 minutes (Casagrande and Zaidman 1999);

• the Earth now appears to be experiencing a sixth mass extinction event that began about 50 000 years ago with the expanding role of humans in the world (Recer 2004). Unlike past events, this mass extinction is being caused by human activities such as transforming the landscape, overexploiting species, pollution, and alien species introductions (Eldredge 2001);

• dams and engineering works have fragmented 60 per cent of the world's large river systems. They have so impeded water flow that the time it takes for a drop of water to reach the sea has tripled;

• human activities are significantly altering the basic chemical cycles upon which all ecosystems depend (Kirby 2000).

Historian J.R. McNeill recently wrote (McNeill 2000): "It is impossible to know whether humankind has entered a genuine ecological crisis. It is clear enough that our current ways are ecologically unsustainable, but we cannot know for how long we can yet sustain

Credit: Noguchi Yoshi/UNEP/Topfoto

them or what might happen if we do." In the past, humanity trod relatively lightly on the Earth, even though civilizations were intensely concentrated in some places such as Mesopotamia and the Nile River valley. Today, however, the evidence from space shows signs of the human presence in almost every corner of the planet.

Global concern about the environment and the fate of the Earth emerged in the 1970s, as did international initiatives to

address those concerns. In roughly the past 30 years, the environment has borne the stresses imposed by a four-fold increase in human population and an eighteen-fold increase in world economic output (UNEP 2002). Not surprisingly, when scientists compare recent satellite images of the Earth's surface with those taken one or several decades ago, the impact people have had on the planet is obvious and often disturbing.

This atlas vividly illustrates some of the changes the human race has brought about on the Earth—both good and bad—over the past 30 years. In doing so, it also serves as an early warning for environmental events that may occur. We hope it will be useful as a basis for developing policy decisions and promoting individual actions to help sustain the Earth and ensure the well-being of its inhabitants.

References

Anon. (n.d.). World population through the years. http://www.neopage.com/know/worldpop.htm on 19 March 2004.

Baldia, M. O. (2000). The origins of agriculture. Version 2.01. http://www.comp-archaeology.org/AgricultureOrigins.htm on 19 March 2004.

Borlaug, N. E. (2000). The Green Revolution revisited and the road ahead. Special 30th Anniversary Lecture, The Norwegian Nobel Institute, Oslo, Norway, September 8, 2000, 23. http://www.nobel.se/peace/articles/borlaug/borlaug- lecture.pdf on 1 August 2004.

Casagrande, J. and Zaidman, Y. (1999). Defining a new balance between humans and the environment. Changemakers. http://www.changemakers.net/journal/99September/index.cfm on 18 March 2004.

Eldredge, N. (2001). The sixth extinction. ActionBioscience Journal. http://www.actionbioscience. org/newfrontiers/eldredge2.html on 19 March 2004.

Grace, J. (n.d.). World Forests and Global Change. University of Edinburgh, The Institute of Ecology & Resource Management, Edinburgh, UK. http://www.ierm. ed.ac.uk/ierm/teaching/slides.pdf on 7 October 2004.

IPC (2003a). Historical estimates of world population. U.S. Census Bureau, Population Division, International Programs Center, Cambridge, UK. http://www.census. gov/ipc/www/worldhis.html on 19 March 2004.

IPC (2003b). Total midyear population for the world: 1950-2050. U.S. Census Bureau, Population Division, International Programs Center, International Data Base, Cambridge, UK. http://www.census.gov/ipc/www/worldpop.html on 19 March 2004.

Kirby, A. (2000). Humans stress ecosystems to the limit. BBC News, UK. http://news.bbc.co.uk/1/hi/sci/tech/926063.stm on 19 March 2004.

Kowalski, W.J. (n.d.). http://www.personal.psu.edu/users/w/x/wxk116/habitat/ on 19 March 2004.

Lund, H.G. and Iremonger, S. (2000). Omissions, commissions, and decisions: the need for integrsted resource assessments. Forest Ecology and Management, 128(1-2): 3-10.

McNeill, J.R. (2000). Something new under the sun – An environmental history of the twentieth century world. W.W. Norton & Company, New York, USA, 421.

Mcrone, J. (2000). The discovery of fire. New Scientist. May 2000. http://www.btinternet.com/~neuronaut/webtwo_features_fire.htm on 18 March 2004.

NASA (2002). Blue Marble: Land Surface, Shallow Water, and Shaded Topography. http://visibleearth.nasa.gov/view_rec.php?vev1id=11656 on 18 August 2004.

Nott, A. (1996). Environmental Degradation. http://www.geocities.com/atlas/env/ on 6 October 2004.

Primal Seeds (n.d.). Agriculture Origins. http://www.primalseeds.org/agricult.htm on 19 March 2004.

Recer, P. (2004). Many species at risk of extinction. Research Study. Associated Press. http://story.news.yahoo.com/news?tmpl=story&u=/ap/wildlife_gone on 19 March 2004.

Sage, R.F. (1995). Was low atmospheric CO2 during the Pleistocene a limiting factor for the origin of agriculture? Global Change Biology, 1:93-106. http://www.greeningearthsociety. org/Articles/origins.htm on 23 March 2004.

Tapestries and More. http://www.tapestries.cc/Imagehtm/gMap.html on 12 May 2004.

UNEP (2002). Global Environment Outlook 3 (GEO3) – Past, present and future perspectives. Earthscan, London, UK, 446. http://www.unep.org/geo/geo3/ on 4 March 2004.

US Global Change Research Program (2004). Our Changing Planet: The U.S. Climate Change Science Program for Fiscal Years 2004 and 2005, 8. http://www.usgcrp.gov/usgcrp/Library/ocp2004-5/ocp2004-5.pdf on 13 October 2004.

Wadley, G. and Martin, A. (1993). The origins of agriculture – a biological perspective and a new hypothesis. Australian Biologist 6: 96 – 105. http://www.veganstraight- edge.org.uk/GW_paper.htm on 19 March 2004.

Wikipedia (n.d.). The free encyclopedia. http://en.wikipedia.org/wiki/Main_Page on 18 March 2004.

Wilford, J. N. (1997). New clues show where people made the great leap to agriculture. The New York Times Company. http://www.spelt.com/origins.html on 19 March 2004.

Williams, G. W. (2001). References on the American Indian use of fire in ecosystems. U.S. Department of Agriculture: Forest Service, Washington, DC, USA. http://www. wildlandfire.com/docs/biblio_indianfire.htm on 15 March 2004.

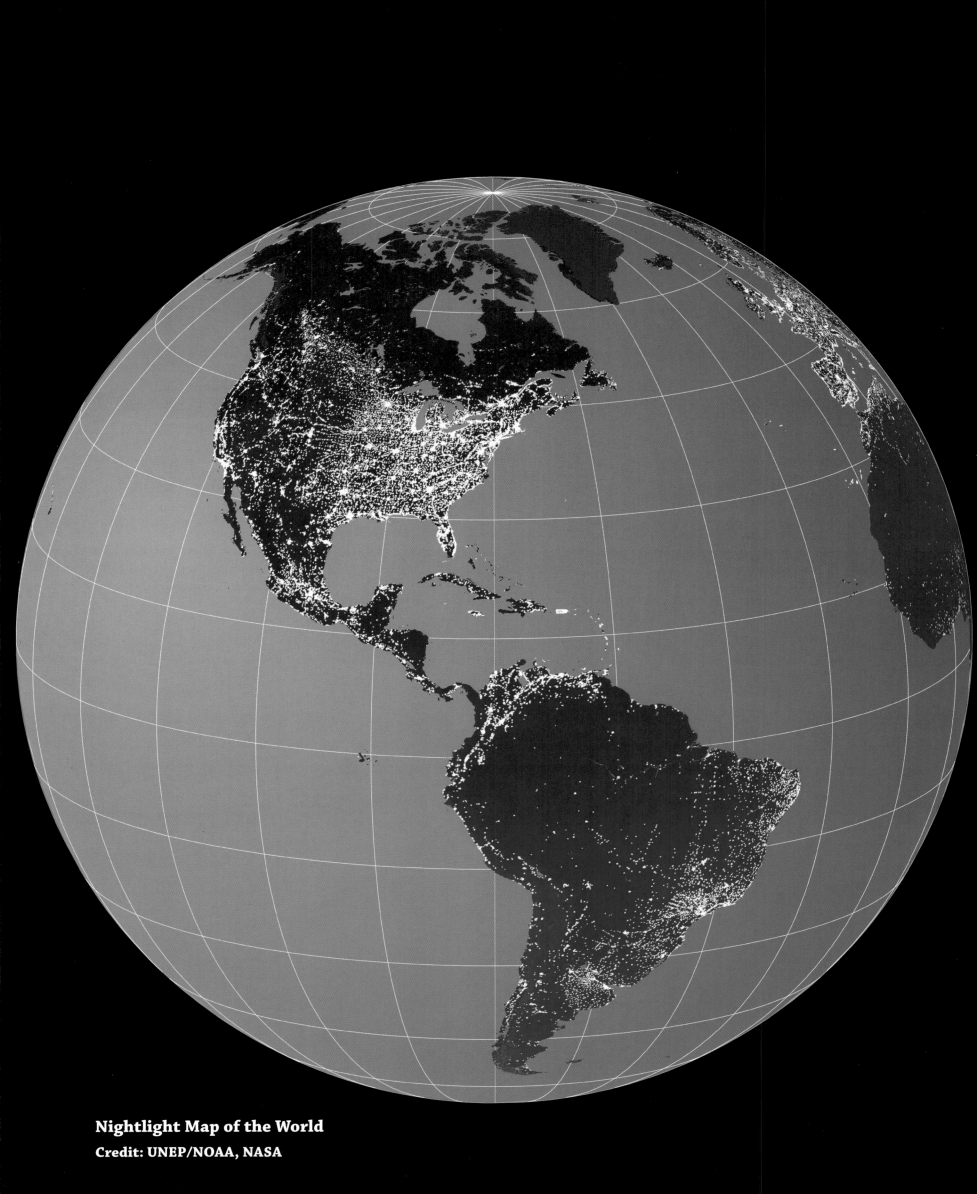

Nightlight Map of the World
Credit: UNEP/NOAA, NASA

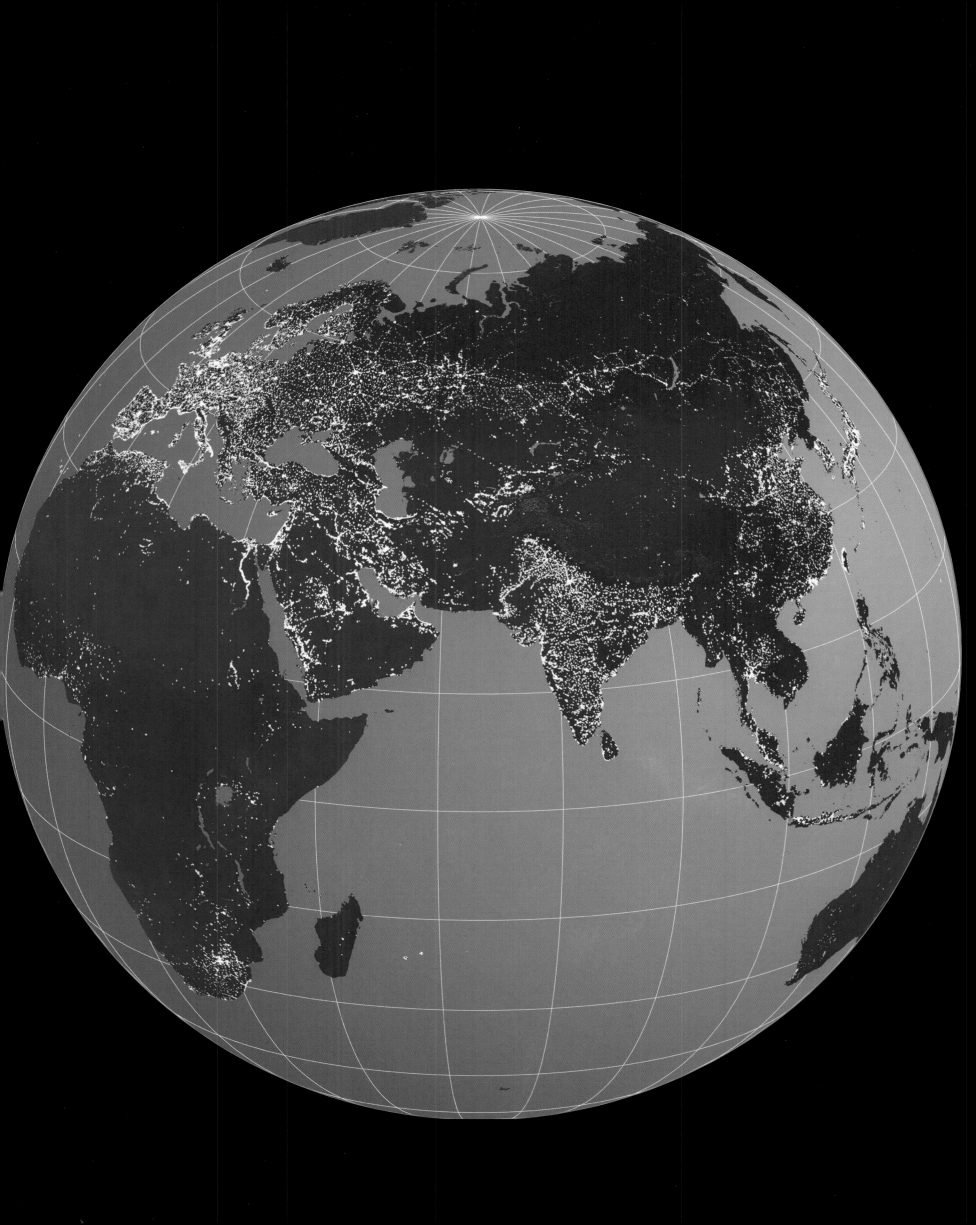

Credit: Busakorn Buranabunpo/UNEP/Topfoto

People and Planet
Human Influences on the Planet

Humans are a prolific and opportunistic species, among the most successful of all the Earth's inhabitants. As the size of the human population has increased, people have spread across the globe into every imaginable habitat. Throughout human history, people have demonstrated an uncanny ability to adapt to and survive in some very harsh places, including, most recently, outer space and the ocean—at least for short periods of time.

As human culture has evolved, people have developed new ways of living in and using their environment, and of helping themselves to all that the Earth has to offer. Their ability to exploit the Earth's seemingly endless resources has been a vital key to the success of the human species.

However, many major advances in human culture—from the cultivation of crops and the development of cities to modern technologies—have tended to insulate people from the very environment that shaped them and upon which they depend. As a group, people have often forgotten that for every action taken there is a reaction, an impact.

The impacts of human activities on the Earth often have both negative and positive components. For example, when people convert forests or grasslands to cropland they improve the means by which to feed their ever-growing numbers. At the same time, they invariably reduce biological diversity in the converted areas. Over time, people have rarely been fully aware of the tremendous change they have wrought on the Earth or that their successes have often been achieved at the expense of other species and the environment.

Since the early 1970s, many excellent texts have been written about the plight of the world (Heywood 1995; Middleton 1997; WRI 2000; Chew 2001; FAO 2001; Harrison and Pearce 2001; IPCC 2001; McNeill 2001; UNEP 2002a). This atlas supplements these works by providing illustrations of both positive and negative human-caused changes that have taken place on the Earth. Satellite images, together with photographs, provide a unique view of how people are impacting the terrestrial environment and what the consequences of environmental change mean in terms of human well-being. The images

Credit: Davoli Silvaho/UNEP/Topfoto

and the changes they illustrate are diverse. But they are united by a common message: environmental change does matter.

The International Conference on Population and Development Programme of Action noted that stabilization of world

Credit: Ed Simpson/UNEP/PhotoSpin

Population Change

Latin America 3% Others 2.5%
Africa 4.5%
US 5%

1900
1,600 million

Europe (including Russia) 25%

Asia 60%

Asia Pacific (including former Soviet Asia) 54%

Europe (including Russia) 14%

2000
6,000 million

Africa 10%

Latin America & Caribbean 8%

Middle East & North Africa 6%

North America 5%

Others 3%

Change in distribution of world population 1900 and 2000. *Source: http://www.newint.org/issue309/facts.html*

population is crucial to achieving sustainable development. Population stabilization is also necessary for managing human impacts on the Earth's environment and resources. In 1999, the Earth's human population reached 6 000 million, having grown during the mid-1990s at a rate of 13 per cent per year, with an average annual addition of 78 million individuals. As of 1999, countries with populations of 100 million or more included China, India, the United States, Indonesia, Brazil, Pakistan, the Russian Federation, Bangladesh, Japan, and Nigeria. According to the medium variant of the United Nations' population estimates and projections, world population will reach 7 200 million by the year 2015. Ninety-eight per cent of the population increase will take place in less-developed regions of the world. Africa will experience, by far, the most rapid rate of growth (Population Division 2000).

The overall impact that humans have on the global environment is proportional to the number of people on the Earth and the average influence of each individual. If that overall impact is to be reduced, addressing both of these factors is essential.

Case Study: Parrot's Beak

Between Sierra Leone and Liberia, there is a small strip of land belonging to Guinea known as the "Parrot's Beak." As civil wars raged in Sierra Leone and Liberia, hundreds of thousands of refugees have fled to relative safety in Guinea, many of them settling in the Parrot's Beak. The United Nations High Commissioner for Refugees (UNHCR) estimates that the refugee population constitutes up to 80 per cent of the local population there (UNEP 2000).

The 1974 image of the Parrot's Beak in Guinea (left) shows the surrounding territory of Liberia and Sierra Leone. Scattered throughout the deep green forest of the Parrot's Beak region are small flecks of light green, where compounds of villages with surrounding agricultural plots are located. Several dark spots in the upper left of the image are most likely burn scars.

The 2002 image (facing page) shows the Parrot's Beak region clearly defined by its light green color surrounded by darker green forest. The light green

Deforestation of indigenous palm trees in the refugee camp has left barren hillsides.

Credit: Unknown/UNEP/UNEP-GRID Geneva

SIERRA LEONE

Jagbwema

Koardu

Koundou

GUINEA

Refugee Occupation Area

Kailahun

Foya

Manowa

0   10   20   Kilometres

24 Jan 1974

LIBERIA

World Biomes

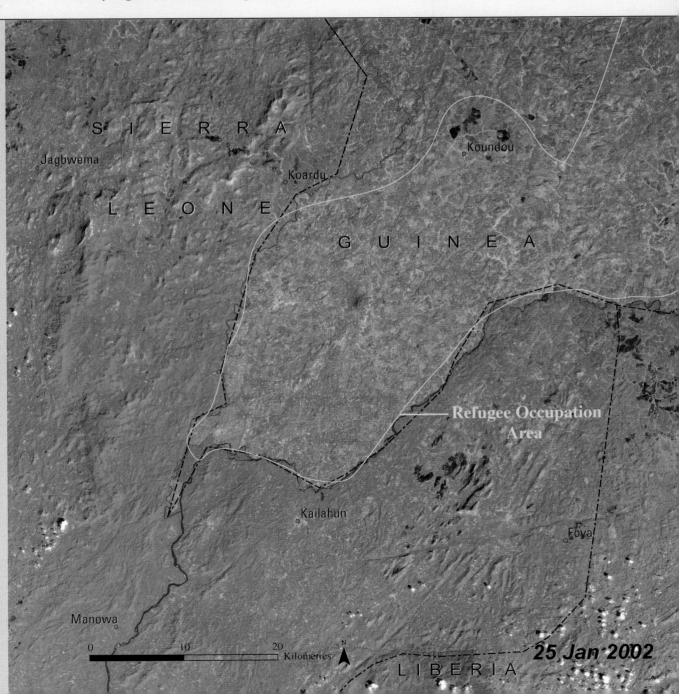

Biomes
- Tropical and Subtropical Moist Broadleaf Forests
- Tropical and Subtropical Dry Broadleaf Forests
- Tropical and Subtropical Coniferous Forests
- Temperate Broadleaf and Mixed Forests
- Temperate Conifer Forests
- Boreal Forests / Taiga
- Tropical and Subtropical Grasslands, Savannas, and Shrublands
- Temperate Grasslands, Savannas, and Shrublands
- Flooded Grasslands and Savannas
- Montane Grasslands and Shrublands
- Tundra
- Mediterranean Forests, Woodlands, and Scrub
- Deserts and Xeric Shrublands
- Mangroves
- Bare Rock / Ice

Source: World Wildlife Fund Terrestrial Ecoregions Dataset.

A biome is a major ecological community of plants and animals with similar life forms and environmental conditions. Some of the Earth's major terrestrial biomes include forests, grasslands, deserts, rain forests, and tundra. Different biomes are the source of different kinds of resources and processes (collectively called ecosystem services) such as water, soil, oil, natural gas and other fuels, minerals and other raw materials, wildlife habitat, erosion control, nutrient cycling, water filtration, food production, and genetic resources. The estimated global value of the Earth's biomes for ecosystem services alone ranges from US $16 trillion to US $54 trillion a year (Costanza et al. 1997).

color is the result of deforestation in the "safe area" where refugees have set up camp. Many of the refugees integrated into local villages, created their own family plots, and expanded the zones of converted forest area until they all merged into the larger defined area. In the upper part of the 2002 image the forest devastation is especially obvious, as areas that were green in the 1974 image now appear gray. Logging interests also moved into the higher elevations of this region, expanding the deforested zone visible in the upper left corner of the image.

Overall impoverishment of the environment of the Parrot's Beak is directly related to the rapidly increasing population in the area, mainly due to immigration, and a growth rate of about three per cent among the indigenous population. Natural resources are being exploited to create more arable land for crops, wood for charcoal, firewood and construction materials, and commercial logging for revenue.

Source: UNEP 2000.

Deforestation is evident on the hills surrounding the refugee camp.

Credit: Unknown/UNEP/UNEP-GRID Geneva

2.1 World Population

A simple definition of world population is the number of people alive on the Earth at any given point in time. World population reached 6 400 million in 2004 and it continues to grow by some 80 million each year (Table 2.1). Since the 1950s, China has been the world's most populous country (Table 2.2). China's population is currently greater than that of some entire world regions (Global Population Profile 2002). By 2050, world population is estimated to reach 7 900 to 10 900 million, when stabilization of the Earth's population is likely to take place. Whether or not world population falls within that range by this middle of this century—rather than exceeding it—will depend upon many of the choices and commitments that people make in the coming years (UNFPA 2001).

Table 2.1 – World population for given points in time
Source: ESA 2003

Year	Population
1970	3 692 492 000
1975	4 068 109 000
1980	4 434 682 000
1985	4 830 979 000
1990	5 263 593 000
1995	5 674 380 000
2000	6 070 581 000
2005	6 453 628 000
2010	6 830 283 000

Credit: Khin Aye Myat/UNEP/Topfoto

The size of any population changes as a result of fluctuations in three fundamental factors: birth rate, death rate, and immigration or emigration. When any or all of these factors deviate from zero, the size of the population will change (Global Population Profile 2002). The primary driving force of population change, whether in an individual country or for the entire world, is change in birth and death rates.

World population is growing more slowly than was expected (Figure 2.1) as a result of aid, family planning programs, and educational and economic programs directed at women. People are also healthier and living longer than they did in the past; average life expectancy has increased while crude birth rate and death rate are following a downward trend (Tables 2.3, 2.4, 2.5 and 2.6).

Most future population growth is likely to be in countries that have relatively large numbers of young people and where large families are still the norm. Furthermore, declining mortality and increased longevity have resulted in, and will continue to lead to, the expansion of older populations. Worldwide,

Figure 2.1: Time to successive billions in world population: 1800-2050.

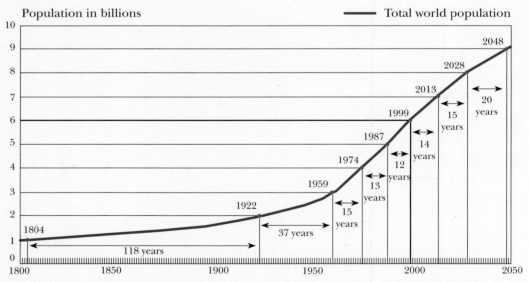

Source: United Nations (1995b); U.S. Census Bureau, International Programs Center, International Data Base and unpublished tables.

Source: Global Population Profile 2002

Table 2.2 – The Top Ten Most Populous Countries: 1950, 2002, 2050

1950	2002	2050
1. China	1. China	1. India
2. India	2. India	2. China
3. United States	3. United States	3. United States
4. Russia	4. Indonesia	4. Indonesia
5. Japan	5. Brazil	5. Nigeria
6. Indonesia	6. Pakistan	6. Bangladesh
7. Germany	7. Russia	7. Pakistan
8. Brazil	8. Bangladesh	8. Brazil
9. United Kingdom	9. Nigeria	9. Congo (Kinshasa)
10. Italy	10. Japan	10. Mexico

Source: U.S. Census Bureau, International Programs Center, International Data Base are unpublished tables.

Table 2.3 – Median age for given points in time *Source: ESA 2003.*

Year	Median age
1970	21.7
1975	22.0
1980	22.7
1985	23.4
1990	24.3
1995	25.3
2000	26.4
2005	27.4
2010	28.4

Table 2.4 – Average life expectancy at birth - Medium variant *Source: ESA 2003.*

Period	Both sexes combined	Male	Female
1970-1975	58.0	56.5	59.5
1975-1980	59.8	58.1	61.5
1980-1985	61.3	59.4	63.2
1985-1990	62.9	60.9	64.8
1990-1995	63.8	61.7	65.9
1995-2000	64.6	62.5	66.9
2000-2005	65.4	63.3	67.6
2005-2010	66.3	64.2	68.4

Table 2.5 – Crude birth rate per 1 000 population - Medium variant *Source: ESA 2003.*

Period	Crude birth rate
1970-1975	30.9
1975-1980	28.1
1980-1985	27.4
1985-1990	26.8
1990-1995	24.5
1995-2000	22.7
2000-2005	21.3
2005-2010	20.4

Table 2.6 – Crude death rate per 1 000 population - Medium variant *Source: ESA 2003.*

Period	Crude death rate
1970-1975	11.6
1975-1980	10.9
1980-1985	10.3
1985-1990	9.7
1990-1995	9.5
1995-2000	9.2
2000-2005	9.1
2005-2010	9.0

the average life expectancy in 1950 was 46 years; in 2050, it is projected to be 76 years (Hunter 2001).

While an increase in life expectancy is a positive development, it presents a new set of challenges. In Europe, for example, where women give birth to an average of 1.4 children, governments are concerned that there will be too few workers in future years to support the growing number of retirees in the population. An aging population strains a nation's social security system and pension plans, and puts pressure on health budgets because of higher health care costs for the elderly. Some governments are also concerned that a shortage of working-age individuals may lead to increased immigration, and that a decline in population may signal a weakening of a country's political and economic clout (Ashford 2004).

One of the main reasons that world population has grown so rapidly over the last 200 years is that mortality rates have declined faster than fertility rates. Improved sanitation, health care, medicine, shelter, and nutrition have all led to dramatic increases in life expectancy. Fertility rates, on the other hand, declined more recently than mortality rates did (UNEP 1999).

There is a striking paradox in global population trends: for more than two decades, many developing countries have experienced a rapid decline in fertility while fertility rates in most highly developed nations have remained very low (Figures 2.3 and 2.4). Yet in the coming years, a massive increase of the world population is almost certain (Heilig 1996).

The Demographic Transition Model (Figure 2.2) shows how a country's population can change as the country develops. However, this model does not take into account migration. Worldwide, migration of people out of rural areas is accelerating, making internal and international

Credit: Ed Simpson/UNEP/PhotoSpin

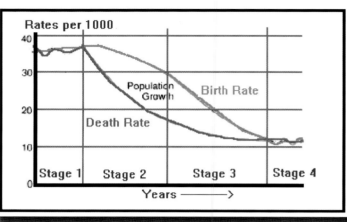

The Demographic Transition

Figure 2.2: The Demographic Transition Model shows how population growth occurs naturally in four stages. Stage 1: Birth rate and death rate are high, limiting both the rate of increase and total population. Stage 2: Birth rate remains relatively high but death rate begins to fall, causing the population to grow rapidly. Stage 3: Declining birth rate and low death rate maintain continued population growth. Stage 4: Both birth rate and death rate are low, slowing population growth, but leaving a large total population. *Source: http://www.geography.learnontheinternet.co.uk/topics/growth*

Figure 2.3: Childbearing trends in major world regions, 1970 and 2004

Total fertility rate (children per woman)

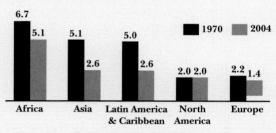

Source: UN Population Division. World Population Prospects: The 2002 Revision (1970 data), and C. Haub, 2004 World Population Data Sheet (2004 data).

migration potentially one of the most important development and policy issues of this century. The migration of labor geographically, out of rural areas, and occupationally, out of farm jobs, is one of the most pervasive features of agricultural transformations and economic growth. Yet in a world of complete and well-functioning markets, there is little or no economic rationale for policies to reduce migration; the movement of labor out of agriculture is both a quintessential feature of agricultural transformations and a prerequisite for efficient and balanced economic growth (Taylor and Martin 2002).

Clearly, human numbers cannot continue to increase indefinitely. The more people there are and the longer they live,

Figure 2.4: Different patterns of fertility decline, 1970-2000

Children per woman

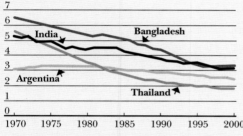

Sources: Registrar General of India; Instituto Nacional de Estadística (Argentina); United Nations Population Division; Institute of Population and Social Research, Mahidol University, Thailand; Demographic and Health Surveys; and Population Reference Bureau estimates.

the more competition there will be for the Earth's limited resources. Unless all nations adopt more sustainable methods of

Population Density Map

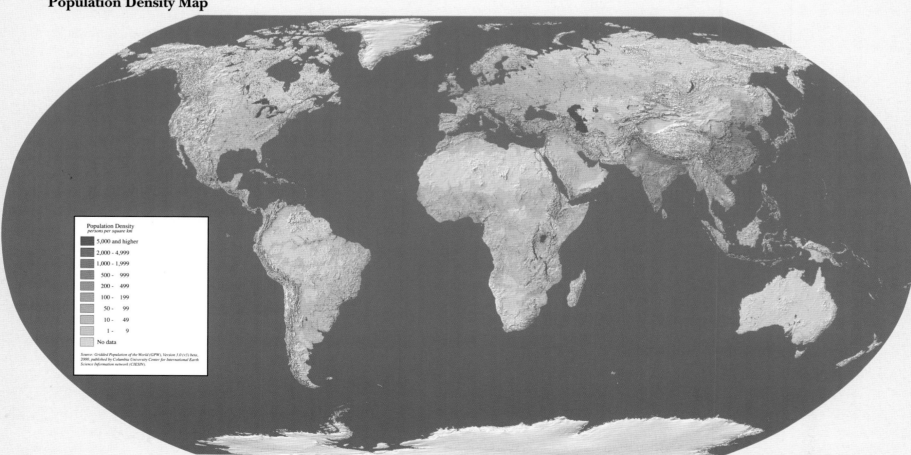

Source: http://beta.sedac.ciesin.columbia.edu/

Global Urban Extent Map

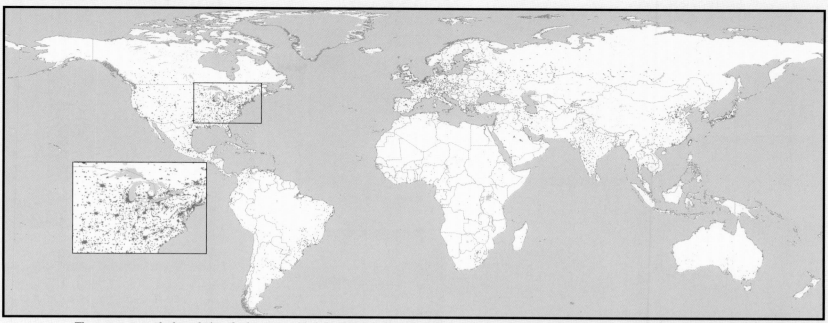

The map portrays the boundaries of urban areas with defined populations of 5 000 persons or more. *Source: Modified from http://beta.sedac.ciesin.columbia.edu/gpw/global.jsp*

production and consumption, the planet's carrying capacity will be exceeded (UNEP 1999).

Natural resources are already severely limited, and there is emerging evidence that natural forces are already starting to control human population numbers through malnutrition and disease (Pimentel et al. 1999).

The environmental challenges that people now face and most likely will continue to face in the future would be less difficult if world population were growing very slowly or not at all. The number of people on the Earth and the rate at which that number increases (Table 2.7) dramatically impact the availability of water, soil, arable land, minerals, fuels, and many other natural resources worldwide. Access to and use of family planning services can help lower fertility rates and delay child-bearing

Case Study: Monitoring Rapid Urban Expansion of Tehran, Iran
1975 and 2000

Tehran is located at the foot of the Alborz Mountains. The city occupies the northern part of the alluvial Tehran Plain, sloping from the mountains to the flat Great Salt Desert. The urban area is bounded by mountains to the north and east making it difficult to differentiate the urban area from the mountainous and desert area that surrounds Tehran.

The population of Tehran has grown three-fold since 1970 when the population was three million. In 1987, the city had grown to more than seven million people and covered an area of 575 km² (230 square miles). Today the city has nine million residents.

The rapid expansion of Tehran, as well as its sharp population growth in recent decades, has had many adverse impacts on the environment. Air and water pollution are major problems in the city. Urban areas are replacing farms and water resources. A major concern is its location on a recognized zone of active faulting with a modest to high seismic risk. Recent planning and construction techniques are designed to improve the resistance to a major earthquake that could threaten the city.

Credit: Saman Salari Sharif/UNEP/UNEP-GRID Geneva

25 Jul 1975

18 Jul 2000

Credit: Ed Simpson/UNEP/PhotoSpin

Population Density (people/km²)

- High (>100)
- Medium (25–100)
- Low (<25)
- Water

Lake Victoria, Kenya

Population growth around Lake Victoria, Kenya, is significantly higher than in the rest of Africa because of the wealth of natural resources and economic benefits the lake region offers. Note the increase in population in a 100-km (62 miles) buffer zone around Lake Victoria between 1960 and 2000. During each decade, population growth within this zone outpaced the continental average.
Source: UNEP/GRID- Sioux Falls

years, thereby helping to slow population growth. Comprehensive population policies are an essential element in a world

development strategy that combines access to reproductive health services, education and economic opportunities, improved energy and natural resource technologies, and more reasonable models of consumption and what constitutes "the good life." Such a strategy has the potential to bring humanity into an enduring balance with the environment and the natural resources upon which people will always depend (Population Fact Sheet 2000).

In addition to the overall global increases in population, the geographic distribution of human population underwent

massive changes during the 20th century (Figure 2.5). For example, between 1900 and 1990, the population of northern South America increased by 214 million, or 681 per cent, compared to the global average population increase of 3 700 million, or 236 per cent (Ramankutty and Olejniczak 2002).

Table 2.7 – World Vital Events Per Time Unit: 2004
(Figures may not add to totals due to rounding)

Time unit	Births	Natural Death	Population Increase
Year	29 358 036	56 150 533	73 207 503
Month	10 779 836	4 679 211	6 100 625
Day	353 437	153 417	200 021
Hour	14 727	6 392	8 334
Minute	245	107	139
Second	4.1	1.8	2.3

Source: http://www.census.gov/cgi-bin/ipc/pcwe

Figure 2.5: Population change in the 20th century
Source: http://www.bioone.org/pdfserv/i0044-7447-031-03-0251.pdf

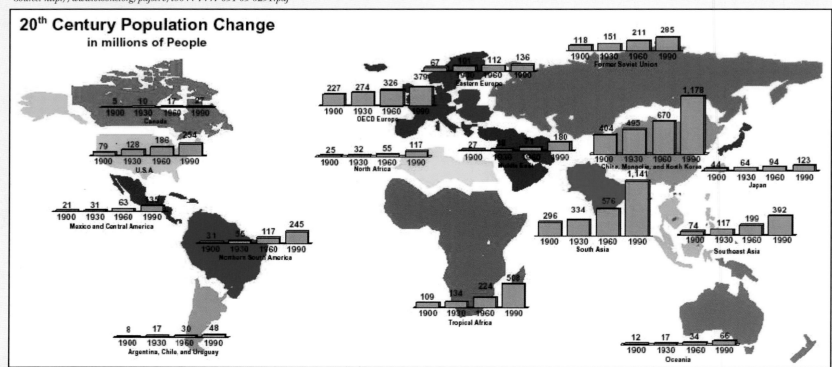

2.2 Culture

Culture encompasses the customary beliefs, social forms, and material traits of a racial, religious, or social group. Culture includes the set of values and institutions that enable a society to develop and maintain its identity. Cultural signatures differ around the globe and often hold to very different ideals and ideas, such as the role of economics as an integrating system of values or the importance of technology and technological change as springboards for human progress. Different cultures also differ in their concepts of justice and fairness and their beliefs about the relationship between people and the natural and spiritual world (UNEP 2002a).

Many of these differences are disappearing as cultures worldwide become increasingly homogeneous. Major steps in this direction occurred in the fifteenth century with European exploration and colonization and in the nineteenth century with the Industrial Revolution. In recent decades the creation of the European Union and spread of globalization has lowered many international barriers and concurrently impacted cultural diversity. Following the collapse of the Eastern Bloc in 1989, capitalism became more pervasive and less nationally limited,

Globally, world-spanning communication networks, and inexpensive air travel have reduced the costs of cross-cultural connections of all kinds, boosting television, tourism, and emigration to new levels. Global financial integration has proceeded at a furious pace, along with the international flow of goods and services as countries become increasingly dependent on one

Credit: Unknown/UNEP/Bigfoto

The World's Biocultural Diversity. People, Languages and Ecosystems

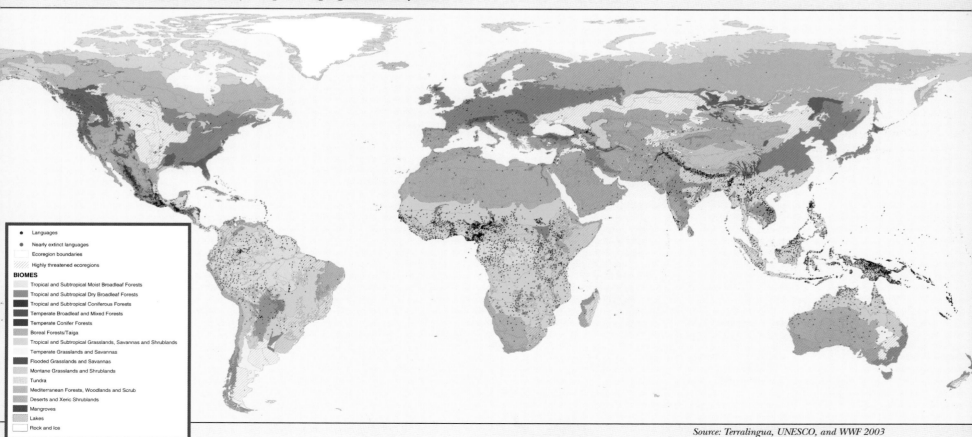

Languages
Nearly extinct languages
Ecoregion boundaries
Highly threatened ecoregions
BIOMES
Tropical and Subtropical Moist Broadleaf Forests
Tropical and Subtropical Dry Broadleaf Forests
Tropical and Subtropical Coniferous Forests
Temperate Broadleaf and Mixed Forests
Temperate Conifer Forests
Boreal Forests/Taiga
Tropical and Subtropical Grasslands, Savannas and Shrublands
Temperate Grasslands and Savannas
Flooded Grasslands and Savannas
Montane Grasslands and Shrublands
Tundra
Mediterranean Forests, Woodlands and Scrub
Deserts and Xeric Shrublands
Mangroves
Lakes
Rock and Ice

Source: Terralingua, UNESCO, and WWF 2003

Languages of the World

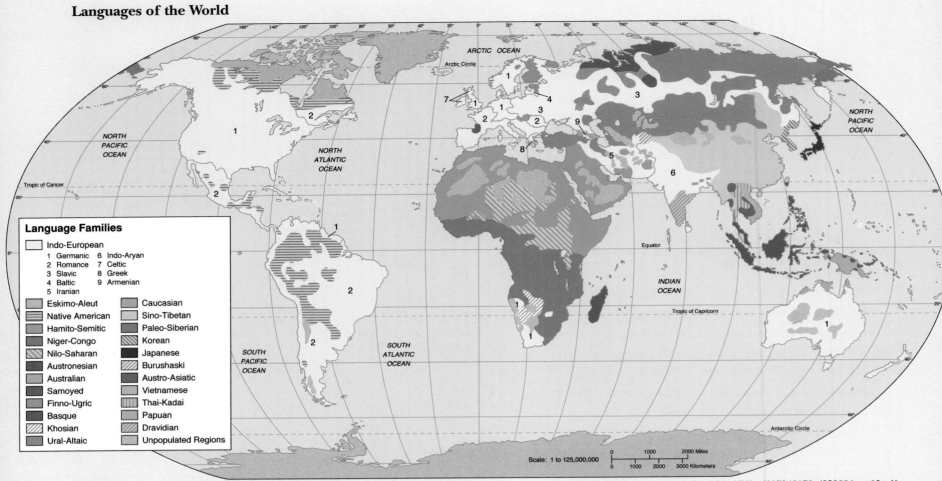

Language Families

- Indo-European
  - 1 Germanic   6 Indo-Aryan
  - 2 Romance   7 Celtic
  - 3 Slavic   8 Greek
  - 4 Baltic   9 Armenian
  - 5 Iranian
- Eskimo-Aleut
- Native American
- Hamito-Semitic
- Niger-Congo
- Nilo-Saharan
- Austronesian
- Australian
- Samoyed
- Finno-Ugric
- Basque
- Khosian
- Ural-Altaic
- Caucasian
- Sino-Tibetan
- Paleo-Siberian
- Korean
- Japanese
- Burushaski
- Austro-Asiatic
- Vietnamese
- Thai-Kadai
- Papuan
- Dravidian
- Unpopulated Regions

Scale: 1 to 125,000,000

Source: http://highered.mcgraw-hill.com/site/dl/free/007248179x/35299/map12.pdf

another for food and basic commodities (Wilk 2000). Explosive development of electronic media has intensified cultural homogenization by promoting the ideals of a handful of cultures over those of many others (Gary and Rubino 2001). The technological expansion of the media, in particular the Internet, is bringing different cultures and civilisations ever closer; while this increases the possibility of dialogue, it can also be perceived as a threat to cultural diversity, In short, current globalization of trade and mass culture, together with unprec¬edented demand for consumer goods, has significantly impacted indigenous cultures around the globe.

Every civilisation and culture is unique and irreplaceable, in that all cultures and civilisations are part of the common legacy of humankind (UN 2000). In many parts of the world, English has become the dominant language, having displaced native tongues and dialects. According to a recent UNEP report (UNEP, 2001) there were 5 000 to 7 000 spoken languages in the world with 4 000 to 5 000 of these classified as indigenous. Thirty-two per cent

WORLD CULTURAL REGIONS

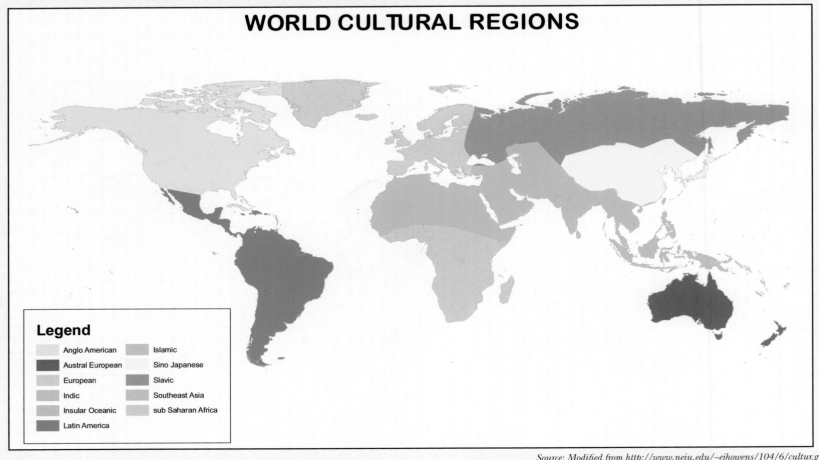

Legend

- Anglo American
- Austral European
- European
- Indic
- Insular Oceanic
- Latin America
- Islamic
- Sino Japanese
- Slavic
- Southeast Asia
- sub Saharan Africa

Source: Modified from http://www.neiu.edu/~ejhowens/104/6/cultur.gif

Credit: Unknown/UNEP/Bigfoto

Table 2.8 – The most common languages in the world

	Language	Approximate number of native speakers (in the year 2000)	Countries with substantial numbers of native speakers
1.	Mandarin Chinese	874 000 000	16
2.	Hindi (India)	366 000 000	17
3.	English	341 000 000	104
4.	Spanish	322-358 000 000	43
5.	Bengali (India and Bangladesh)	207 000 000	9
6.	Portuguese	176 000 000	33
7.	Russian	167 000 000	30
8.	Japanese	125 000 000	26
9.	German (standard)	100 000 000	40
10.	Korean	78 000 000	31
11.	French	77 000 000	53
12.	Wu Chinese	77 000 000	1
13.	Javanese	75 000 000	4
14.	Yue Chinese	71 000 000	20
15.	Telegu (India)	69 000 000	7

Note: These statistics are only rough approximations in most cases.
(Source: The World Almanac and Book of Facts, 2003)

of the world's spoken languages are found in Asia, 30 per cent in Africa, 19 per cent in the Pacific, 15 per cent in the Americas, and three per cent in Europe. More than 2 500 languages are in danger of immediate extinction, while 234 have already died out and many more were losing their connection to the modern world. Some researchers estimate that over the next century 90 per cent of the world's languages will have become extinct or virtually extinct. More than 350 languages already have fewer than 50 speakers (Table 2.8). Such rare lan¬guages are more likely to decline or disap¬pear than those that are more common (Sutherland 2003). The disappearance of any language represents an irreparable loss for the heritage of all humankind (Wurm 1970). The loss has been likened to the extinction of a species—an unfortunate cultural analog to the alarming events now occuring in the biological world. In fact,

Religions of the World

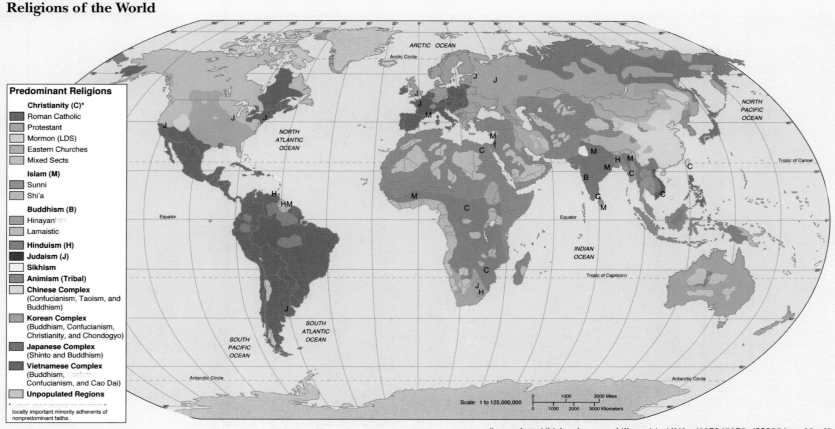

Predominant Religions

Christianity (C)*
- Roman Catholic
- Protestant
- Mormon (LDS)
- Eastern Churches
- Mixed Sects

Islam (M)
- Sunni
- Shi'a

Buddhism (B)
- Hinayanistic
- Lamaistic

Hinduism (H)

Judaism (J)

Sikhism

Animism (Tribal)

Chinese Complex (Confucianism, Taoism, and Buddhism)

Korean Complex (Buddhism, Confucianism, Christianity, and Chondogyo)

Japanese Complex (Shinto and Buddhism)

Vietnamese Complex (Buddhism, Confucianism, and Cao Dai)

Unpopulated Regions

locally important minority adherents of nonpredominant faiths.

Source: http://highered.mcgraw-hill.com/site/dl/free/007248179x/35299/map11.pdf

A natural wonder formed by natural processes, Rainbow Bridge (far left) straddles a tributary of the Colorado River in southern Utah in the United States. Two contemporary bridges, one from Sydney, Australia (left), and the other from London, England (below left), echo the natural form of Rainbow Bridge, but are the obvious byproducts of modern culture.

the number of "living" languages spoken on the Earth is dwindling faster than the planet's biodiversity.

Culture is an aspect and a means of development. Much has been said about the expansion of Western culture to the detriment of others. It is clear that many individuals aspire to Western lifestyles, while others associate Western cultural values with selfish individualism and excessive consumption. The spread of Western culture is both a cause and an effect of economic globalization, aided by the far-reaching penetration of information technologies and electronic media. At the same time, there have been nationalist and religious reactions against that culture, sometimes resulting in terrorist activities and in open warfare within or between nations (UNEP 2002a).

The World has some 6 000 communities. The international migration rate is growing every year and the number of migrants has doubled since the 1970s. While the reasons for migration vary, it is safe to say that we live in an increasingly heterogeneous society. Difference naturally leads to diversity of vision, values, beliefs, practices and expression, which all deserve equal respect and dignity (UNESCO 2003). While highlighting the role of culture in development, there is also a need to emphasize the role of culture in promoting peace (UNESCO n.d).

Just as biodiversity enriches our natural environment and is essential for its protection, cultural diversity is a treasure of humanity and a prerequisite for human development (UN 2000). Cultural Diversity presupposes respect of fundamental freedoms, namely freedom of thought, conscience and religion, freedom of opinion and expression, and freedom to participate in the cultural life of one's choice. Cultural Diversity is not just a natural fact that we need simply recognize and respect. It is about plurality of knowledge, wisdom and energy, all of which contribute to improving and moving the World forward (UNESCO n.d.).

Variety in all aspects of life has been a source of wonder and celebration for countless centuries, and the loss of that variety is an unfortunate prospect (Gary and Rubino 2001).

2.3 Land Use and Degradation

Growing crops, clearing land, planting trees, draining a wetland—these and many other activities fall into the broad category of land use, or how people use land. Land-use intensity is the extent to which land is used. It is an indication of the amount and degree of development in an area, and a reflection of the effects generated by that development (Planning Department 2001).

As a measure of activity, land-use intensity can range from very low (for example, a pristine wilderness area) to intermediate (a managed forest ecosystem) to very high (urban and industrial settings) (Lebel and Steffen 1998). From a global change perspective (Figure 2.6), land-use intensity is an important characteristic in assessing change and its impact (Berka et al. 1995). Land-use intensity is determined by the spatial requirements of a land-use activity, relationship to open space, requirements for infrastructure (transportation routes, water, sewer, electricity, and communications), and environmental impact. Parameters for measuring land-use intensity typically include:

- type of land-use activity, such as agriculture, grazing, wood production, or residential, commercial or industrial usage,
- duration of use,
- number of people, animals, plants, structures, or machines that occupy the land during a given period, and
- amount of land involved.

Credit: Andre Louzas/UNEP/Topfoto

Figure 2.6: This series of illustrations depicts global land-use change, particularly the expansion of cropland and grazing land, between 1700 and 1990. *Credit: Klein Goldewijk, K., 2001. Source: NASA 2002, http://www.gsfc.nasa.gov/topstory/20020926landcover.html*

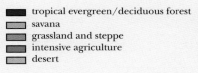
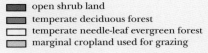

- tropical evergreen/deciduous forest
- savana
- grassland and steppe
- intensive agriculture
- desert
- open shrub land
- temperate deciduous forest
- temperate needle-leaf evergreen forest
- marginal cropland used for grazing

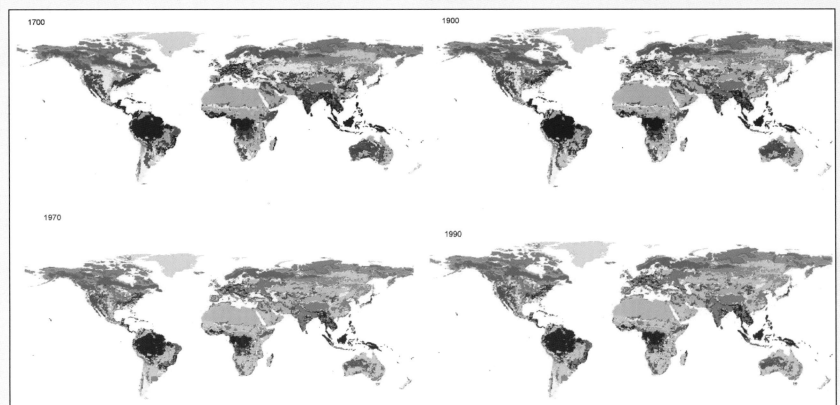

Also important in assessing land-use intensity is to examine the relative imperviousness of the landscape. Impervious surfaces, such as paved roads, inhibit or entirely block the absorption of water by underlying soil (Forney et al. 2001). Once paved or otherwise made impervious, land is not easily reclaimed. As environmentalist Rupert Cutler noted (Brown 2001), "Asphalt is the land's last crop."

Land-use intensity trends are usually expressed through changes in inputs, management, or number of harvests over a given period of time. Only changes within the same land-use category and on the same area (change of intensity)—as opposed to changes from one type of land use to another (for example, forest to cropland)—are taken into account when assessing trends (van Lynden et al. n.d.; FAO 2002).

The Agro-Ecological Zones (AEZ) methodology (Figure 2.7) is a system developed by the Food and Agriculture Organization of the United Nations (FAO) with the collaboration of the International Institute for Applied Systems Analysis (IIASA), that enables rational land use planning on

Figure 2.7: Conceptual framework of the Agro-Ecological Zones methodology

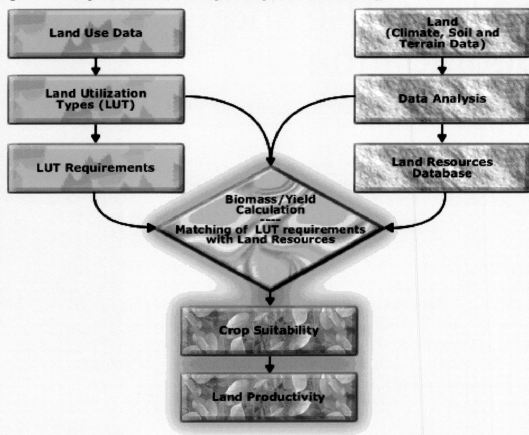

Source: FAO 2000, http://www.fao.org/ag/agl/agll/gaez/index.htm

the basis of an inventory of land resources and evaluation of biophysical limitations and potentials. This methodology utilizes a land resources inventory to assess, for

specified management conditions and levels of inputs, all feasible agricultural land-use options and to quantify expected production of cropping activities relevant

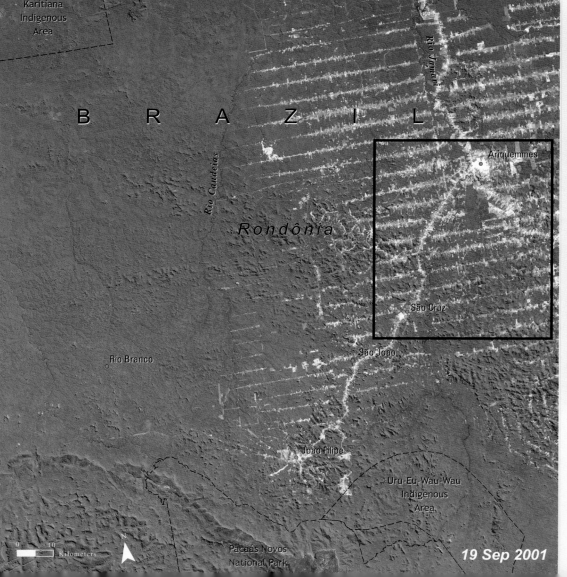

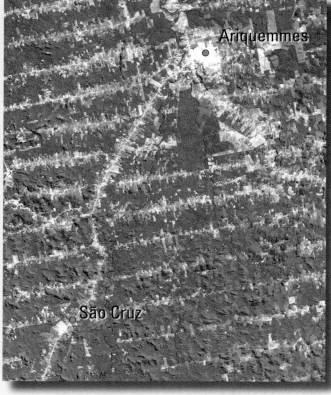

Figure 2.8: A satellite image reveals a typical "feather" or "fishbone" pattern of deforestation in Brazil. The pattern follows the construction of a new road through the rain forest. Roads provide easy access for mechanized logging to clear cut forest sections. Clear cut sections can then be turned into agricultural fields as roads provide easy access to local markets.
Source: UNEP/GRID–Sioux Falls

Credit: Xintian Pan/UNEP/Topfoto

in the specific agro-ecological context. The characterization of land resources includes components of climate, soils and landform, which are basic for the supply of water, energy, nutrients and physical support to plants (FAO 2000).

Worldwide, the effect people are having on the Earth is substantial and growing. Satellite images reveal in startling detail the signs of human impact on the landscape. From the herringbone patterns of deforestation etched into once-undisturbed rain forests (Figure 2.8) to the patchwork patterns of agricultural fields and concrete splotches of urban sprawl, the evidence that people have become a powerful force capable of reshaping the Earth's environment is everywhere.

Scientists estimate that between one-third and one-half of the Earth's land surface has been transformed by human activities (Figure 2.9) (Herring n.d.). The activity that has had the greatest impact on the global landscape is agriculture. Twelve per cent of the world's land surface—an area equivalent to that of the South American continent—is under permanent cultivation (Ramankutty and Foley 1999; Devitt 2001).

Over the next 30 years, the annual rate of growth in global crop production is expected to decrease. However, the Food and Agriculture Organization of the United Nations predicts that production will still exceed demand, despite the world's growing population. By 2030, 75 per cent of the

projected global crop production will occur in developing countries, compared to 50 per cent in the early 1960s. Increases in production will be achieved by improving plant yields and through more intensive land-use activities, including multi-cropping or high-cropping intensities (UCS 2004). In light of these projections, continued support of agricultural research and policies in developing countries is vital.

Nearly one-third of the world's cropland—1 500 million hectares—has been abandoned during the past 40 years because erosion has made it unproductive (Pimentel et al. 1995). Restoring soil lost by erosion is a slow process; it takes roughly 500 years for a mere 2.5 cm (1 inch) of soil to form under agricultural

Global Soil Degradation Map

Source: http://www.isric.nl/

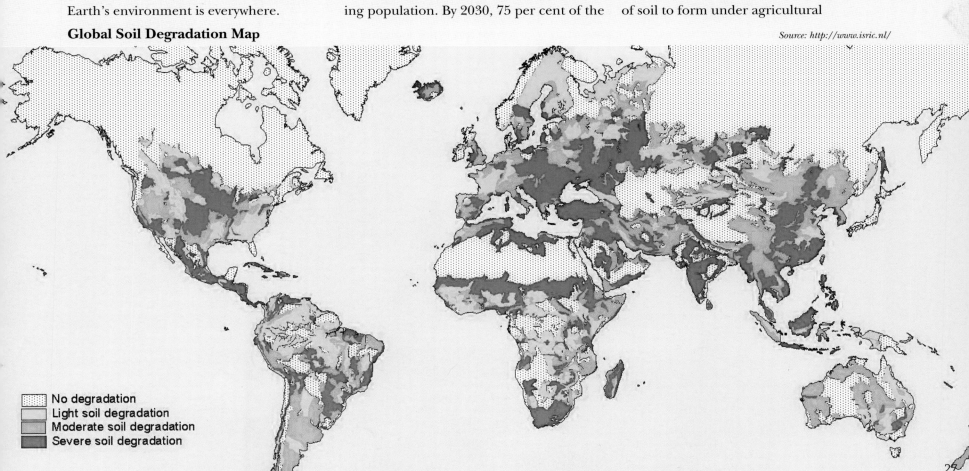

- No degradation
- Light soil degradation
- Moderate soil degradation
- Severe soil degradation

Figure 2.9: Human-induced land degradation (severe and very severe) as percentage of total land area

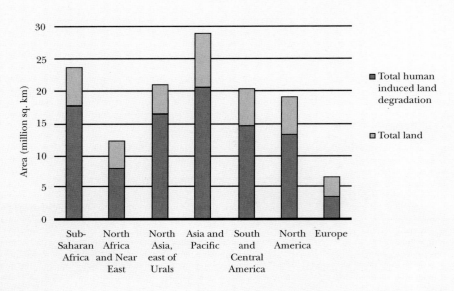

Legend:
- Total human induced land degradation
- Total land

Source: World Atlas of Desertification (UNEP 1992)

conditions (Pimentel et al. 1996). Thus the approach to replacing eroded agricultural lands typically has been to clear more and more areas of grassland or forest and convert them to cropland. The ever-growing need for agricultural land accounts for 60 to 80 per cent of the world's deforestation (Figure 2.9).

Despite such "replacement" strategies, the amount of available cropland worldwide has declined to 0.27 hectare (0.67 acres) per person (Pimentel et al. 1996). It is possible to feed one adult on a plant diet grown on about 0.2 hectares (0.5 acres) of land (Knee 2003)—and this land-per-person minimum is roughly what will be available when world population reaches 8 000 million—but only if crop yields now being achieved

in developed countries are achieved worldwide. To do so requires that most countries' inputs of fertilizer, and probably pesticides, rise to match those of North America and Europe. Furthermore, any mechanization of crop production will entail additional energy consumption. Increased mechanization is likely given the mass migration from rural areas to cities currently underway on all continents. While agriculture accounts for only about two per cent of energy consumption in North America and Europe, it accounts for roughly ten per cent of energy consumption in the rest of the world (Knee 2003).

The shortage of cropland, together with falling productivity, is a significant factor contributing to global food shortages and

associated human malnutrition. Political unrest, economic insecurity, and unequal food distribution patterns also contribute to food shortages worldwide (Pimentel et al. 1996).

In addition to agriculture, the global trend toward urbanization is another key factor bringing change to the landscape. Historically, forests and grasslands have been converted to cropland. Increasingly, cropland is being converted to urban areas (Ramankutty and Foley 1999; Devitt 2001). Millions of hectares of cropland in the industrial world have been paved to create roads and parking lots. The average car requires 0.07 hectares (0.17 acres) of paved land for roads and parking space (Brown 2001).

If farmers worldwide fail to meet the challenge of increasing yields on existing cropland, or they cannot access the tools necessary to achieve increased yields, the only alternative will be to clear the world's remaining forests and grasslands (Green 2001). Yet indications are that the world does not have enough forests to fulfill all the current and future demands being placed on them (Nilsson 1996).

As natural forests are exhausted or come under protection, the demand for wood and wood products will be increasingly satisfied by tree farms. Between 1980 and 1995, forest plantations in developed countries increased from 45-60 million hectares (111-147 million acres) to 80-100 million hectares (198-247 million acres).

Credit: Choosak Khemtai/UNEP/Topfoto

Credit: Paiboon Patta/UNEP/Topfoto

In the developing world, the area in forest plantations doubled from roughly 40 million to about 81 million hectares (99-200 million acres) over the same period. More than 80 per cent of forest plantations in the developing world are found in Asia, where demand for paper and other wood products continues to grow rapidly. Forest plantations now cover more than 187 million hectares (462 million acres) worldwide. That accounts for less than five per cent of the Earth's total forested area, but 20 per cent of current global wood production (Larsen 2003).

Land Degradation and Desertification

By the beginning of the twenty-first century, unprecedented global environmental changes had reached sufficient proportions to impinge upon human health—simultaneously and often interactively. These changes include the processes of land degradation and desertification (Menne and Berollini 2000).

Land degradation is the decline in the potential of land resources to meet human economic, social, and environmental functions needs (Africa Mountain Forum n.d.). Desertification is soil degradation in arid regions, often to such an extent that it is impossible to make the soil productive again (Table 2.9). Desertification is the result of complex interactions between unpredictable climatic variations and unsustainable land use practices by communities who, in their struggle to survive, overexploit agricultural, forest, and water resources (CIDA 2001).

Over 3 600 million hectares (8 896 million acres)—25 per cent of the Earth's land area—are affected by land degradation. Desertification occurs to some extent on 30 per cent of irrigated lands, 47 per cent of rain-fed agricultural lands, and 73 per cent of rangelands (Figure 2.10). Annually, an estimated 1.5 to 2.5 million hectares (3.7 to 6 million acres) of irrigated land, 3.5 to 4.0 million hectares (8.6 to 9 million acres) of rain-fed agricultural land, and about 35 million hectares (86 million acres) of rangeland lose all or part of their productivity due to land degradation processes (Watson et al. 1998).

Desertification and drought are problems of global dimension that directly affect more than 900 million people in 100 countries, some of which are among the least developed nations in the world (Watson et al. 1998). The consequences of desertification include (UNEP 2002a):

- reduction of the land's natural resilience to recover from climatic disturbances;
- reduction of soil productivity;
- damaged vegetation cover, such that edible plants are easily replaced by non-edible ones;
- increased downstream flooding, reduced water quality, sedimentation in rivers and lakes, and siltation of reservoirs and navigation channels;
- aggravated health problems due to wind-blown dust, including eye infections, respiratory illnesses, allergies, and mental stress;
- undermined food production; and
- loss of livelihoods forcing affected people to migrate.

Desertification results from mismanagement of land and thus deals with two interlocking, complex systems: the natural

Table 2.9 – Degree of soil degradation by subcontinental regions (per cent of total area)

	None	Light	Moderate	Strong	Extreme
Africa	83	6	6	4	0.2
Asia	82	7	5	3	<0.1
Australiasia	88	11	0.5	0.2	<0.1
Europe	77	6	15	1	0.3
North America	93	1	5	1	0
South America	86	6	6	1	0
World:					
Per centage	85	6	7	2	<0.1
Area ('000 km²)	110 483	7 490	9 106	2 956	92

Source: World Atlas of Desertification (UNEP 1992)

ecosystem and the human social system (Eswaran et al. 1998). While much desertification is attributed to poor land-use practices, hotter and drier conditions brought about by potential global warming would extend the area prone to desertification northwards to encompass areas currently not at risk. In addition, the rate of desertification would increase due to increases in erosion, salinization, fire hazard, and reductions in soil quality. As a result, the process of desertification is likely to become irreversible (Karas n.d.).

Worldwide, an estimated 6 to 27 million hectares (15 to 67 million acres) of land are lost each year to desertification. Seventy per cent of the world's dry land is degraded enough to be vulnerable to desertification (Anon 2002). The amount of land susceptible to desertification (areas known as tension zones) also is increasing. Currently, 7.1 million km^2 (2.7 million square miles) of land face low risk of human-induced desertification, 8.6 million km^2 (3.4 million square miles) are at moderate risk, 15.6 million km^2 (6.2 million square miles) are at high risk, and 11.9 million km^2 (4.6 million square miles) are at very high risk. Tension zones result from:

- excessive and continuous soil erosion resulting from overuse and improper use of lands, especially marginal and sloping lands;

- nutrient depletion and/or soil acidification due to inadequate replenishment of nutrients or soil pollution from excessive use of organic and inorganic agrichemicals;

- reduced water-holding capacity of soils due to reduced soil volume and reduced organic matter content, both of which are a consequence of erosion and reduced infiltration due to crusting and compaction;

- salinization and water-logging from over-irrigation without adequate drainage; and

- unavailability of water stemming from decreased supply of aquifers and drainage bodies.

The following negative effects are highest in the tension zones (Eswaran et al. 1998):

- systematic reduction in crop performance, leading to failure in rain-fed and irrigated systems;

- reduction in land cover and biomass production in rangelands, with an accompanying reduction in quality of feed for livestock;

- reduction of available woody plants for fuel and increased distances to harvest them;

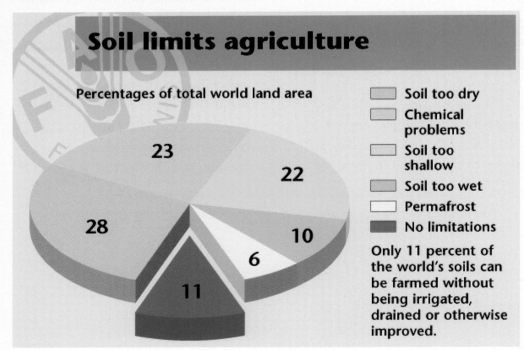

Soil limits agriculture

Percentages of total world land area

- Soil too dry
- Chemical problems
- Soil too shallow
- Soil too wet
- Permafrost
- No limitations

23
22
28
10
6
11

Only 11 percent of the world's soils can be farmed without being irrigated, drained or otherwise improved.

Figure 2.10: Soils are classified according to the proportions of different sized particles they contain. As seen in this figure, the largest percentage of world land area unsuitable for agriculture is land that is too dry. *Source: FAO 2000, http://www.fao.org/desertification/default.asp?lang=en*

This sub-scene of an ASTER satellite image shows sand dunes covering an area roughly 12 km x 15 km (8 x 9 miles) in the Thar Desert of northwestern India and eastern Pakistan. The dunes here shift constantly, taking on new shapes. Approximately 800 km (497 miles) long and 490 km (305 miles) wide, the Thar Desert is bounded on the south by a salt marsh known as the Rann of Kutch, and on the west by the Indus River plain. The desert's terrain is primarily rolling sand hills, with scattered outcroppings of shrub and rock. *Source: NASA 2004, http://asterweb.jpl. nasa.gov/gallery/gallery.htm?name=Thar*

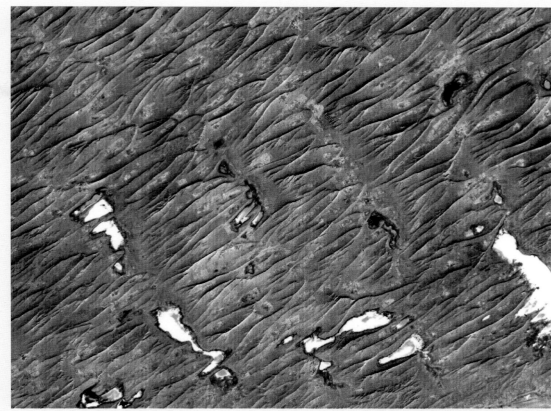

Credit: NASA/GSFC/METI/ERSDAC/JAROS, and U.S./Japan ASTER Science Team.

- significant reduction in water from overland flows or aquifers and a concomitant reduction in water quality;

- encroachment of sand and crop damage by sand-blasting and wind erosion; and

- increased gully and sheet erosion by torrential rain.

Ultimately, desertification processes impact about 2 600 million people, or 44 per cent of the world's population (Eswaran et al. 1998).

Case Study: Mt. Kenya–Diversity in Ecosystems

Christian Lambrechts

Mount Kenya is located on the equator 180 kilometres north of Nairobi. It is a solitary mountain of volcanic origin with the base diameter of about 120 km (75 miles). Its broad cone shape reaches an altitude of 5 199 m (17 057 ft) with deeply incised U-shaped valleys in the upper parts. Forest vegetation covers the major part of the mountain, with a total area around 220 000 hectares (548 574 acres). The forests are critical and invaluable national assets that must be protected.

High diversity in ecosystems and species

The wide range in altitude clines—from 1 200 to 3 400 m (3 900 to 11 000 ft)—and rainfall clines from—from 900 mm/year (35 in/year) in the north to 2 300 mm/year (91 in/year) in the south-eastern slopes—contributes to the highly diverse mosaic patterns of Mount Kenya forests. Mount Kenya adds value to the nation by providing tourism potential and local cultural and economic benefits. It also provides important environmental services to the nation such as a water catchment area of the

Tana River where 50 per cent of Kenya's total electricity output is generated.

Forest conservation initiative

Following a 1999 aerial survey, the entire forest belt of Mount Kenya was gazetted as National Reserve and placed under the management of Kenya Wildlife Services in the year 2000. In 2002, a study was carried out to assess the effectiveness of the new management practices put in place in 2000. The study revealed significant improvement in the state of conservation of the forests.

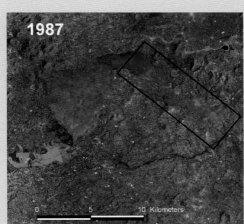

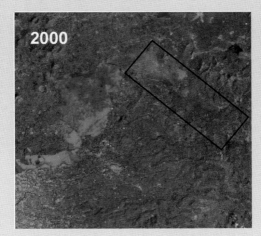

Forest is shown in red on these images. Note the changes in forest cover in the boxes.
Source: UNEP/GRID–Nairobi

2.4 Ecoregions and Ecosystems

An ecosystem is an organic community of plants and animals viewed within its physical environment (habitat); the ecosystem results from the interaction between soil and climate. It is a dynamic complex of plant, animal, and microorganism communities and their non-living environment interacting as a functional unit (UNEP-WCMC 2003).

An ecoregion is a cartographical delineation of a relatively large unit of land or water containing a geographically distinct assemblage of species, natural communities, and environmental conditions. An ecoregion is often defined by similarity of climate, landform, soil, surface form, potential natural vegetation, hydrology, and other ecologically relevant variables. Ecoregions contain multiple landscapes with different spatial patterns of ecosystems.

The ecoregion concept is one of the most important in landscape ecology, both for management and understanding (Hargrove and Hoffman 1999). Ecoregion classifications are based on particular environmental conditions and designed for specific purposes, and no single set of ecoregions would be appropriate for all potential uses (Wikipedia n.d.).

The environment of an ecoregion in terms of climate, resource endowments, and socioeconomic

Credit: John R Jones/UNEP/Topfoto

GLOBAL ECOLOGICAL ZONES

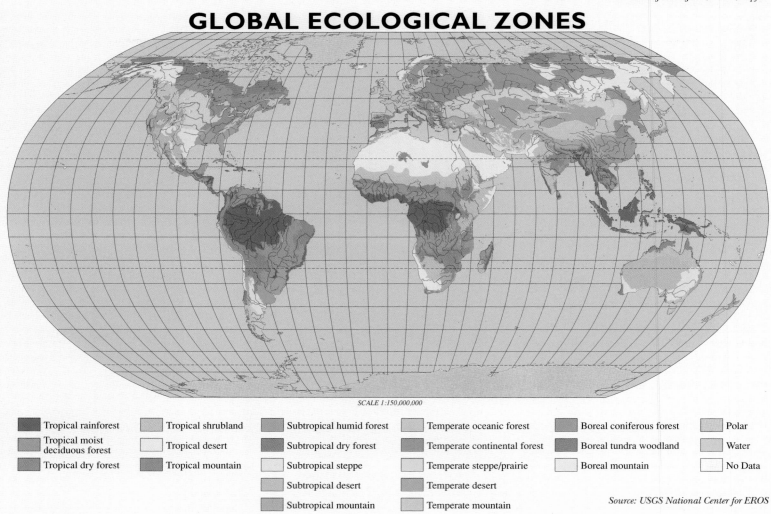

SCALE 1:150,000,000

Tropical rainforest	Tropical shrubland	Subtropical humid forest	Temperate oceanic forest	Boreal coniferous forest	Polar
Tropical moist deciduous forest	Tropical desert	Subtropical dry forest	Temperate continental forest	Boreal tundra woodland	Water
Tropical dry forest	Tropical mountain	Subtropical steppe	Temperate steppe/prairie	Boreal mountain	No Data
		Subtropical desert	Temperate desert		
		Subtropical mountain	Temperate mountain		

Source: USGS National Center for EROS

Credit: Philip De Mancz/UNEP/Topfoto

conditions is homogeneous. Specific advantages of using an ecoregion approach for planning and decision-making include:

- easier identification of production capabilities and constraints;
- better targeting of prospective technologies;
- improved assessment of responses to new technologies; and
- wider adoption and larger impact of research outputs (Saxena et al. 2001).

Trends

An increase in average global temperature has the potential to bring about dramatic change in ecosystems. Some species may be forced out of their habitats (possibly to extinction) because of changing conditions. Other species may flourish and spread. Few, if any, terrestrial ecoregions on the Earth are expected to remain unaffected by significant global warming.

Since 1970, there has been a 30 per cent decline in the world's living

Table 2.10 – Loss of biodiversity from 2000 to 2003—expressed as changes in species numbers—in animals and plants classified as critically endangered, endangered, and vulnerable

Group	Critically Endangered			Endangered			Vulnerable		
	2000	2002	2003	2000	2002	2003	2000	2002	2003
Mammals	180	181	184	340	339	337	610	617	609
Birds	182	182	182	321	326	331	680	684	681
Reptiles	56	55	57	74	79	78	161	159	158
Amphibians	25	30	30	38	37	37	83	90	90
Fishes	156	157	162	144	143	144	452	442	444
Insects	45	46	46	118	118	118	392	393	389
Mollusks	222	222	250	237	236	243	479	481	474
Plants	1 014	1 046	1 276	1 266	1 291	1 634	3 311	3 377	3 864

Source: http://www.redlist.org/info/tables/table2.html

things and the downward trend is continuing at one per cent or more per year (Collins 2000; UNEP 1997). Table 2.10 shows an increase in the number of endangered and vulnerable species between the years 2000–2003. Modification of landscapes, loss of native species, introduction of exotic species, monoculture-focused agriculture, soil enhancement, irrigation, and land degradation have all tended to "simplify" ecosystems, leading to a reduction in biodiversity. In aquatic environments, eutrophication and habitat destruction have had a similar effect (Tilman et al. 2001). As ecosystems become simpler, so do ecoregions.

Ecoregion and ecosystem fragmentation also contributes to a decline in biodiversity and threatens many species. Globally, over half of the temperate broadleaf and mixed forests and nearly one quarter of the tropical rain forests have been fragmented or removed (Wade et al. 2003).

Impacts

Simplification of ecosystems and ecoregions results in species extinctions and a loss of natural resources (Tilman et al. 2001). Climate change and the way in which ecological communities respond to it have enormous conservation implications. These include developing awareness of the transience of native ranges and plant associations and the significance of population declines and increases, as well as the need to develop targets and references for restoration, and strategies for dealing with global warming (Millar 2003). For example, changes in the potential distribution of tree and shrub taxa in North America in response to projected climate change are expected to be far-reaching and complex; growing ranges for various species will shift not only northward and upward in elevation but in all directions (Shafer et al. 2001). Some models predict that more than 80 per cent of the world's

Credit: Ron Levy/UNEP/Topfoto

ecoregions will suffer extinctions as a result of global warming (Malcolm et al. 2002). Ecoregions expected to be most dramatically altered by climate change include the boreal forests of the North-

ern Hemisphere, the fynbos of Southern Africa, and the Terai-Duar savanna and grasslands of northeastern India (Malcolm et al. 2002).

Credit: Gary Wilson/UNEP/NRCS

2.5 Biodiversity, Invasive Species, and Protected Areas

Biological diversity, or "biodiversity," refers to the variety of life on the Earth in all its forms. There are three levels of biodiversity: biodiversity of a landscape or ecosystem, species biodiversity, and genetic biodiversity (IUCN, UNEP, and WWF 1991). These three levels are intimately connected. For example, genetic diversity is often the key to survival for a species, equipping it with the necessary resources to adapt to changing environmental conditions. Species diversity, in turn, is typically a measure of ecosystem health (Rosenzweig 1999).

We have just begun to identify and fully understand the diverse living things that currently inhabit the Earth. Scientists have discovered and described roughly 1.75 million species to date. That number is expected to increase substantially when all marine organisms, arthropods, bacteria, and viruses are eventually added to the list. Tragically, however, humans are destroying this great diversity at an alarming rate. Rates of human-induced species extinction are estimated to be 50 to 100 times the natural background rate; this could

Credit: Gyde Lund/UNEP

Credit: William M. Ciesla/UNEP/Invasive.org

increase to 1 000 to 10 000 times the natural rate in the next 25 years (Lund et al. 2003; Pellew 1996).

Why are so many species becoming extinct? Human activities over the last three centuries have significantly transformed the Earth's environment, primarily through the conversion of natural ecosystems to agriculture (Ramankutty and Foley 1999). It is estimated that cropland expanded from 3-4 million km^2 (1.2-1.5 million square miles) in 1700 to 15-18 million km^2 (5.8-6.9 million square miles) in 1990, primarily at the expense of forests. At the same time, grazing lands expanded from 5 million km^2 (1.9 million square miles) in 1700 to 31 million km^2 (12 million square

miles) in 1990, largely via the conversion of native grasslands (Goldewijk and Ramankutty 2001). In addition to agriculture-driven landscape transformations, the move to monoculture-based forms of agriculture has contributed to declining biodiversity.

Wild plants and animals are a major source of food. Billions of people still harvest wild or "bush" food around the world. Between one-fifth and one-half of all food consumed by poor people in developing countries is gathered rather than cultivated. On a global scale, ocean fish caught in the wild account for 16 per cent of the human diet (Harrison and Pearce 2001).

Wild plants are also a major source of medicine, and the loss of biological diver-

sity has serious implications in terms of human health. Of the 150 most frequently prescribed drugs, more than half are derived from or patterned after chemical compounds found in plants (Brehm 2003). Moreover, plants are an important source of fuel. Nearly 15 per cent of the world's energy is derived from the burning of plant materials (De Leo and Levin 1997).

Worldwide, people eat only a small fraction of the 70 000 plants known to be edible or to have edible parts (Wilson 1989). But retaining biodiversity is still vital for the food supply, since most food crops constantly require an infusion of "wild" genes to maintain their resistance to ever-evolving pests (Harrison and Pearce 2001).

Photos left to right: A family of elephants in Africa (Credit: Gyde Lund/UNEP). A clear-cut section of forest (Credit: Steven Poe/UNEP/Topfoto). Yellowstone National Park (Credit: Gyde Lund/UNEP). People often peel the loose bark off birches for souvenirs. Danforth, ME USA (Credit: Randy Cyr/UNEP/Forestryimages.org).

Despite our dependence on biodiversity, it has been estimated that 27 000 species are lost every year—roughly three per hour. Other estimates put the number much higher. The greatest loss of biodiversity is currently taking place in wet tropical regions where rain forest ecosystems are being altered dramatically. But loss of biodiversity is also evident in drier regions, due to desertification. Major contributors to species extinctions and loss of biodiversity worldwide include:

- human population growth;
- unsustainable patterns of consumption such as over-harvesting of plant and animal resources;
- poor agricultural practices;
- increased production of wastes and pollutants;
- urban development; and
- international conflict (UNEP 2002b).

Loss of biodiversity occurs hand-in-hand with habitat loss, and habitat loss is generally greatest where human population density is highest (Harrison 1997). One type of habitat loss is fragmentation. Fragmentation occurs where a once-continuous ecosystem is broken up into many small, poorly connected patches of land, which happens when blocks of trees are removed from a forest. A change in land cover typically accompanies fragmentation. Six categories of fragmentation have been identified (interior, perforated, edge, transitional, patch, and undetermined) depending on how a given area of land is broken up (Riitters et al. 2000). Fragmentation may be human-induced or due to natural causes such as fire, floods, or wind. Fragmentation may create more diverse landscapes than were originally present, and while it may destroy habitats of some species, it can also create habitats for others.

Left to right: Junipers near near Paulina, Oregon. Without natural fires to control their spread, junipers can become invasive in rangelands (Credit: Gyde Lund/UNEP). Women herding goats (Credit: Unknown/UNEP/Topfoto). Cattle grazing in a bog (Credit: Rubai Wang/UNEP/Topfoto). Kudzu taking over the land and trees in the southeastern United States (Credit: Gyde Lund/UNEP).

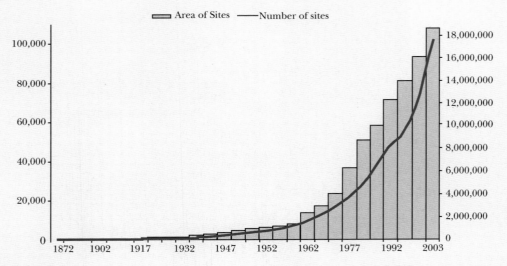

Cumulative Growth in Protected Areas by 5 Year Increments:
1872 – 2003

Legend: Area of Sites — Number of sites

Figure 2.11: The number and extent of the world's protected lands increased significantly during the period from 1872 to 2003. The greatest increase has occurred over the past few decades. In 2003, the total number of protected sites surpassed 100 000, while total area increased to more than 18 million km^2 (7 million square miles). *Source: Chape et al. 2003*

Invasive Species

Most plants and animals exist in places in which they did not originate. They moved or were introduced into new areas over time. While rooted plants cannot themselves move from place to place, the dispersion of their seeds by wind, water, and animals has enabled them to spread into many new habitats.

An introduced species is one whose existence in a given region is due to some type of human activity. That activity may enable the species to cross natural geographic barriers or it may transform conditions in an area as to be in some way favorable to the species' growth and spread. Introduced species are also called alien, or exotic, species.

Many introduced species have been actively transported by people to new areas for specific purposes and have played important and beneficial roles in human history. Most modern agricultural crops were introduced into the regions they now inhabit. For instance, corn (maize) is thought to have originated in Mexico some 7 000 years ago. Today it is found worldwide. Wheat probably originated in the Middle East. Currently, wheat is grown on more land area worldwide than any other crop and is a close third to rice and corn in total world production.

Many modern domesticated animals were also new species introductions at some point in their history. Modern domestic cattle evolved from a single early ancestor, the auroch. Cattle were domesticated between 10 000 and 15 000 years ago near the boundary of Europe and Asia or Southwest Asia. Cattle are now widely distributed throughout the world. The total world cattle population in the late

Case Study: Lake Maracaibo, Venezuela

Lake Maracaibo in northwestern Venezuela is the largest natural lake in South America at 13 330 km^2 (5 146 square miles). At its widest point, it is more than 125 km (78 miles) wide. The lake itself lies in the Maracaibo basin, which is semi-arid in the north, but averages over 1 200 mm (47 in) of annual rainfall in the south. It has been suffering from a serious problem of invasive duckweed, a tiny aquatic plant that grows in freshwater. This first image (left), taken by the Aqua MODIS satellite on 17 December 2003, shows the lake during the winter months, when duckweed is absent from the lake's waters, and the silvery sunglint is absent. In summer the weed blooms. The true-colour image from 26 June 2004 (middle) shows strands of duckweed curling through the lake, floating at the surface, or slightly submerged in the brackish water. A closer look in August 2004 (right) reveals the stranglehold the duckweed has on port areas, especially along the important oil shipping routes in the neck of Lake Maracaibo. Fish and the fishing industry suffer as thick green mats block photosynthesis and alter fish habitats. The weed also adheres to boats, affects cooling systems, and obstructs travel. In September 2004, Venezuela's Ministry of Environment and Natural Resources reported that it had reduced the duckweed area by 75 per cent, using duckweed harvesting machines from the United States. The ministry is investigating using the harvested weed as animal fodder.

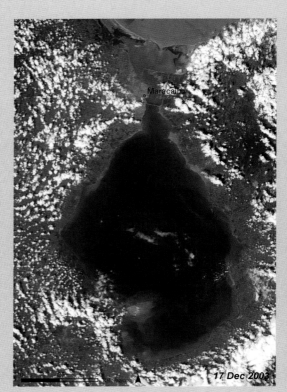

17 Dec 2003

26 Jun 2004

9 Aug 2004

Credit: (Left, right images) NASA; (middle) LPDAAC – USGS National Center for EROS

Table 2.11 – Growth of protected areas of the world in 1994 and 2004 (in per cent)

Ratio	1994	2004
World	7.8	9.5
Developed Countries	11.3	14.1
Commonwealth Independent States (CIS)	2.8	3.0
CIS-ASIA	3.6	3.8
CIS-Europe	2.6	2.8
Developing	7.6	9.1
Northern Africa	3.5	3.9
Sub-Saharan Africa	8.1	8.3
Latin America & the Caribbean	8.0	9.9
Eastern Asia	8.2	14.2
Southern Asia	4.5	5.1
South-eastern Asia	4.8	5.9
Western Asia	21.4	22.0
Oceania	1.0	1.1
Least Developed Countries (LDCs)	7.7	7.9
Landlocked Developing Countries (LLDCs)	8.4	9.6
Small Island Developing States (SIDs)	1.6	2.8

There is considerable variation in the total area protected between regions, ranging from 1.1 per cent in the developing countries of Oceania to 22.0 per cent in Western Asia. The percentage coverage in both Western and Eastern Asia (14.2 per cent) exceeds the coverage of all developed countries (14.1 per cent) representing a significant commitment by these regions to conservation. However, the constraints imposed upon the data by the criterion for a date of establishment within the Millennium Development Goals (MDG) reporting period suggest that all figures should be treated cautiously.

Source: UNEP-WCMC 2005

1980s was estimated to be nearly 1.3 billion. Chickens—the world's most abundant domesticated bird—are generally believed to have descended from jungle fowl in Southeast Asia. They were subsequently introduced into almost every country and region of the world.

In stark contrast to such positive species introductions are those where introduced exotic species have become invasive. Scientifically speaking, invasive species are those organisms that are unwanted and have a tendency to spread. Invasive species harm, or have the potential to harm, a given ecosystem or peoples' health or economic well-being (Clinton 1999). Historically, some invasive exotic species have been intentionally introduced into new settings; the introduction of the common starling into the United States and the rabbit into Australia are two classic examples. Introductions of other invasive species have occurred by accident, such as that of zebra mussels into the American Great Lakes as a result of shipping activities.

Native or indigenous species are those that occur naturally in an area or habitat. Invasive species often out-compete and displace native species because the invaders have no natural enemies and can spread easily and quickly. Both managed and natural ecosystems throughout the world are under siege from increasing numbers of harmful invasive species. These include disease organisms, agricultural weeds, and destructive insects and small mammals that threaten economic productivity, ecological stability, and biodiversity. On a local scale, such invasions decrease diversity of native flora and fauna. Globally, they contribute to making the biosphere more homogeneous and less resilient.

Natural biodiversity helps to maintain ecological resilience in the face of varying environmental conditions (Holling et al. 1995). Invasion by exotic species lessens ecological resilience and can transform ecosystems in unpredictable ways that may have negative consequences for people. This problem is growing in severity and geographic extent as global trade and international travel expand, as markets are liberalized and deregulated, as ecosystems are further altered and fragmented, and as global climate continues to change (Brandt 2003; Dalmazzone 2000).

Invasions by alien species are set to worsen in the next few decades if the world continues to warm as most scientists predict it will. Longer growing seasons spawned by global warming may give invasive weedy plants time to flower and set seeds where previously they could only spread asexually. This new-found ability could allow the weeds to adapt to new environments more quickly, and better resist attack by insects. Higher levels of carbon dioxide in the atmosphere may also favor plants that can utilize extra carbon dioxide and grow faster. One such example is cheatgrass, an introduced species that now dominates vast areas of the American West (Holmes 1998). In other parts of the world, invasive exotic plant species make up 4 to 44 per cent of the total number of species in ecosystems (Lövei 1997).

Invasive exotic species are one of the most significant drivers of environmental change worldwide. They also contribute to social instability and economic hardship, and place constraints on sustainable development, economic growth, and environmental conservation. Worldwide, the annual economic impact of invasive species on agriculture, biodiversity, fisheries, forests, and industry is enormous. The World Conservation Union (IUCN) estimates that the global economic costs of invasive exotic species are about US$400 billion annually (UNEP 2002a). Alien invaders cost 140 billion dollars a year in the USA alone (McGrath 2005). Less easily measured costs also include unemployment, impacts on infrastructure, shortages of food and water, environmental degradation, increases in the rate and severity of natural disasters, and human illness and death. Invasive exotic species represent a growing problem, and one that is here to stay—at least for the foreseeable future (Brandt 2003).

Protected and Wilderness Areas

Wilderness areas are those areas of land that are relatively untouched by human activities. To qualify as wilderness, an area must have 70 per cent or more of its original vegetation intact, cover at least 10 000 km^2 (3 861 square miles), and be inhabited by fewer than five people per km^2 (12 people per square mile).

Wilderness areas are major storehouses of biodiversity. They also provide critical ecosystem services to the planet, including watershed maintenance, pollination, and carbon sequestration. Wilderness areas currently cover nearly half the Earth's terrestrial surface (Mittermeier et al. 2003). While that represents a significant amount of land area, however, most wilderness areas are not protected, and are therefore at risk.

World Environmental Hotspots as identified by Conservation International.

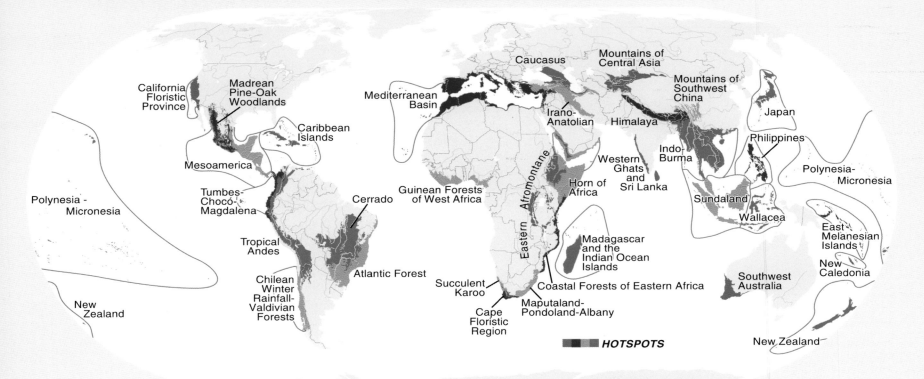

Source: www.conservation.org

One of the most effective means for conserving wilderness is through the development of protected areas. A protected area is an area of land (or water) especially dedicated to the protection and maintenance of biological diversity, along with natural and associated cultural resources, that is managed through legal or other effective means (IUCN 1994). Protected areas are managed for a wide variety of purposes, including:

- scientific research;
- wilderness protection and preservation of species and ecosystems;
- maintenance of environmental services;
- protection of specific natural and cultural features;
- tourism and recreation;
- education;
- sustainable use of resources from natural ecosystems; and
- maintenance of cultural and traditional attributes (Green and Paine 1997).

The 2003 United Nations List of Protected Areas—compiled by UNEP and the IUCN and released during the Fifth World Parks Congress in Durban, South Africa—reveals that there are now 102 102 protected areas, together representing a total land area roughly equivalent to China and Canada combined, or more than 12 per cent of the Earth's surface (Chape et al. 2003). That total exceeds the ten per cent called for in the Caracas Action Plan formulated at the Fourth World Parks Congress held in 1992 in Caracas, Venezuela. Between 10 and 30 per cent of some of the planet's vital natural features, such as Amazonian rain forests and tropical savannah grasslands, are classified as protected areas. However, ecoregional and habitat representation remains uneven.

Currently, almost half of the world's protected areas are found in regions where agriculture and logging are primary land-use strategies. All indications are that food

and timber production will need to increase in coming decades to keep up with population growth and increasing demand for wood and wood products. Thus, establishing additional protected areas, while helping to preserve biological diversity, will take land out of production and put more stress on lands elsewhere (Sohngen et al. 1999). Balancing the need to protect wild species and conserve habitat while at the same time increasing agricultural and timber production represents a tremendous challenge (McNeely and Scherr 2001).

Of the world's protected areas, the vast majority—91.3 per cent—are found in terrestrial ecosystems. Fewer than ten per cent of the world's lakes and less than 0.5 per cent of the world's seas and oceans lie within protected areas (SBSTTA 2003). Recognizing that the world's marine environment remains largely unprotected, the Fifth World Parks Congress put forth the Durban Action Plan. The Plan calls for the establishment of at least 20-30 per cent marine protected areas worldwide by 2012. The Plan also calls for the conservation of all globally threatened or endangered species by 2010.

In the coming years, further development of global networks of protected areas will need to focus on four areas (Green and Paine 1997):

- consolidating existing networks by addressing major gaps;
- physically linking protected areas to one another so they function more effectively as networks;

Credit: Harriett O'Mahony/UNEP/Topfoto

- expanding networks by forming or strengthening links with other sectors, notably the private sector;
- and improving the effectiveness with which protected areas are managed.

Protected areas are often considered a kind of sacrifice, a financial burden rather than an asset. Yet establishing, maintaining, and expanding protected areas is a fundamental approach to safe-guarding the environment and conserving biological diversity.

Protected areas are also important in other respects, such as helping to maintain freshwater resources. Protected areas may also hold the cures to some of the world's most devastating diseases in the form of unique chemical compounds and as-yet-undiscovered genetic material.

A recent analysis published by Conservation International identified nine additional sites as areas of extraordinarily high biological diversity, popularly known as hotspots, taking the count of hotspots to 34. These 34 regions worldwide are where 75 per cent of the planet's most threatened mammals, birds, and amphibians survive. In these 34 hotspots, estimated 50 per cent of all vascular plants and 42 per cent of terrestrial vertebrates exist. Therefore it is critical to protect and preserve these areas (Conservation International 2005).

Credit: Christian Lambrechts

Case Study: Kameng and Sonitpur Elephant Reserves, Arunachal Pradesh, India

S. P. S. Kushwaha and Rabul Hazarika

The Kameng and Sonitpur Elephant Reserves in northeastern India are comprised of transborder subtropical evergreen to tropical moist deciduous forests of Arunachal Pradesh and Assam. The reserves are facing deforestation and habitat loss in recent years. This study attempts to investigate the loss of habitat in these reserves using temporal satellite imagery of periods 1994, 1999 and 2002. The on-screen visual interpretation of the three-period imagery revealed alarming and continuous habitat loss from 1994 to 2002. The overall habitat loss was found to be 344 km^2 (133 square miles) between 1994 and 2002. The average annual rate of deforestation worked out to be 1.38 per cent, which is much bigger than the national average. The rate of deforestation was highest between 1999 and 2002. The study indicated that at this rate much of the forests in the study area would be depleted within the next few years. It also showed that moist deciduous forests, which possess the highest biodiversity in Assam, are facing maximum deforestation. High deforestation has resulted in high man-elephant conflicts.

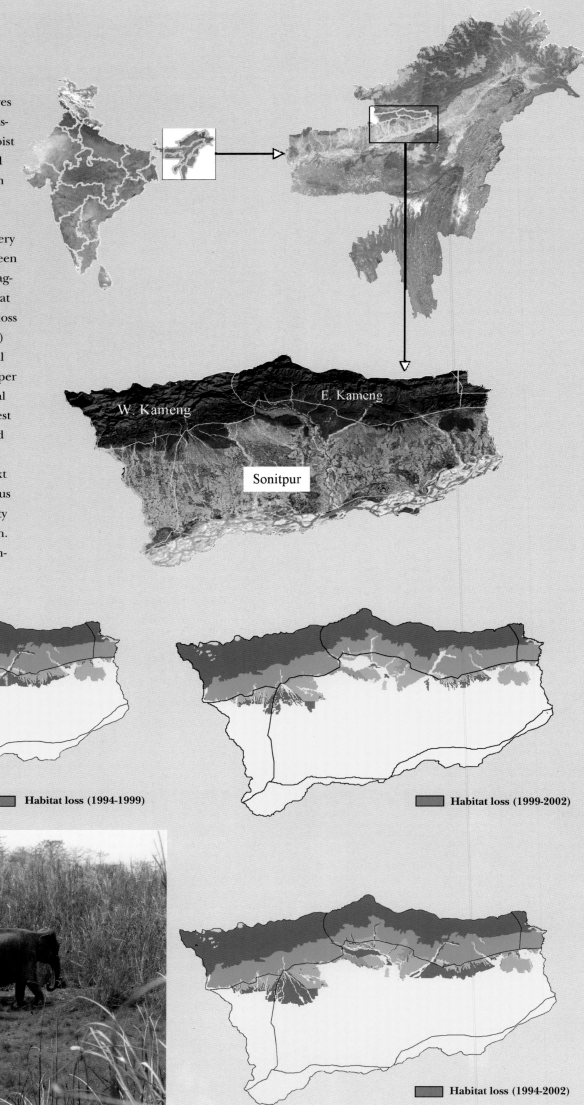

W. Kameng

E. Kameng

Sonitpur

Habitat loss (1994-1999)

Habitat loss (1999-2002)

Habitat loss (1994-2002)

Credit: Dr. Bibhab Talukdar/UNEP

2.6 Energy Consumption and Resource Extraction

Energy is measured by its capacity to do work (potential energy) or the conversion of this capacity to motion (kinetic energy). Most of the world's convertible energy comes from fossil fuels that are burned to produce heat that is then converted to mechanical energy or other means in order to accomplish tasks (EIA 2004a).

Energy is essential for the fulfillment of many basic human needs, such as generating electricity, heating and cooling living spaces, cooking food, forging steel, and powering engines for many forms of transportation (Harrison and Pearce 2001). Energy use is closely tied to human health and well-being. Worldwide, roughly 2 000 million people do not have access to electricity. Countries in which energy use is low tend to have high infant mortality rates, low literacy rates, and low life expectancies. It is through the utilization of convertible energy sources that the modern world has transcended its agrarian roots and fostered the energy-driven societies that characterize it today. Generating the power to sustain these societies has entailed extracting massive amounts of natural resources from the planet. Extraction is the process of obtain-

Credit: Darren Defner/UNEP/Topfoto

Nightlight Map of the World

Credit: NOAA/UNEP/NASA

Global Natural Resources

Affluent countries consume vast quantities of global natural resources, but contribute proportionately less to the extraction of many raw materials. This imbalance is due, in part, to domestic attitudes and policies intended to protect the environment. Ironically, developed nations are often better equipped to extract resources in an environmentally prudent manner than the major suppliers. Thus, although citizens of affluent countries may imagine that preservationist domestic policies are conserving resources and protecting nature, heavy consumption rates necessitate resource extraction elsewhere and oftentimes with weak environmental oversight. A major consequence of this "illusion of natural resource preservation" is greater global environmental degradation than would arise if consumption were reduced and a larger portion of production was shared by affluent countries. Clearly, environmental policy needs to consider the global distribution and consequences of natural resource extraction (Berlik et al. 2002).

Credit: Jon P Bonetti /UNEP/Topfoto

ing a useful substance from a raw material (NCR&LB 2003). Such raw materials may include fossil fuels, metals, minerals, water, and biomass, including animals and raw materials from plants and crops (EEA 2001).

Resources can be divided into those that are renewable, such as plant and animal material, and those that are not, including coal, oil, and minerals. The Earth has a finite supply of non-renewable resources. Even renewable resources, however, are exhaustible if they are used faster than they can be replenished.

Total world energy consumption has risen almost 70 per cent since 1971 (WRI 1998). It is expected to increase by 58 per

cent between 2001 and 2025, from 404 quadrillion British thermal units (Btu) in 2001 to 640 quadrillion Btu in 2025 (EIA 2003; EIA 2004b). While a slow, steady increase in energy consumption is expected in industrialized nations, where most energy use currently takes place, a meteoric increase in consumption is anticipated in the developing world during that period (Tilford 2000).

The tempo at which energy resources are being used to fuel modern societies is rapidly depleting supplies of non-renewable resources and far exceeding the rate at which renewable resources can be naturally renewed (Ernst 2002). In many least–developed countries, for instance,

burning biomass in the form of wood (largely a non-renewable resource) generates 70 to 90 per cent of the energy needed and disregard for the fate of non-renewable resources is prevalent. In the United States, for example, 72 per cent of the country's electricity is generated using non-renewable resources. Only about 10 per cent comes from renewable resources, with nuclear power providing the rest.

Currently, 85 per cent of world energy consumption comes from the burning of fossil fuels—oil, coal, and natural gas (BP 2003). Although no immediate shortages of these non-renewable energy resources exist, supplies are finite and will not last forever. What took millions of years to

Case Study: Blackout in United States and Canada 14 August 2003

On 14 August 2003, parts of the northeastern United States and southeastern Canada experienced widespread power blackouts. Among the major urban agglomerations affected by the electrical power outage were the cities of New York City, Albany, and Buffalo in New York, Cleveland and Columbus in Ohio, Detroit in Michigan, and Ottawa and Toronto in Ontario, Canada. Other U.S. states, including New Jersey, Vermont, Pennsylvania, Connecticut, and Massachusetts, were also affected. The blackout resulted in the shutting down of nuclear power plants in New York and Ohio, and air traffic was slowed as flights into affected airports were halted. Approximately 50 million people were affected by the outage. The change in the nighttime city lights is apparent in this pair of Defense Meteorological Satellite Program (DMSP) satellite images. The top image was acquired on 14 August, about 20 hours before the blackout, and the bottom image shows the same area on 15 August, roughly

seven hours after the blackout. In the bottom scene, notice how the lights in Detroit, Cleveland, Columbus, Toronto, and Ottawa are either missing or visibly reduced. Previous major blackouts include the 9 November 1965 outage caused by a faulty relay at a power plant in Ontario, which affected a large swath of land stretching from Toronto to New York. Another one followed on 14 July 1977, the result of a lightning strike, affecting New York City. The power supply in nine western states was also affected in August 1996 as a result of a high demand for electricity, a heat wave, and sagging electrical power lines.

Source: NASA 2002, http://earthobservatory.nasa. gov/NaturalHazards/natural_hazards_v2.php3?img_ id=11628; GlobalSecurity.org 2003, http://www. globalsecurity.org/eye/blackout_2003.htm

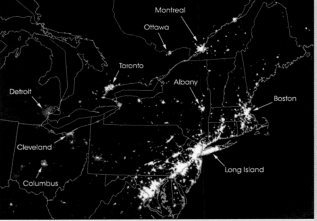

14 August 2003 before power failure

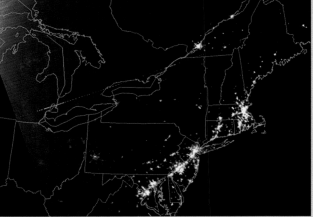

15 August 2003 after power failure

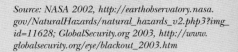

Credit Image courtesy Chris Elvidge, U.S. Air Force http://earthobservatory. nasa.gov/NaturalHazards/Archive/Aug2003/NE_US_OLS2003227_lrg.jpg

Case Study: SWERA, the Solar and Wind Energy Resource Assessment

The Solar and Wind Energy Resource Assessment (SWERA)—co-financed by the Global Environment Facility (GEF)—is a project to assist 13 developing countries identify optimal locations for potential solar and wind energy production. SWERA assists by creating a global archive of information gathered through a network of international and national agencies. These agencies collect and analyse data on solar and wind energy resources, energy demand, and electrification. Using inputs derived from satellite and surface observations, SWERA partners model wind and solar energy resource potential and produce maps of wind power density and monthly average and daily total solar radiation of a given area. This information can then be used to facilitate investments and create policies in the participating countries for developing solar and wind energy.

Source: UNEP/GRID–Sioux Falls

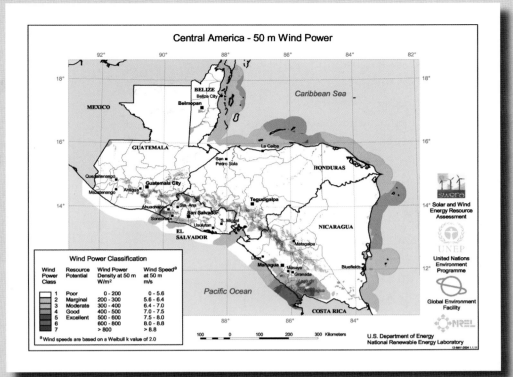

Credit: NREL

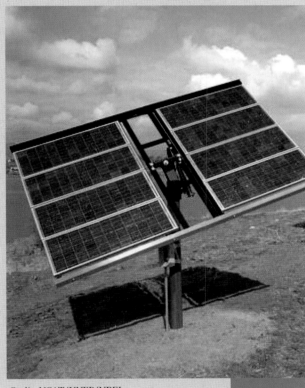

Credit: NCAT/UNEP/NREL

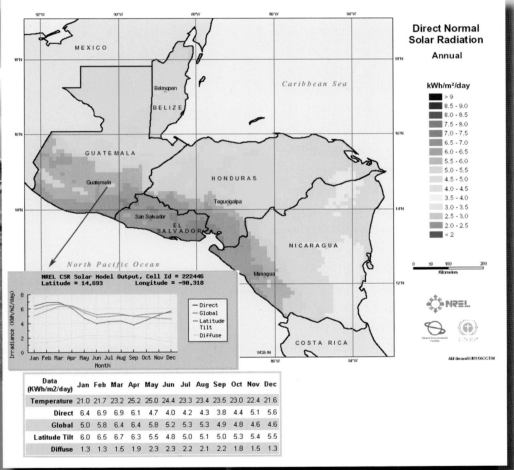

Data (KWh/m2/day)	Jan	Feb	Mar	Apr	May	Jun	Jul	Aug	Sep	Oct	Nov	Dec
Temperature	21.0	21.7	23.2	25.2	25.0	24.4	23.3	23.4	23.5	23.0	22.4	21.6
Direct	6.4	6.9	6.9	6.1	4.7	4.0	4.2	4.3	3.8	4.4	5.1	5.6
Global	5.0	5.8	6.4	6.4	5.8	5.2	5.3	4.9	4.8	4.6	4.6	4.6
Latitude Tilt	6.0	6.5	6.7	6.3	5.5	4.8	5.0	5.1	5.0	5.3	5.4	5.5
Diffuse	1.3	1.3	1.5	1.9	2.3	2.3	2.2	2.1	2.2	1.8	1.5	1.3

Credit: Energy Northwest/UNEP/NREL

produce will be consumed in the time frame of a century or two (Hawken 1994; Tilford 2000).

This rapid, large-scale consumption of a fuel source that took millennia to produce has generated unforeseen complications, several with global implications. Burning fossil fuels produces atmospheric pollutants such as oxides of sulfur and nitrogen and unburned hydrocarbons. Fossil fuel combustion also adds one of the most prevalent greenhouse gases—carbon dioxide—to the atmosphere. As a result, world energy use has emerged at the center of the climate change debate. World carbon dioxide emissions are projected to rise from 23 900 million metric tonnes in 2001 to 27 700 million metric tonnes in 2010 and 37 100 million metric tonnes in 2025 (EIA 2004b). The Earth's atmosphere and biosphere will not remain unchanged by the combustion of such enormous amounts of these fuels. The relatively sudden release of massive amounts of carbon has the potential not only to disrupt the Earth's heat balance and climate, but other parts of the global carbon cycle as well, and in unpredictable ways.

The Earth cannot sustain existing levels of resource consumption. Furthermore, resource extraction methods are often environmentally destructive. The impact of environmental degradation hits those who are poorest the hardest. Many of the world's energy sources and other natural resources come from developing countries. Extracting and harvesting these resources can result in air, soil, and water pollution. It also generates waste; the amount of waste associated with extracting minerals, for instance, can be enormous. Disposing of wastes in environmentally friendly ways is a daunting, if not impossible, task in many developing nations.

A major challenge for the 21st century is to develop methods of generating and using energy that meet the needs of the population while protecting the planet (Harrison and Pearce 2001). Yet most of the world is still without energy policies that direct or restrict consumption patterns.

Only through conservation and resource recovery strategies can we hope to reach a sustainable balance between available resources and their consumption. The utiliza-

Sources of Energy

Coal

Coal is the world's largest source of fuel for electricity production. The byproducts of coal combustion are also a major source of environmental damage.

Credit: Unknown/UNEP/Freefoto.com

Oil

Although used primarily in the production of transportation fuels, oil is also used for generating electricity, for heating, and in the production of chemical compounds and synthetic materials.

Credit: Unknown/UNEP/Freefoto.com

Natural Gas

Compared to coal and oil, natural gas is a relatively clean-burning fossil fuel. It is used primarily for heating and for powering many industrial processes. Increasingly, natural gas is burned to drive turbines used in the production of electricity.

Credit: Unknown/UNEP/Freefoto.com

Biomass

Plant and animal material, or biomass, is a rich source of carbon compounds. When burned to release energy, biomass does not add additional carbon to the natural carbon cycle as do fossil fuels. Fast-growing plants, such as switchgrass, willow, and poplar can be harvested and used as "energy crops." Biomass wastes, including forest residues, lumber and paper mill waste, crop wastes, garbage, and landfill and sewage gas, can be used for heating, as transportation fuels, and to produce electricity, while at the same time reducing environmental burdens. According to the World Bank, 50 to 60 per cent of the energy used in developing countries in Asia, and 70 to 90 per cent used in developing countries in Africa, comes from wood or other biomass; half the world cooks with wood.

Credit: Volker Quaschning/UNEP

Sources:
www.freefoto.com, http://www.topfoto.co.uk/, Prof. Dr.-Ing. habil. Volker Quaschning http://www.volker-quaschning.de

Credit: Unknown/UNEP/Freefoto.com

Nuclear

Nuclear power harnesses the heat of radioactive materials to produce steam for electricity generation. The use of nuclear power is expected to decline as aging plants are taken out of operation.

Credit: Volker Quaschning/UNEP

Hydroelectric

Hydroelectric power uses the force of moving water to produce electricity. A large part of the world's electricity is produced in hydroelectric plants. Many of these plants, however, are associated with large dams that disrupt habitats and displace people. Smaller "run of the river" hydroelectric plants have less environmental impact.

Credit: Lupidi/UNEP/Topfoto

Geothermal

Geothermal energy is energy contained in intense heat that continually flows outward from deep within the Earth. Geothermal energy is typically used to heat water, which is then used to heat buildings directly or to drive turbines to produce electricity.

Credit: Anthony Karbowski/UNEP/Topfoto

Solar

Solar energy—power from the sun—is readily available and inexhaustible. Humans have used sunlight for heating and drying for thousands of years. Converting the power of sunlight into usable energy forms, such as electricity, is not without cost, but the sunlight itself is free. Solar cells, or photovoltaics, are devices used to transform sunlight into electric current.

Credit: Sanjay Singh/UNEP/Topfoto

Wind

Wind power is an ancient energy source that has moved into the modern era. Aerodynamically designed wind turbines can produce electricity more cheaply than coal-burning power plants.

tion of non-renewable resources is theoretically not sustainable. But if used wisely, some non-renewable resources can be conserved and recycled for a very long time. To recycle means to make new products from old ones. Recycling materials such as paper, aluminum, and glass saves energy, reduces pollution, and conserves natural resources (EPA 2003). Every nation must assume responsibility for recycling its own wastes. Some industrialized Western countries "dispose" of their electronic wastes by shipping them to Asia—an increasingly common practice (FOEE 2004).

To insure that renewable resources can be replenished at a sustainable rate, people must switch to more environmentally friendly energy sources and employ new technologies that can help make such sustainability a reality (Ernst 2002). Indeed, new technologies—coupled with effective and efficient use of existing technologies—are essential to increasing the capabilities of countries to achieve sustainable development on many fronts, as well as sustaining the world's economy, protecting the environment, and alleviating poverty and human suffering (Hay and Noonan 2000).

Most changes in the Earth, including changes brought about by increasing energy consumption, can be observed through such tools as remote sensing. Remote sensing is the collection of information about an object without being in physical contact with the object. Aircraft and satellites are the common platforms from which remote sensing observations are made. Satellite imagery, a crucial component of this publication, is especially useful for studying changes in our Earth's environment. Most satellite imagery is collected using multispectral scanners, which record light intensities in different wavelengths in the spectrum from infrared through visible light through ultraviolet light. Satellite imagery is useful because of its stable nature (same resolution, same time, same data characteristics). Aided by the global positioning system (GPS), these satellites know their orbital position precisely. Thus, satellite imagery is ideally suited for applications requiring large-area coverage, such as agricultural monitoring, regional mapping, environmental assessment, and infrastructure planning (Krouse et al. 2000).

GERMANY

Annaberg-
Buchholz

48°

Most

Jirkov

Chomutov

CZECH

REPUBLIC

Kadan

Žatec

0   5   10
Kilometres

N

29 Apr 1975

Black Triangle

MINING
THE BLACK TRIANGLE, EUROPE

The so-called Black Triangle is an area bordered by Germany, Poland, and the Czech Republic and is the site of extensive surface coal mining operations. In the 1975 satellite image above, the gray areas are surface mines located primarily in the Czech Republic. Air-borne pollutants from coal extraction activities tended to become trapped by the mountainous terrain to the northeast and were concentrated in the area around the mines.

GERMANY

Annaberg-Buchholz

Most

Jirkov

Chomutov

C Z E C H

R E P U B L I C

Kadan

Žatec

0   5   10
Kilometres

N

3 May 2000

eventually causing severe deforestation along the border between the Czech Republic and Germany. In the 2000 image, this deforestation is very obvious, appearing as large brownish patches. Interestingly, the 2000 image also reveals somewhat improved vegetation cover—a slight "greening" of the landscape—as compared to conditions in 1975. Some of this improvement may be attributable to actions taken by the three countries bordering the

Black Triangle to reduce pollutants produced by the mining operations. The implementation of anti-pollution technologies, including circulating fluid-ized-bed boilers, clean coal technology, and nitrous oxide emission burners, appears to have reversed some, albeit not all, of the environmental damage experienced by the region as a result of the mines.

Alba

R O M A N I A

Mediaş

Tirnava Mare

Copşa Micã

Sibiu

In one year up to 67 000 tonnes of sulfur dioxide, 500 tonnes of lead, 400 tonnes of zinc and 4 tonnes of cadmium can be released by the city's two active smelters. The affected area is huge: in excess of 180 000 hectares (445 000 acres) of land are affected by air pollution and 150 000 hectares (371 000 acres) of agricultural land are untenable. 31 000 hectares (77 000 acres) of forest are also unacceptably polluted.

Credit: Lorant Czaran/UNEP

18 Sep 1986

0   3   6
Kilometres

N

MINING
Copşa Micã, Romania

Copşa Micã is a large industrial city located in the very center of Romania and is classified as an "environmental disaster area." The environmentally damaged area covers hundreds of square kilometres of land. The main industries in Copşa Micã are non-ferrous metalworking and chemical processing plants, and their effect on the environment has been devastating. Air pollution by heavy metals is 600 times

•Copsa Mica

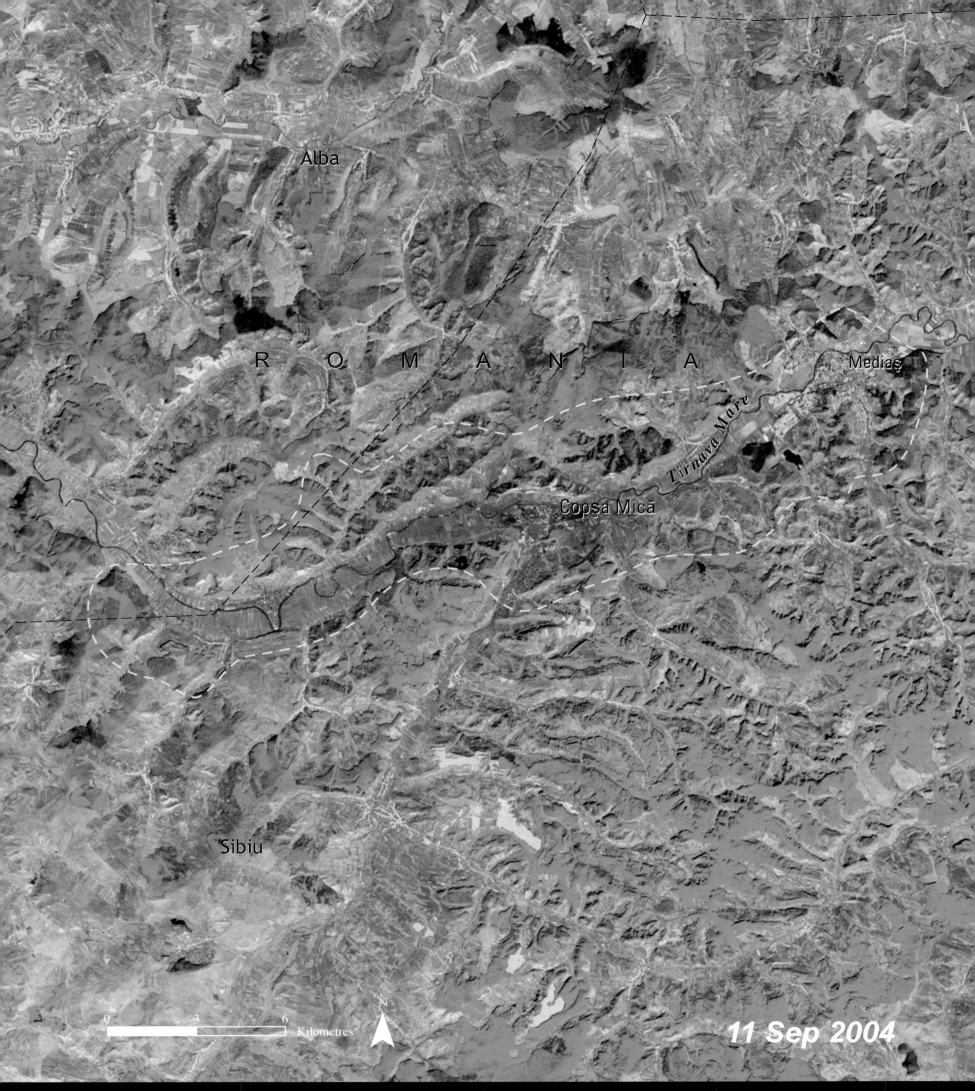

Alba

R   O   M   A   N   I   A

Mediaș

Tîrnava Mare

Copșa Mică

Sibiu

0   3   6
|———————|———————| Kilometres

N

11 Sep 2004

the allowed levels. To make matters worse, a lead-smelting facility emitted fumes containing sulfur dioxide, lead, cadmium, and zinc on the town and surrounding area for 50 km² (19 square miles). The entire town and much of the surrounding area were covered with a blanket of black soot daily until the facilities were forced to close in 1993.

In 1989 Copșa Mică was exposed as one of the most polluted places in Europe. It has the highest infant mortality rate in Europe, 30.2 per cent of children suffer reduced "lung function" and 10 per cent of the total population of 20 000 suffer "neurobehavioral problems." The soil and the local food chain probably will remain contaminated for at least another three decades.

C H I L E

0   1 2 3   6   Kilometres

N

27 Oct 1989

C H I L E

12 Dec 2003

0   3   6 Kilometres

N

impoundment, which appears on the 1989 image as a white patch in the lower left corner. Impoundments of this type help reduce water consumption and enhance water conservation, two areas where mining activities typically fall short. The Escondida Mine also minimizes the impact of its operation on the environment by means of a 170-km-long (106 miles) underground pipeline that carries copper concentrate slurry from the mine to the port of Coloso. This underground scheme is efficient and ecologically sound, as the copper travels downhill without disrupting the environment. The 2003 image shows how the Escondida Mine has grown and expanded while at the same time continues to minimize negative impacts from its mining operations on the environment.

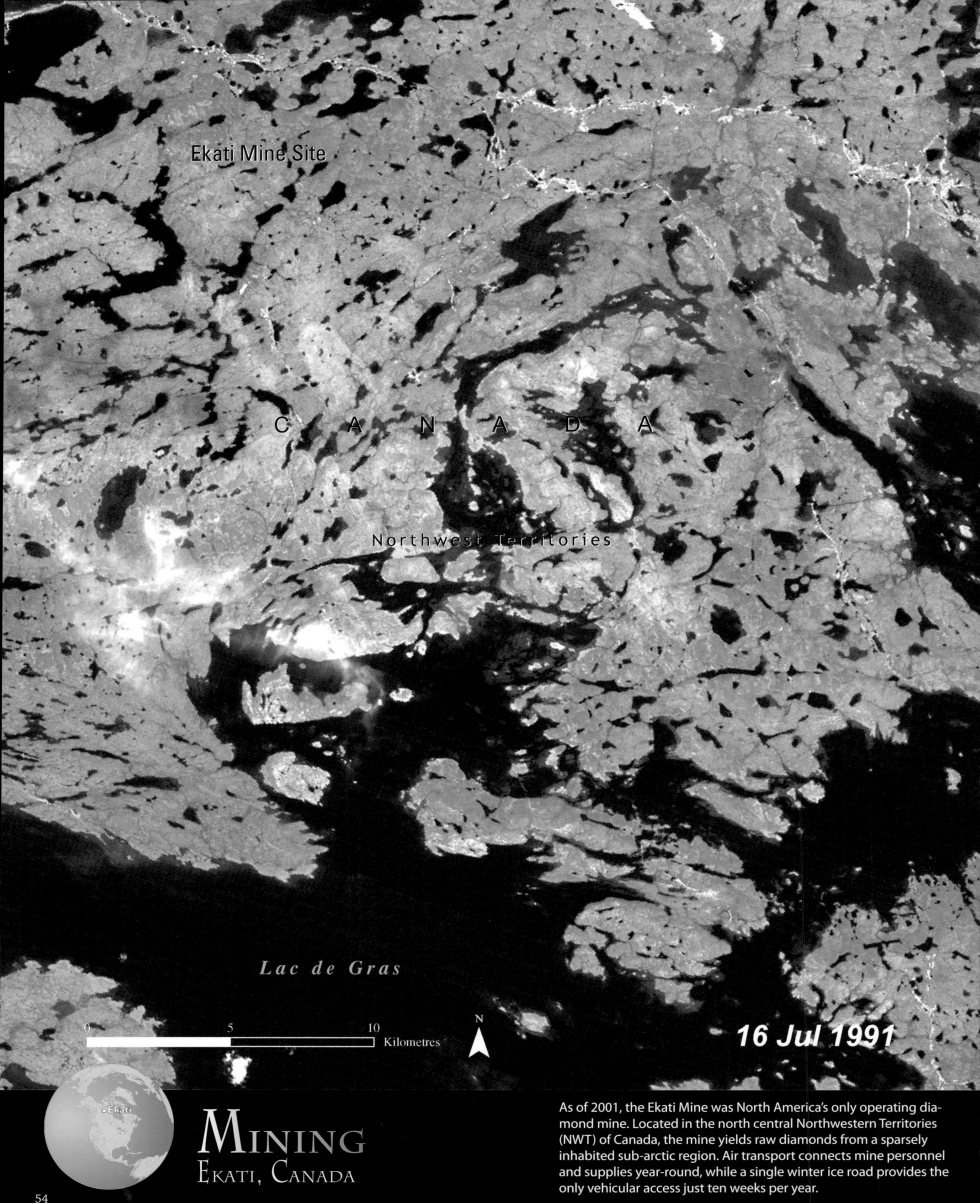

Ekati Mine Site

C A N A D A

Northwest Territories

Lac de Gras

0   5   10
Kilometres

N

16 Jul 1991

• Ekati

MINING
EKATI, CANADA

As of 2001, the Ekati Mine was North America's only operating diamond mine. Located in the north central Northwestern Territories
(NWT) of Canada, the mine yields raw diamonds from a sparsely
inhabited sub-arctic region. Air transport connects mine personnel
and supplies year-round, while a single winter ice road provides the
only vehicular access just ten weeks per year.

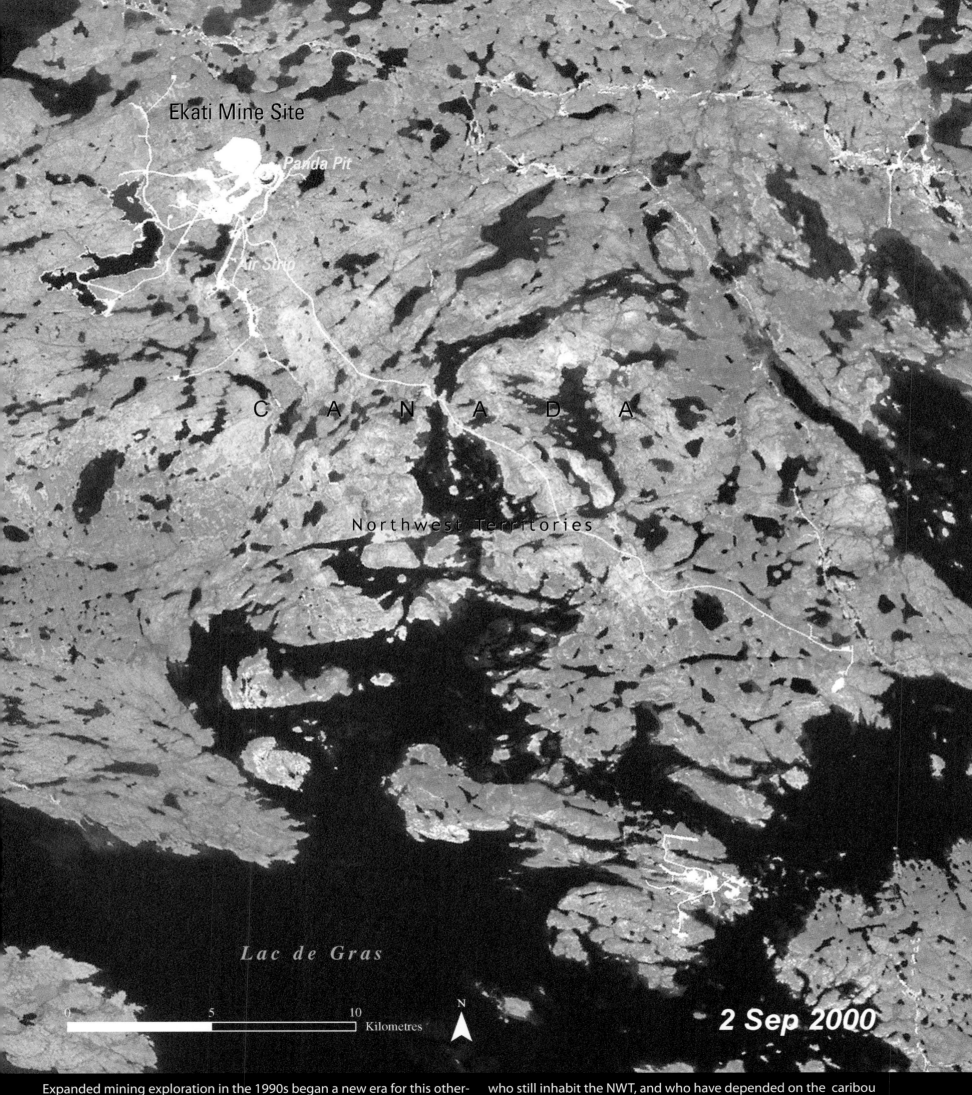

Ekati Mine Site

Panda Pit

Air Strip

C A N A D A

Northwest Territories

Lac de Gras

0   5   10
                          Kilometres

N

2 Sep 2000

Expanded mining exploration in the 1990s began a new era for this otherwise undeveloped region. Wildlife officials have collared and tracked caribou,
in a herd ranging from 350 000 to half a million, to monitor their movement
and behavior in proximity to the mines. Historical information about the
herds comes from Dogrib and Inuit knowledge obtained from elder natives,

who still inhabit the NWT, and who have depended on the caribou
for centuries.

These two images compare the same area, pre-mining and after mine operations have commenced. The white patch in the northwest portion of the
2000 image represents the mine and the associated infrastructure.

PAPUA NEW GUINEA

Ok Mani

Tabubil

Ok Tedi

0   2   4 Kilometres

N

5 Jun 1990

PAPUA
NEW GUINEA

Ok Mani

Tabubil

Ok Tedi

26 May 2004

0   2   4   Kilometres

N

pair of satellite images reveals the tremendous environmental impact the mine has had in 20 years. The uncontrolled discharge of 70 million tonnes of waste rock and mine tailings annually has spread more than 1 000 km (621 miles) down the Ok Tedi and Fly rivers, raising river beds and causing flood-

biodiversity. In the 1990 image, both the mine and the township of Tabu-bil—developed east of the river in support of the mine—are clearly visible. Lighter patches of green show disturbance of the original forest cover from subsistence agriculture, road clearing, and other infrastructure development.

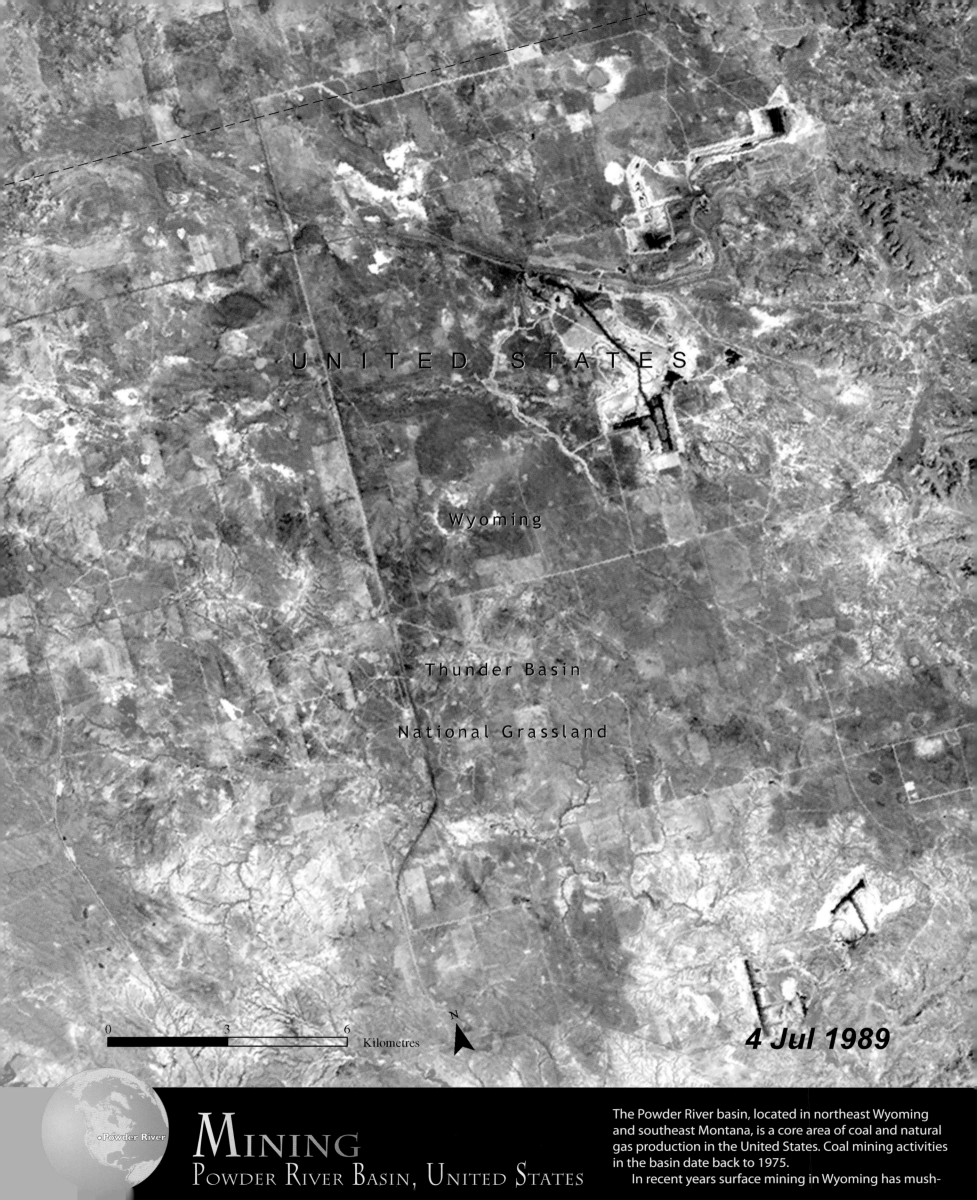

UNITED STATES

Wyoming

Thunder Basin

National Grassland

0   3   6   Kilometres

N

4 Jul 1989

○Powder River

MINING
POWDER RIVER BASIN, UNITED STATES

The Powder River basin, located in northeast Wyoming and southeast Montana, is a core area of coal and natural gas production in the United States. Coal mining activities in the basin date back to 1975.

In recent years surface mining in Wyoming has mush-

UNITED STATES

Wyoming

Thunder Basin

National Grassland

0   3   6
Kilometres

N

29 Jul 2001

at almost 300 000 tonnes per year. Similarly, coal bed methane gas development in this region is unrivaled in the U.S.

The images show areas under coal exploitation. Notable is the remarkable increase in mining operations in the 2001 image. Reasons for the expansion

deposits, and the depletion of high-grade deposits in other major coal mining areas such as West Virginia and Kentucky.

The open cast coal mines appear as white-purplish while the reddish brown areas are bare ground. Vegetated areas are green.

AUSTRALIA

Queensland

Albatross
Bay

Weipa

26 Jul 1973

0   5   10
                      Kilometres

N

Weipa mine

MINING
WEIPA BAUXITE MINE, AUSTRALIA

Mining of bauxite (aluminum ore) began at Weipa, on the Cape York
Pennisula in Queensland, Australia, in 1963. The mine produces 8.5
metric tonnes of ore annually, making it one of the world's largest
open-cut bauxite mines.

Under current mining practices, vegetation is cleared and the
topsoil is removed and either stockpiled for later use or immediately

A U S T R A L I A

Queensland

Albatross
Bay

Weipa

The total lease covers an area of approximately 2 590 km² (1 000 square miles) of which 68 km² (26 square miles) have been mined. Approximately four km² (1.5 square miles) of the mined land is revegetated each year, and over 50 km² (19 square miles) of land has been revegetated to date.

0   5   10
                    Kilometres

N

14 May 2002

replaced on previous mined areas. After topsoil removal, the bauxite is removed, resulting in a lowering of the entire landscape to a depth equivalent to the thickness of the orebody, often several meters. If the topsoil can be returned to a mined-out area after only a short time, it still contains most of the original soil fungi, bacteria and micro fauna. In addition, the seeds from the original plant community are likely to be viable. On slopes, rigorous

soil conservation measures are implemented, and the area is then normally planted with suitable native species so that it gradually reverts to bushland. Some of the profits generated by the mining operation are being placed in a trust for cultural protection, development and long-term investments to compensate for the disruption of local Aboriginal inhabitants and their environment.

References

Africa Mountain Forum (n.d.). Definitions of terms used. Land Use Change and Intensification. Introduction. http://anmf.web1000.com/publications/LUCIdocuments%5CLUCIintroduction.html on 2 April 2004.

Anon. (2002). Desertification. Nature Gallery – Global Trends. Pacific Island Travel. http://www.pacificislandtravel.com/nature_gallery/desertification.html on 1 April 2004.

Ashford, L. (2004). World Population Highlights, BRIDGE, Population Reference Bureau. http://www.prb.org/Template.cfm?Section=PRB&template=/ContentManagement/ContentDisplay.cfm&ContentID=11267 23 September 2004.

Berka, C., McCallum, D., Wernick, B. (1995). Land use impacts on water quality: case studies in three watersheds. Presented at The Lower Fraser Basin in Transition: A Symposium and Workshop, May 4, 1995. http://www.ire.ubc.ca/ecoresearch/publica3.html on 1 May 2004.

Berlik, M. M., Kittredge, D. B., Foster, D. R. (2002). The illusion of preservation - a global environmental argument for the local production of natural resources. Harvard Forest Paper No. 26. Harvard University, Harvard Forest, Petersham, Massachusetts, USA, 23. http://harvardforest.fas.harvard.edu/publications/pdfs/illusion.pdf on 8 November 2004.

BP (2003). BP statistical review of world energy. London: British Petroleum, 44. http://www.bp.com/liveassets/bp_internet/globalbp/STAGING/global_assets/downloads/B/BP_statistical_review_of_world_energy_2003_print_version.pdf on 18 April 2004.

Brandt, K. (2003). Invasive alien species and the Global Invasive Species Programme. The Phycological Society of Southern Africa – Newsletter #54, December 2003: 7-10. http://www.botany.uwc.ac.za/pssa/articles/features/no54.htm on 22 March 2004.

Brehm, D. (2003). Nobelists sound alarm on global environmental trends. MIT News March 7, 2003. http://web.mit.edu/newsoffice/nr/2003/nobels.html on 22 March 2003.

Brown, L. R. (2001). Paving the Planet: Cars and Crops Competing for Land. Earth Policy Institute Alert 12: 14 February 2001. http://www.earth-policy.org/Alerts/Alert12_printable.htm on 2 May 2004.

Chape, S., Blyth, S., Fish, L., Fox, P., Spalding, M. (compilers) (2003). 2003 United Nations List of Protected Areas. IUCN, Gland, Switzerland and Cambridge, UK and UNEP-WCMC, Cambridge, UK. ix, 44. http://www.iucn.org/themes/wcpa/wpc2003/pdfs/unlistpa2003.pdf on 20 March 2004.

Chew, S. C. (2001). World Ecological Degradation: Accumulation, Urbanization, and Deforestation 3000 B.C. – A.D. 2000. Rowman & Littlefield. 216.

CIDA (2001). Canadian International Development Agency: Canada combats desertification. http://www.acdi-cida.gc.ca/cida_ind.nsf/vall/05633A876C55D09C852569A60080FA70?OpenDocument on 2 April 2004.

CIESIN (2000). Center for International Earth Science Information Network, Gridded Population of the World (GPW), Version 3.0 beta., Columbia University, Palisades, New York, USA. http://beta.sedac.ciesin.columbia.edu/gpw/global.jsp on 5 April 2005.

Clinton, W. J. (1999). Invasive species. Presidential Documents, Executive Order 13112 of 3 February 1999. Federal Register 64(25): 6183-6186, Monday, 8 February 1999. http://frwebgate.access.gpo.gov/cgi-bin/getdoc.cgi?dbname=1999_register&docid=fr08fe99-168.pdf on 21 March 2004.

Collins, M. (2000). Globalizing solutions. Biological diversity. http://www.ourplanet.com/imgversn/105/collins.html on 15 May 2004.

Conservation International (2005). Hotspots Revisited: Nine new biodiversity hotspots are added to list of high priority conservation targets by John Tidwell. www.conservation.org on 14 February 2005.

Costanza, R., d'Arge, R., de Groot, R., Farberk, S., Grasso, M., Hannon, B., Limburg, K., Naeem, S., O'Neill, R. V., Paruelo, J., Raskin, R. G., Suttonkk, P. van. den., Belt, M. (1997). The value of the world's ecosystem services and natural capital. Nature, 387:253-259. http://www.esd.ornl.gov/benefits_conference/nature_paper.pdf on 16 July 2004.

Dalmazzone, S. (2000). Chapter 2 - Economic factors affecting vulnerability to biological invasions. In: Perrings C., Williamson M. and Dalmazzone S. (eds.) The Economics of Biological Invasions. Cheltenham, Elgar, 17-30. http://www.soc.uoc.gr/calendar/2000EAERE/papers/PDF/D6-Dalmazzone.pdf on 21 March 2004.

De Leo, G. A. and Levin, S. (1997). The multifaceted aspects of ecosystem integrity. Conservation Ecology 1(1): 3. http://www.consecol.org/vol1/iss1/art3 on 21 March 2004.

Devitt, T. (2001). World land database charts troubling course. NEWS@UW-MADISON, University of Wisconsin, Madison, Wisconsin, USA. http://www.news.wisc.edu/story.php?get=6315 on 2 May 2004.

EEA (2001). Total material requirement of the European Union. Technical report No 55. European Environment Agency, Copenhagen, Denmark. http://reports.eea.eu.int/Technical_report_No_55/en/ on 8 November 2004.

EIA (2003). International Energy Outlook 2003. World Energy Consumption. U.S. Department of Energy, Energy Information Administration, Washington, DC, USA. http://www.eia.doe.gov/oiaf/archive/ieo03/world.html on 19 April 2004.

EIA (2004a). Energy glossary. U.S. Department of Energy, Energy Information Administration, Washington, DC, USA. http://www.eia.doe.gov/glossary/glossary_e.htm on 18 April 2004.

EIA (2004b) International Energy Outlook 2004. U.S. Department of Energy, Energy Information Administration, Washington, DC, USA. http://www.eia.doe.gov/oiaf/ieo/highlights.html on 19 April 2004.

EPA (2003). Glossary. U.S. Environmental Protection Agency, Office of Solid Waste, Terminology Reference System, Washington, DC, USA. http://oaspub.epa.gov/trs/trs_proc_qry.navigate_term?p_term_id=8880&p_term_cd=TERM on 9 November 2004.

Ernst, W. G. (2002). Global equity, sustained resource consumption through efficient extraction-conservation-recycling, and development of cheap, inexhaustible energy. The Geographic Society of America Annual Meeting, 27-30 October 2002, Denver, Colorado, USA. http://gsa.confex.com/gsa/2002AM/finalprogram/abstract_37598.htm on 8 November 2004.

ESA (2003). World Population Prospects: The 2002 Revision. Population Division of the Department of Economic and Social Affairs of the United Nations Secretariat. http://esa.un.org/unpp/index.asp?Panel=1 on 25 April 2004; http://esa.un.org/unpp/index.asp?panel=2 on 25 April 2004.

Eswaran, H., Reich, P., Beinroth, F. (1998). Global desertification tension zones. Proc. Of International Soil Conservation Organization Conference, Purdue University, Indiana, USA. http://www.nrcs.usda.gov/technical/worldsoils/papers/tensionzone-paper.html on 2 April 2002.

FAO (2000). Global Agro-Ecological Zones (Global-AEZ). http://www.fao.org/ag/agl/agll/gaez/index.htm on 8 August 2004.

FAO (2001). Forest Resources Assessment 2000. Main Report FAO Forestry Paper 140, Rome, Italy. http://www.fao.org/forestry/foris/webview/forestry2/index.jsp?siteId=101&langId=1 on 25 February 2004.

FAO (2002). Land resources information systems in the Near East. Regional Workshop Cairo, 3-7 September 2001. Food and Agriculture Organization of the United Nations, Rome, Italy. http://www.fao.org/DOCREP/005/Y4357E/y4357e20.htm on 1 May 2004; http://www.fao.org/desertification/default.asp?lang=en on 1 May 2004.

FOEE (2004). Reducing resource use. Friends of the Earth Europe's response to the European Commission communication towards a Thematic Strategy on the Sustainable Use of Natural Resources, 10. http://www.foeeurope.org/publications/reducing_resource_use_feb2004.pdf on 9 November 2004.

Forestry Images: Forest Health, Natural Resources & Silviculture Images. http://www.forestryimages.org on 24 November 2004.

Forney, W., Richards, L., Adams, K. D., Minor, T. B., Rowe, T. G., Smith, J. L., Raumann, C. G. (2001). Land Use Change and Effects on Water Quality and Ecosystem Health in the Lake Tahoe Basin, Nevada and California. Open-File Report 01-418, 29. U.S. Geological Survey, Menlo Park, California, USA. http://pubs.usgs.gov/of/of01-418/of01-418.pdf on 18 November 2004.

Gary, J. and Rubino, C. (2001). Facts about the world's languages. H. W. Wilson Co., Bronx, New York, USA, 896. http://www.hwwilson.com/print/factslangs_introduction.htm on 25 April 2004.

Global Population Profile 2002 (2004). International Population Report. U.S. Agency for International Development and U.S. Department of Commerce Publication. http://www.census.gov/ipc/prod/wp02/wp-02.pdf on 27 October 2004.

Global Security.org (2003). Great Northeast Power Blackout of 2003. http://www.globalsecurity.org/eye/blackout_2003.htm on 22 February 2005

Goldewijk, K.K. and Ramankutty, N. (2001). Land cover change over the last three centuries due to human activities: the availability of new global data sets. Submitted to Geojournal in November 2001. http://www.sage.wisc.edu/pubs/Abstracts/goldewijkGEOJ.html on 20 March 2004.

Green, M. J.B. and Paine, J. (1997). State of the world's protected areas at the end of the Twentieth Century. Paper presented at IUCN World Commission on Protected Areas Symposium on "Protected Areas in the 21st Century: From Islands to Networks" Albany, Australia, 24-29th November 1997, 35. http://www.unep-wcmc.org/protected_areas/albany.pdf on 20 March 2004.

Green, R. (2001). Production agriculture is only hope for world's remaining forests. Perspicacity & Paradigms Online. http://www.perspicacityonline.com/101/environment10101.htm on 1 May 2004.

Hargrove, W. W. and Hoffman, F. M. (1999). Locating and characterizing the borders between ecoregions using multivariate geographic clustering. http://research.esd.ornl.gov/~hnw/borders/ on 16 May 2004.

Harrison, P. (1997). Population and sustainable development: five years after Rio. United Nations Population Fund (UNFPA), New York City, New York ii, 38. http://www.ncseonline.org/PopPlanet/ePopulationReports/rest/abstracts/134981.pdf on 20 March 2004.

Harrison, P. and Pearce, F. (2001). AAAS Atlas of Population and Environment. American Association for the Advancement of Science, Washington, D.C., USA 215. http://atlas.aaas.org/ on 2 March 2004.

Hawken, P. (1994). The Ecology of Commerce: A Declaration of Sustainability. New York: Harper Business, 250.

Hay, J. E., Noonan, M. (2000). Anticipating the Environmental Effects of Technology -A manual for decision-makers, planners and other technology stakeholders. United Nations Environment Programme, Division of Technology, Industry and Economics, Consumption and Production Unit, Paris, France. http://www.unep.or.jp/ietc/Publications/Integrative/EnTA/AEET/index.asp on 9 November 2004.

Heilig, G. K. (1996). World Population Prospects: Analyzing the 1996 UN Population Projections. IIASA Electronic Working Paper WP-96-146. International Institute for Applied Systems Analysis (IIASA), IIASA LUC-Project, Laxenburg, Austria. http://www.iiasa.ac.at/Research/LUC/Papers/gkh1/chap1.htm on 10 November 2004.

Herring, D. (n.d.). Changing Global Land Surface. TERRA –The EOS Flagship. http://terra.nasa.gov/FactSheets/LandSurface/ on 2 May 2004.

Heywood, V.H. (ed.) (1995). Global Biodiversity Assessment. United Nations Environment Programme, Cambridge University Press, Cambridge, UK, 1140.

Holling C. S., Schindler, D. W., Walker, B. W., Roughgarden, J. (1995). Biodiversity in the functioning of ecosystems: an ecological primer and synthesis. In: Perrings C. A., K.- G. Maler, C. Folke, C. S. Holling, and B.- 0. Jansson (eds.), Biodiversity Loss: Ecological and Economic Issues, Cambridge University Press, Cambridge, UK, 44-83.

Holmes, B. (1998). The coming plagues – non-native species on the move due to global warming. New Scientist (weekly), London, UK, 18 April 1998. http://www.sare.org/htdocs/hypermail/html-home/29-html/0117.html on 23 March 2004.

Howenstine, E. (2004). Northeastern Illinois University, Department of Geography, USA. http://www.neiu.edu/~ejhowens/104/6/cultur.gif on 26 June 2004.

Hunter, L. M. (2001). The environmental implications of population dynamics, RAND Publications, Santa Monica, California, USA, 98. http://www.rand.org/publications/MR/MR1191/ on 25 April 2004.

ISRIC (2004). International Soil and Reference Information Centre: Global Assessment of Human-induced Soil Degradation, Soil Degradation Status Assessment, AR Wageningen, The Netherlands. http://www.isric.nl/ on 26 June 2004.

IPCC (2001). Climate Change 2001: The Scientific Basis. Contribution of Working Group I to the Third Assessment Report of the Intergovernmental Panel on Climate Change [Houghton, J.T.,Y. Ding, D.J. Griggs, M. Noguer, P.J. van der Linden, X. Dai, K. Maskell, and C.A. Johnson (eds.)]. Cambridge University Press, Cambridge, United Kingdom and New York, NY, USA, 881.

Isaac Newton School Geography Department, Internet Geography, UK. http://www.geography.learnontheinternet.co.uk/topics/popn1.html#growth on 14 October 2004.

IUCN (1994). Guidelines for protected areas management categories. IUCN: Publication CNPPA with assistance of WCMC. Gland, Switzerland and Cambridge, UK, 261. http://www.unep-wcmc.org/protected_areas/categories/eng/ on 20 March 2004.

IUCN, UNEP and WWF (1991). Caring for the Earth. A strategy for sustainable living. Gland, Switzerland, 228.

Karas, J. (n.d.). Climate change and the Mediterranean Region. Desertification. Greenpeace.org, 34. http://archive.greenpeace.org/climate/climatecountdown/documents/desert.pdf on 1 April 2004.

Klein G. K., 2001: Estimating global land use change over the past 300 years: The HYDE Database. Global Biogeochemical Cycles, 15, 417-433. http://www.gsfc.nasa.gov/topstory/20020926landcover.html on 2 May 2004.

Knee, M. (2003). Domestic and Global Trends. Studies in Quality, Ethics, and the Global Environment, Ohio State University, Ohio,USA. http://hcs.osu.edu/hcs600/sm1067.htm on 1 May 2004; http://hcs.osu.edu:16080/hcs600/sm1011.htm on 1 May 2004.

Krouse, A. J., Ferry, M. M., Crowsey, R. C. (2000). Satellite imagery: the space odyssey arrives in the courtroom. The Defense 42(6):18-23. http://www.crowsey.com/spacearticle.htm on January 20, 2005.

Kushwaha, S. P. S., Hazarika, Rubul (2004). Assessment of habitat loss in Kameng and Sonitpur Elephant Reserves. Current Science, Volume 87, No. 10, 25 November 2004.

Larsen, J. (2003). Shrinking forests. Peopleandplanet.net. People and forests. http://www.peopleandplanet.net/doc.php?id=1481 on 1 May 2004.

Lebel, L. and Steffen, W. (eds.) (1998). Global Environmental Change and Sustainable Development in Southeast Asia: Science Plan for a SARCS Integrated Study. Chiang Mai, Thailand: Southeast Asian Regional Committee for START (SARCS). http://www.sarcs.org/issp/isspindx.htm on 1 May 2004.

Lotfi, S. (2001). Remote Sensing for Urban Growth in Northern Iran, Department of Urban Planning, University of Mazandaran, Babolsar, Iran, 9 November 2001.

Lövei, G. L. (1997). Global change through invasion. Nature 388: 627-628. http://www.nature.com/cgi-taf/DynaPage.taf?file=/nature/journal/v388/n6643/index.html on 22 March 2004.

Lund, H. G., Dallmeier, F., Alonso, A. (2004). Biodiversity: Biodiversity in forests. MS 146, 33-40. In: Encyclopedia of Forest Sciences. J. Burley, J. Evans and J.A. Youngquist (eds.). Elsevier/Academic Press. ISBN 0-12-145160-7.

Malcolm, J.R., Liu, C., Miller, L.B., Allnutt, T., Hansen, L. (2002). Habitats at risk: Global warming and species loss in globally significant terrestrial ecosystems, WWF, 2002. http://www.undp.org.vn/mlist/envirovlc/022002/post54.htm on 16 May 2004.

McGrath, S. (2005). Attack of the alien invaders. National Geographic 207(3): 92-117.

McGraw-Hill Higher Education: http://highered.mcgraw-hill.com/sites/dl/free/007248179x/35299/map12.pdf on 10 August 2004; http://highered.mcgraw-hill.com/sites/dl/free/007248179x/35299/map11.pdf on 10 August 2004.

McNeely, J. A., Scherr, S. J. (2001). Common ground - common future: how ecoagriculture can help feed the world and save wild biodiversity. IUCN Publication, Gland, Switzerland, 23. http://www.futureharvest.org/pdf/biodiversity_report.pdf on 20 March 2004.

McNeill, J.R. (2001). Something new under the sun - An environmental history of the twentieth-century world. New York: W.W. Norton & Company, 421.

Menne, B. and Bertollini, R. (2000). The health impacts of desertification and drought. Down to Earth: Newsletter of the UNCCD, No. 14, 4-7. http://www.unccd.int/publicinfo/newsletter/no14/news14eng.pdf.

Middleton, N. et al. eds. (1997). World Atlas of Desertification. Second Edition. United Nations Environment Program 0340691662, 192.

Millar, C. I. (2003). Climate change as an ecosystem architect: Implications to rare plant ecology, conservation, and restoration, 37. http://www.fs.fed.us/psw/programs/snrc/staff/millar/MillarCNPS.pdf on 15 May 2004.

Mittermeier, R. A., Mittermeier, C. G., Gil, P. R., Pilgrim, J., Fonseca, G., Brooks, T., Konstant, W. R. (2003). Wilderness: Earth's Last Wild Places. University of Chicago Press. Distributed for the Conservation International, 576. http://www.mindfully.org/Heritage/Earth-Half-Wilderness4dec0.htm on 21 March 2004.

NASA (2002). Landcover Changes May Rival Greenhouse Gases As Cause Of Climate Change, NASA-GSFC News Release, 1October 2002. http://www.gsfc.nasa.gov/topstory/20020926landcover.html on 13 July 2004.

NASA (2003). Blackout in US and Canada. http://earthobservatory.nasa.gov/NaturalHazards/natural_hazards_v2.php3?img_id=11628 on 18 July 2004.

NASA (2004). Thar Desert. NASA/GSFC/METI/ERSDAC/JAROS, and U.S./Japan ASTER Science Team. http://asterweb.jpl.nasa.gov/gallery/gallery.htm?name=Thar on 25 August 2004.

NCR&LB (2003). Construction Terms. National Contractor Referrals and License Bureau. http://www.contractorreferral.com/glossary/ on 8 November 2004.

New Internationalist Magazine (n.d). http://www.newint.org/issue309/Images/population.gif on 16 July 2004.

Nilsson, S. (1996). Do We Have Enough Forests? IUFRO Occasional Paper No. 5, International Union of Forestry Research Organizations, Vienna, Austria. http://www.actionbioscience.org/environment/nilsson.html on 2 May 2004. http://iufro.boku.ac.at/iufro/publications/dowehave.htm on 2 May 2004.

O'Neil, D. (2004). An introduction to the processes and consequences of culture change. Tutorial. Palomar College, Behavioral Sciences Department, San Marcos, California, USA. http://anthro.palomar.edu/change/change_1.htm on 15 July 2004; http://anthro.palomar.edu/language/language_1.htm on 15 July 2004.

Pellew, R. (1996). The Global Biodiversity Assessment. Our Planet 7.5. http://www.ourplanet.com/imgversn/75/pellew.html on 22 March 2004.

Pimentel, D., Harvey, C., Resosudarmo, P., Sinclair, K., Kurz, D., McNair, M., Crist, S., Sphpritz, L., Fitton, L., Saffouri, R., Blair, R. (1995). Environmental and economic costs of soil erosion and conservation benefits. Science 267: 1117-1123.

Pimentel, D., Huang, X., Cordova, A., Pimentel, M. (1996). Impact of population growth on food supplies and environment. Presented at AAAS Annual Meeting, Baltimore, Maryland, USA, 9 February 1996. http://dieoff.org/page57.htm on 2 May 2004.

Pimentel, D., B., O., Mullaney, K.E., Calabrese, J., Nelson, W.F., Yao, X. (1999). Will Limits of the Earth's Resources Control Human Numbers? Cornell University, New York, USA. http://www.dieoff.com/page174.pdf on 17 November 2004.

Planning Department (2001). Chapter 12 Glossary. Raleigh Comprehensive Plan, City of Raleigh Planning Department, Raleigh, North Carolina, USA. http://www.raleigh-nc.org/planning/CP/Text/Chapter_12/Glossary.pdf on 1 May 2004.

Population Division (2000). Charting the Progress of Populations. Department of Economic and Social Affairs, United Nations, New York, USA, 95. http://www.un.org/esa/population/pubsarchive/chart/contents.htm on 10 November 2004.

Population Fact Sheet (2000). People in the Balance: Population and Natural Resources at the Turn of the Millennium, Engelman, R. with Cincotta, R. P., Dye, B., Gardner- Outlaw, T. and Wisnewski, J.: Population Action International, Washington, DC, USA. www.populationaction.org on 14 October 2004.

Ramankutty, N. and Foley, J. A. (1999). Global Potential Vegetation Data. Climate, People, and Environment Program, University of Wisconsin, Madison, Wisconsin, U.S.A. http://cpep.meteor.wisc.edu/pages/available.html on 2 March 2004.

Ramankutty, N., Foley, J. A., Olejniczak N. J. (2002). People on the Land: Changes in Global Population and Croplands during the 20th Century, Ambio Vol. 31 No. 3, May 2002, Royal Swedish Academy of Sciences 2002, 251-257. http://www.bioone.org/pdfserv/i0044-7447-031-03-0251.pdf on 16 August 2004.

Rau, J. L. (2003). Atlas of Urban Geology: Environmental and urban geology of selected cities in Central, Southwest and South Asia, United Nations Economic and Social Commission for Asia and the Pacific.

Riitters, K., Wickham, J., O'Neill, R., Jones, B., Smith, E. (2000). Global-scale patterns of forest fragmentation. Conservation Ecology 4(2): 3. http://www.consecol.org/vol4/iss2/art3 on 21 September 2004.

Rosenzweig, M.L. (1999). Heeding the basic warning in biodiversity's basic law. Science 284:276-277. http://www.sciencemag.org/cgi/content/full/284/5412/276 on 21 March 2004.

Saxena, R., Pal, S., Joshi, P.K. (2001). Delineation and characterization of agro-ecoregions, 3. National Centre for Agricultural Economics and Policy Research, New Delhi, India. http://www.icar.org.in/ncap/publications/pmenotes/pmenotes6.pdf on 15 May 2004.

SBSTTA (2003). Status and trends of, and threats to, protected areas. UNEP/CBD/SBSTTA/9/5/Rev.1. Ninth meeting of the Convention on Biological Diversity, Subsidiary Body on Scientific, Technical and Technological Advice. Montreal, 10-14 November 2003 Item 4.1, 19. www.biodiv.org/doc/meetings/sbstta/sbstta-09/official/sbstta-09-05-rev1-en.doc on 5 May 2004.

Shafer, S. L., Bartlein, P. J., Thompson, R. S. (2001). Potential changes in the distributions of Western North America tree and shrub taxa under future climate scenarios. Ecosystems 4: 200–215. http://www.usgcrp.gov/usgcrp/Library/nationalassessment/forests/Ecosystems4%20Shafer.pdf on 16 May 2004.

Sohngen, B., Mendelsohn, R., Sedjo, R. (1999). Forest management, conservation, and global timber markets. American Journal of Agricultural Economics, 81(1): Abstract 1-10. http://www.acs.ohio-state.edu/units/research/archive/timber.htm on 24 February 2004.

Southern Oregon University, Geography Department, Oregon, USA. http://www.sou.edu/Geography/JONES/GEOG111.112/atlas/Map14.GIF on 14 October 2004.

Sutherland, W.J. (2003). Parallel extinction risk and global distribution of languages and species. Nature 423: 276-279. http://www.nature.com/cgi-taf/DynaPage.taf?file=/nature/journal/v423/n6937/abs/nature01607_fs.html on 25 April 2004.

Taylor, E.J., Martin, P.L. (2002). Human Capital: Migration and Rural Population Change (Chapter for Handbook of Agricultural Economics, Bruce L. Gardner and Gordan C. Rausser, eds.), Elsevier Science, New York, USA. http://www.iga.ucdavis.edu/human.pdf on 14 October 2004.

Terralingua, UNESCO, and WWF (2003). From the poster 'The World's Biocultural Diversity: People, Languages, and Ecosystems', a companion to the booklet 'Sharing a World of Difference: The Earth's Linguistic, Cultural, and Biological Diversity', produced by Terralingua in collaboration with UNESCO and WWF and published by UNESCO Publishing, Paris.

The World Almanac and Book of Facts (2003).

Tilford, D. (2000). Why Consumption Matters. Center for a New American Dream. http://www.newdream.org/core/whyconsumptionmatters9.html#_edn177 on 18 April 2004.

Tilman, D., Fargione, J., Wolff, B., D'Antonio, C., Dobson, A., Howarth, R., Schindler, D., Schlesinger, W. H., Simberloff, D., Swackhamer, D. (2001). Forecasting agriculturally driven global environmental change. Science 292: 281-284. http://pangea.stanford.edu/courses/GES56Q/Tilman2001-Agriculture%20and%20Env%20Change.pdf on 5 May 2004.

UCS (2003). Clean Energy: The Sources of Energy, Union of Concerned Scientists, Cambridge, Massachusetts, USA. http://www.ucsusa.org/clean_energy/renewable_energy/page.cfm?pageID=77 on 19 April 2004.

UCS (2004). FAO Report: Enough food in the future–without genetically engineered crops. Food and Environment. Backgrounder. Union of Concerned Scientists, Cambridge, Massachusetts, USA. http://www.ucsusa.org/food_and_environment/biotechnology/page.cfm?pageID=331 on 3 May 2004.

UN (2000). Dialogue among Civilizations and Cultures: Resolution adopted without a vote by the 103rd Inter-Parliamentary Conference, Amman, Jordan, 5 May 2000. http://www.dialoguecentre.org/PDF/culture.pdf on 20 July 2005.

UNEP (1992). The World Atlas of Desertification, First Edition.

UNEP (1997). Global Environment Outlook 2000. United Nations Environment Programme. Oxford University Press. http://www.grida.no/geo1/ on 15 May 2004.

UNEP (1999). Taking Action – An Environmental Guide for you and your Community. http://www.nyo.unep.org/action/default.htm on 24 November 2004.

UNEP (2000). In cooperation with UNHCS and UNHCR, Environmental impact of refugees in Guinea, March 2000, Morten R. (2001). Aid and safety for Guinea's refugees, The Lancet, London, 7 April 2001, Volume 357, Issue 9262, 1123; Anonymous (2002). Surveillance of mortality during a refugee crisis—Guinea, January-May 2001, JAMA, 9 January 92002, Volume 287, Issue 2, Chicago, 182. http://www.grid.unep.ch/guinea/ on 8 August 2004; www.ids.ac.uk/ids/env/GuineaBiodiversity.pdf on 8 August 2004.

UNEP (2001). Globalization: Threat to World's Cultural, Linguistic and Biological Diversity, 8 February 2001. http://www.unep.org/GC/GC21/NR%20_18.doc on 20 July 2005.

UNEP (2002a). Global Environment Outlook 3 (GEO3) – Past, present and future perspectives. London: Earthscan, 446. http://www.unep.org/geo/geo3/ on 4 March 2004.

UNEP (2002b). Children of the New Millennium: environmental impact on health. Nairobi, United Nations Environment Programme. http://www.unep.org/ceh/ch04.html Accessed on 1 April 2004.

UNEP-WCMC (2003). A Survey of Global and Regional Marine Environmental Assessments and Related Scientific Activities. United Nations Environment Programme and World Conservation Monitoring Centre, Cambridge, UK. Miscellaneous Pagination. http://www.unep-wcmc.org/marine/GMA/ on 15 May 2004.

UNEP-WCMC (2005). Millennium Development Goals: Goal 7, Target 9 - Indicator 26: Ratio of area protected to maintain biological diversity to surface area.

UNESCO (2003). Dialogue among Civilizations: The International Ministerial-Conference on Dialogue among Civilizations: Quest for New Perspectives, New Delhi, India, 9-10 July 2003. http://unesdoc.unesco.org/images/0013/001343/134394e.pdf on 20 July 2005.

UNESCO (n.d.): What is Cultural Diversity? http://portal.unesco.org/culture/en/ev.php-URL_ID=13031&URL_DO=DO_TOPIC&URL_SECTION=201.html on 20 July 2005.

UNFPA (2001). The State of World Population 2001 - Footprints and Milestones: Population and Environmental Change. The United Nations Population Fund. http://www.unfpa.org/swp/2001/english/contents.html on 1 August 2004.

US Census Bureau (2004). World Vital Events. http://www.census.gov/cgi-bin/ipc/pcwe on 14 October 2004.

van Lynden, G., Schwilch, G., Liniger, H. (n.d.). A standardised method for assessment of soil degradation and soil conservation: the WOCAT mapping methodology, 7. http://www.wocat.org/ftp/ISCOmap.pdf on 1 May 2004.

Volker Quaschning. Personal Communication. http://www.volker-quaschning.de/fotos/re-spanien/index_e.html on 14 October 2004.

Wade, T. G., Riitters, K. H., Wickham, J. D., Jones, K. B. (2003). Distribution and causes of global forest fragmentation. Conservation Ecology 7(2): 7. http://www.consecol.org/vol7/iss2/art7 on 15 May 2004.

Watson, R. T., Dixon, J. A. Hamburg, S. P., Janetos, A. C., Moss, R. H. (1998). Protecting our planet - securing our future -linkages among global environmental issues and human needs, MEA Resources: Fact Sheets, Desertification and Land Degradation. http://www.gdrc.org/uem/mea/factsheets/fs5.html on 1 April 2004.

Wilk, R. (2000). Globalization, consumer culture, and the environment. Course Outline ER290-6 Spring 2000 cc#26251. Bloomington, Indiana University, Indiana, USA. http://www.indiana.edu/~wanthro/290-00.htm on 15 July 2004.

Wilson, E. O. (1989). Threats to biodiversity. Scientific American 261, September 1989, 108–112.

WRI (1998). 1998-99 World Resources: A Guide to the Global Environment. The World Resources Institute, The United Nations Environment Programme, The United Nations Development Programme, The World Bank. New York: Oxford University Press, 384. http://www.wri.org/wri/wr-98-99/ on 18 April 2004.

WRI (2000). World Resources 2000-2001: People and ecosystems: The fraying web of life. World Resources Institute in cooperation with United Nations, Washington, DC, USA.

Wurm, S. A. (1970). Atlas of the world's languages in danger of disappearing. UNESCO Publishing, 90. http://portal.unesco.org/culture/en/ev.php@URL_ID=2229&URL_DO=DO_TOPIC&URL_SECTION=201.html on 25 April 2004.

Site References

Black Triangle

Common Report on Air Quality in the Black Triangle Region (2000). Report prepared by the JAMS Working Group and published with the assistance of the PHARE Black Triangle Project. http://www.env.cz/envdn.nsf/0/1f9cf50ae801b07fc1256b5a00309f85/$FILE/Trojuhel.pdf on 14 December 2004.

NASA's Earth Observatory, Changes in Biochemical Cycles. www.eos-ids.sr.unh.edu/ids-cycles.html on 15 December 2004.

Czech Power Company CEZ, a.s. http://www.cez.cz/eng/ on 14 December 2004.

Ardö, J., Lambert, N. J., Henzlik, V. and Rock, B. N. (1997). Satellite Based Estimations of Coniferous Forest Cover Changes, Krusne Hory, Czech Republic, 1972-1989. Ambio, Volume 26, No. 3., 158-166.

Strub-Aeschbacher, N. (2002). Presentation - Black Triangle. UNEP/DEWA/GRID-Geneva.

Cape York Bauxite Mining

State of the Environment Queensland (1999). Underlying Pressures Weather and Climate. http://www.env.qld.gov.au/environment/science/environment/soe3_pre.pdf on 14 December 2004.

Australian Academy of Technological Sciences and Engineering (2000). Aluminum Technology in Australia 1788-1988. Australian Science and Technology Heritage Centre. http://www.austehc.unimelb.edu.au/tia/873.html on 31 December 2002.

Tomago Aluminum (2002). The Aluminum Industry in Australia. Tomago Aluminum. http://www.tomago.com.au/alum/industry.html on 31 December 2002.

Hick, P. and Ong, C. (2001). Remote-sensing for the assessment and management of the Comalco Andoom Bauxite Mine. CSIRO Australia. http://www.per.dem.csiro.au/research/EGG/weipa/index.html on 31 December 2002.

Industry Science Resources (2002). Developing Opportunities for Aboriginal People. Backing Australia's Ability. http://www.innovation.gov.au/resources/indigenouspartnerships/ on 31 December 2002.

Comalco (2002). Queensland, Australia. CaseStudies/RioTinto/index.html on 31 December 2002. http://www.comalco.com.au/05_operations/02_weipa.htm on 31 December 2002.

Copsa Mica_Romania

Thompson, J. (1991). East Europe's Dark Dawn, The Iron Curtain Rises to Reveal a Land tarnished by Pollution. National Geographic Magazine, 36-69 Volume 179, No. 6 June 1991.

Czaran, L. (2004). Personal Communication.

Escondida Mine, Chile

SPG Media Limited. (2004). http://www.mining-technology.com/ on 3 March 2005.

NASA's Earth Observatory. http://earthobservatory.nasa.gov/Newsroom/NewImages/images.php3?img_id=4492 on 3 March 2005.

NASA/GSFC/METI/ERSDAC/JAROS, and U.S./Japan ASTER Science Team. http://asterweb.jpl.nasa.gov/gallery/gallery.htm?name=Escondida on 3 March 2005.

Etaki Mine, Canada

The Canada Centre for Remote Sensing, Natural Resources Canada. Tour Canada from Space. http://www.ccrs.nrcan.gc.ca/ccrs/learn/tour/43/43nwt_e.html on 3 March 2005.

Gunn, A., Dragon J., Boulanger, J. (2001). Seasonal movements of satellite-collared caribou from the Bathurst herd. Wildlife and Fisheries Division, Resources, Wildlife and Economic Development, Government of the Northwest Territories, Yellowknife, Canada.

BHP Billiton (2003). Ekati Diamond Mine Facts. BHP Billiton Diamonds Inc. Issue 6. www.bhpbilliton.com on 3 March 2005.

Ok Tedi Mine, Papua New Guinea

Claassen, D. (2004). Personal Communication. October 2004.

About the Environment (2001). Ok Tedi Mining Limited. http://www.oktedi.com/resources/pages/aboutEnvironment.php on 3 March 2005.

Powder River Basin, Wyoming

Wyoming Oil and Gas Conservation Commission (n.d.). Map of coalbed distribution. http://wogcc.state.wy.us/CoalBedTwpRge.htm?RequestTimeout=3500 on 3 March 2005; Statistics. http://wogcc.state.wy.us/coalbedchart.cfm?RequestTimeout=3500 on 3 March 2005.

BLM White Paper on the San Juan Basin. http://oil-gas.state.co.us/blm_sjb.htm on 3 March 2005.

Powder River Basin Resource Council. http://www.powderriverbasin.org/ on 3 March 2005.

Conserving Wyoming Heritage. http://www.wyomingoutdoorcouncil.org/programs/cbm/Conserving_WY_heritage.pdf on 3 March 2005.

Powder River CBM Information Council. http://www.cbmwyo.org/ on 3 March 2005.

USGS Central Energy Resources Team Report. http://greenwood.cr.usgs.gov/energy/CBmethane/OF00-372/index.html on 3 March 2005.

Northern Plains Research Council, Coal Bed Methane – comprehensive list of impacts, includes white paper on proposed wastewater reinjection. http://www.nprcmt.org/Issues/CBM/CBM%20index.asp on 3 March 2005; http://www.nprcmt.org/media/2001/op-ed-TomSchneider-CBM.asp on 3 March 2005; http://www.escribe.com/culture/native_news/m664 on 3 March 2005.

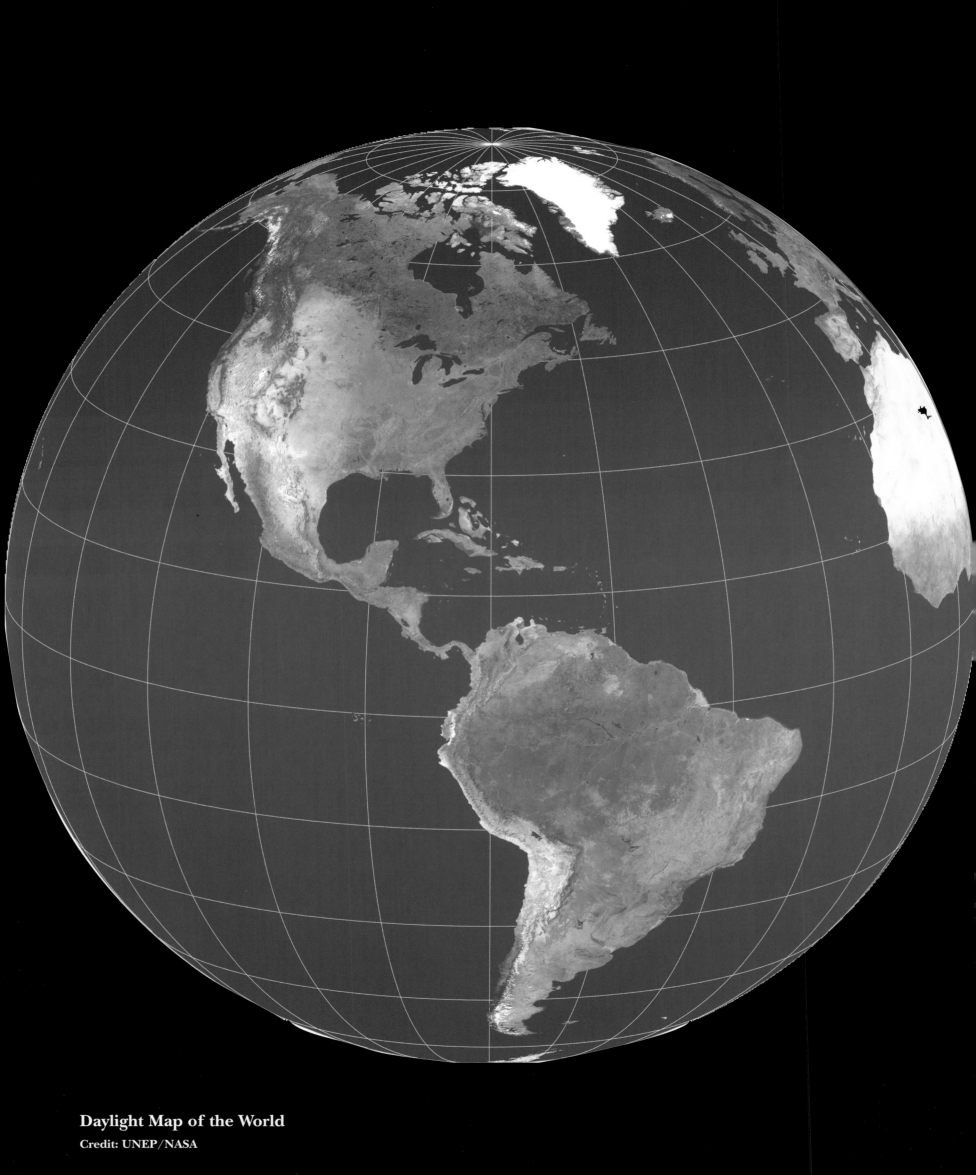

Daylight Map of the World
Credit: UNEP/NASA

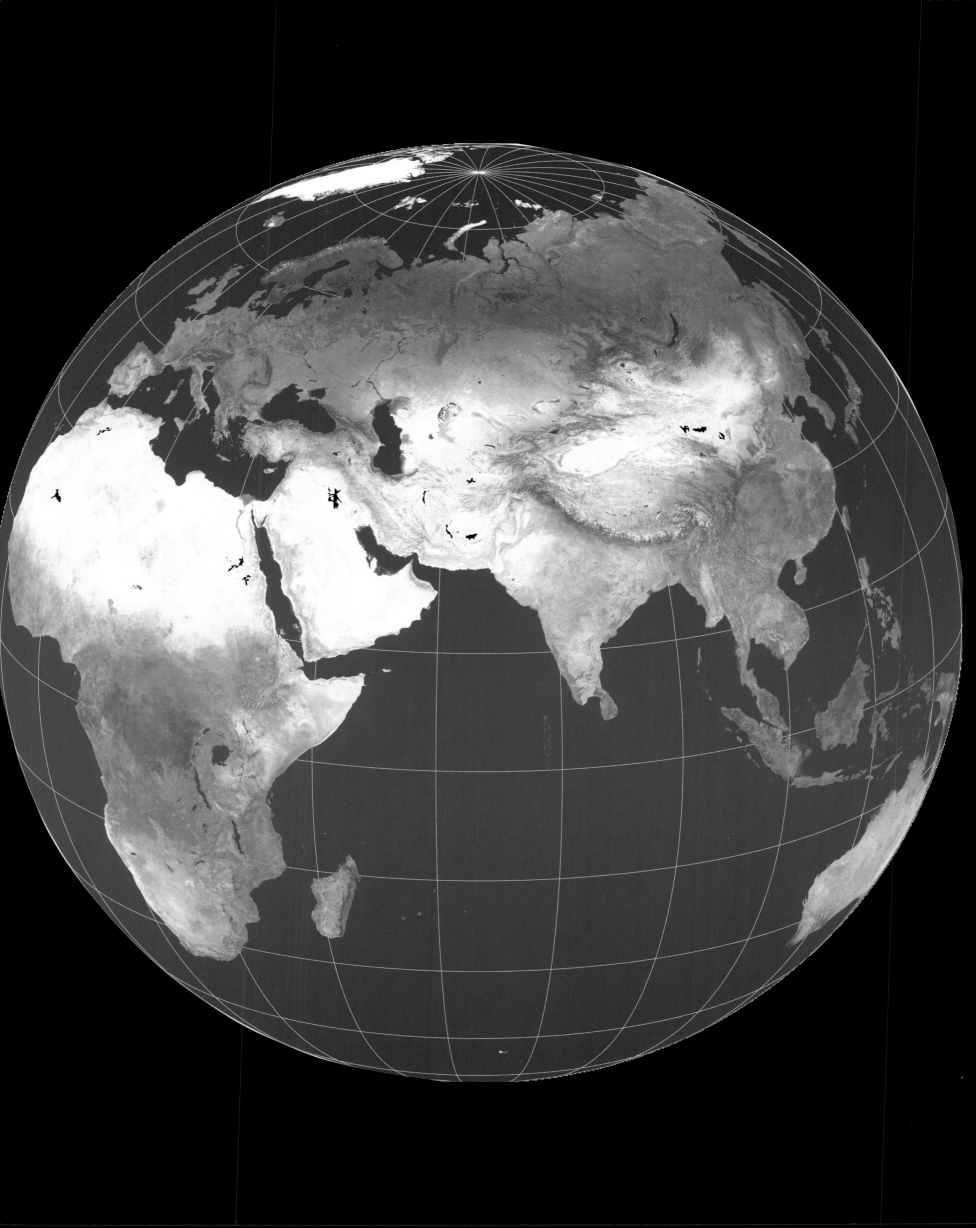

Human Impacts on the Planet
Visualising Change over Time

3

Human interactions with the environment leave many traces. For much of human history, human impact on the Earth's surface has been relatively minor. In the last several hundred years, however, that impact has grown tremendously. Change brought about by human activities can now be objectively measured; it can even be seen from space. A study by the National Aeronautics and Space Administration (NASA 2003a) known as The Human Footprint (Figure 3.1) is a quantitative analysis of human influence across the globe that illustrates the impact of people and their activities on the Earth.

Evidence of change is not always visible on the landscape. Change also occurs in the atmosphere, in the soil, and in the oceans and other water bodies. In these environments, evidence of change can still be "seen," however, by detecting and measuring things such as rising average global temperatures, the concentrations of certain gases in the atmosphere, and various chemical contaminants in water.

Change alone is not the only problem. It is the degree to which human activities are changing the Earth that is also cause

Map of the Human Footprint

Figure 3.1: The Human Footprint is a quantitative analysis of human influence on the Earth's surface. In this map, human impact is rated on a scale from 0 (minimum) to 100 (maximum) for each terrestrial biome. The color green indicates areas of minimal impact while purple indicates areas of major impact. *Credit: Scott, Michon 2003. The Human Footprint. NASA: Socioeconomic Data and Applications Center. Source: http://earthobservatory.nasa.gov/Newsroom/ NewImages/Images/human_footprint.gif*

for growing concern. For instance, the results of a recent ten-year study concerning the ecological effects of industrialized fishing in the world's oceans reveals that large predatory fish species including tuna, marlin, sharks, cod, and halibut have declined by an estimated 90 per cent from pre-industrial levels (Myers and Warm

2003). Furthermore, the average size of surviving individuals among these species is only one-fifth to one-half what it was previously.

The composition of the Earth's atmosphere is also undergoing rapid change. Since life began on Earth, changes in climate have ordered the distribution of

Credit: Apollo Mbabaz/UNEP/Topfoto

organisms and their behavior. Today, increases in atmospheric concentrations of greenhouse gases are expected to cause more rapid changes in the Earth's climate than have been experienced for millennia (Figure 3.2). At least some of the increase globally is due to human activity, and certainly, local impacts such as urban heat islands have profound effects on regional climatic conditions. As shown in Figure 3.2, waste generation and disposal is one of the ways in which humans contribute greenhouse gases to the atmosphere.

An emerging global impact issue is that of electronic or E-waste—a collective term for discarded electronic devices. Over the past decade, E-waste has become one of the world's fastest growing waste streams and—due to the presence of lead, mer-

cury, brominated flame retardants, and other hazardous substances—one of the most toxic. The disposal of computer waste in particular is becoming a difficult issue as millions of computers and other electronic devices are rapidly becoming obsolete as each year the industry produces ever-greater quantities of less-expensive equipment. There are an estimated 300 million obsolete computers in the United States, with fewer that ten per cent destined for recy-

cling each year. Even when a computer is sold to a secondhand parts dealer, however, there is a good chance it will end up in a dump in the developing world (Figure 3.3).

The Earth's forests are also under pressure. Tropical forests are now being subjected to the same heavy exploitation as were temperate forests a few generations ago. Pressures from logging, mining, hydropower, and a hunger for land are

Figure 3.2: The disposal and treatment of waste can produce emissions of several greenhouse gases that contribute to global climate change. Even the recycling of waste produces some emissions, although these are offset by the reduction in fossil fuels that would be required to obtain new raw materials. Both waste prevention and recycling help address global climate change by decreasing greenhouse gas emissions and saving energy (Environmental Protection Agency). *Source: http://www.grid.unep.ch/waste/html_file/42-43_climate_change.html*

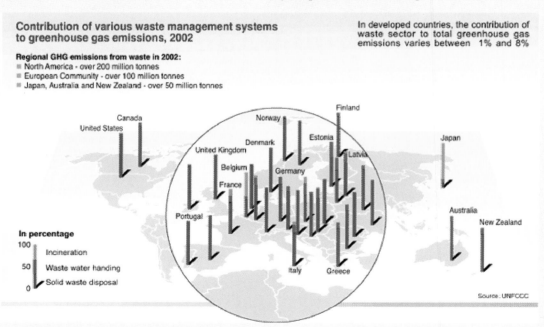

leading to large areas of forest being converted to serve other purposes. The integrity of forest ecosystems is being affected as the timber and paper industries remove vast areas of mature tropical and temperate forests. As a consequence, forest ecosystems lose their ability to support complex biodiversity and thousands of plant and animal species disappear forever.

Several globally significant environmental trends that have occurred between 1980 and 2000 may also be contributing to loss of forest ecosystems, including global warming (the two warmest decades on record are the 1980s and 1990s), three intense El Niño events, changes in cloudiness and monsoon dynamics, and a 9.3 per cent increase in atmospheric CO_2. Although these factors, along with others, are thought to exert their influence globally, their relative roles are still unclear.

An observed decline in tropical cloud cover is probably one of the more important recent climatic changes, although none of the existing climate models can accurately simulate this effect. It is known that continued reductions in tropical cloud cover, if accompanied by reduced rainfall, will have profound implications for tropical ecosystems in terms of water stress, productivity, ecological community composition, and disturbance patterns.

Images of Change

Various types of ground-based instruments, together with *in situ* surveys and analyses, can measure many of the changes being brought about on the Earth as a result of human activities. But such changes can also be observed—in more detail and with a "big picture" perspective—from space by Earth-orbiting satellites that gather images of the Earth's surface at regular intervals. The Landsat series of Earth-observing satellites has compiled a data record of the planet's land surfaces that spans the past thirty years and continues today.

By comparing two images of the same area taken ten, twenty, or even thirty years apart, it is often easy to see human-induced changes in a particular landscape. Few places remain on our planet that do not show at least some impact from human activities.

The focus of this chapter is a set of specific case studies in which satellite images, taken at different times, are paired so as to reveal changes and human impacts on the atmosphere, oceans and coastal zones, freshwater ecosystems and wetlands, forests, croplands, grasslands, urban areas, and the tundra regions.

The changes that we see in pairs of satellite images should make us cautious. Some are positive changes. But many more are negative. These images could be seen as warning signs. At the least they should provide us with food for thought and prompt us to ask pointed questions: How can we be more protective of our environment? How can we use the environment in ways that will not reduce the ability of the Earth to support us in perpetuity?

Figure 3.3: The high tech boom has been accompanied by E-waste, which represents the largest and fastest-growing type of manufacturing waste product. Recycling E-waste involves major producers and users, and the shipping of obsolete equipment and other products to Asia, Eastern Europe, and Africa where recyclers, such as the people in this photo, are exposed to toxic substances. *Source: http://www.grid.unep.ch/waste/html_file/36-37_ewaste.html*

Credit: Unknown/UNEP/Basel Action Network

3.1 Atmosphere

The Earth's atmosphere is a collection of gases, vapor, and particulates that together form a blanket of "air" that surrounds the planet. The atmosphere extends over 560 km (348 miles) from the surface of the Earth out toward space, and can be roughly divided into five major layers or sections (Figure 3.4). The primary components of the atmosphere are three gases: nitrogen (N_2, 78 per cent), oxygen (O_2, 21 per cent), and argon (Ar, 1 per cent). Other components, present in smaller amounts, include water vapor (H_2O, 0-7 per cent), ozone (O_3, 0-0.01 per cent), and carbon dioxide (CO_2, 0.01-0.1 per cent) (Phillips 1995).

The Earth's atmosphere plays many vital roles essential to sustaining life on the planet. The air we breathe circulates through its lowest level. The chemical elements carbon, nitrogen, oxygen, and hydrogen, which are constituents of all living things, are cycled and recycled in the atmosphere. Organisms convert these elements into carbohydrates, proteins, and other chemical compounds. The atmosphere also shields life on the planet's surface from harmful solar radiation, and—for the most part—from the threat of meteorites, which typically burn up as they go through the atmosphere (UNEP 1999a).

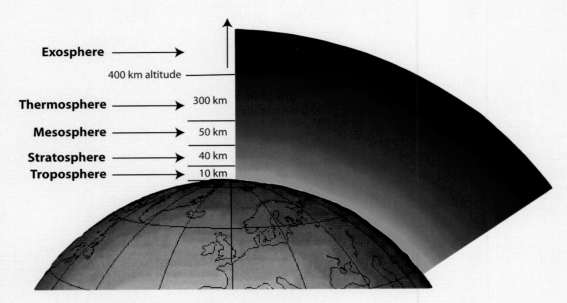

Figure 3.4: The Earth's atmosphere *Credit: Used with permission from Centre for Atmospheric Science, University of Cambridge, UK. Source: http://www.atm.ch.cam.ac.uk/tour/atmosphere.html*

Human activities impact the Earth's atmosphere in many ways. Some activities produce a quite direct effect, such as generating and releasing pollutants that foul the air, and adding carbon dioxide and other greenhouse gases to the atmosphere that induce global warming and climate change. Other human impacts, such as water pollution, land degradation, and human-induced loss of biodiversity, can indirectly affect the atmosphere, as well as the water and land.

In this section, four major issues that involve human impacts on the atmosphere—ozone depletion, global warming, climate change, and air pollution—are addressed.

Ozone Depletion

Ozone is a relatively unstable molecule, made up of three oxygen atoms (O_3). In the atmosphere, ozone is formed naturally in the stratosphere. It is concentrated there as an "ozone layer" that acts as a protective shield against harmful ultraviolet (UV) radiation coming from the Sun. The loss of stratospheric ozone allows more UV radiation to reach the Earth's surface, where it can cause skin cancer and cataracts in people and negatively affect other living things as well.

Ozone is also found in the troposphere, the layer of the Earth's atmosphere that is closest to the planet's surface. Ozone can

Credit: Image Analysis Laboratory/UNEP/NASA Johnson Space Center

be formed naturally in the troposphere—for example, by lightning. However, tropospheric ozone is also a byproduct of certain human activities. Vehicle exhaust contributes large quantities of ozone to the troposphere each year.

Depending on where ozone resides, it can protect or harm life on the Earth (Figure 3.5). In the stratosphere, ozone is "good" as it shields life on the surface from harmful solar radiation. In the troposphere, ozone can be "bad" as it becomes a type of air pollution. Changes in the amount of ozone in either the stratosphere or the troposphere can have serious consequences for life on the Earth. For several decades, "bad" tropospheric ozone has been increasing in the air we breathe, while "good" stratospheric ozone has been decreasing, gradually eroding the Earth's protective ozone shield (Thompson 1996).

Since the late 1970s, scientists have detected a slow but steady decline in the amount of ozone in the stratospheric ozone layer. This ozone destruction results from the presence of certain types of chemicals in the atmosphere, especially chlorofluorocarbons (CFCs) and other chlorine- and bromine-containing compounds, coupled with fluctuations in stratospheric temperature.

In polar regions, particularly the area of the atmosphere that overlies Antarctica, ozone depletion is so great that an "ozone hole" forms in the stratosphere there every

Stratosphere:

In this region, ozone is "good." It protects us from the sun's harmful ultraviolet radiation.

Troposphere:

In this region, ozone is "bad." It can damage lung tissure and plants.

Earth

Mesosphere

Ozone in the Earth's Atmosphere

Figure 3.5: Ozone in the stratosphere forms the protective ozone layer that shields the Earth's surface from harmful solar radiation. Ozone in the troposphere, the lowest part of the atmosphere, can be a form of air pollution. *Source: http://www.atmos.umd.edu/~owen/CHPI/IMAGES/ozonefig1.html*

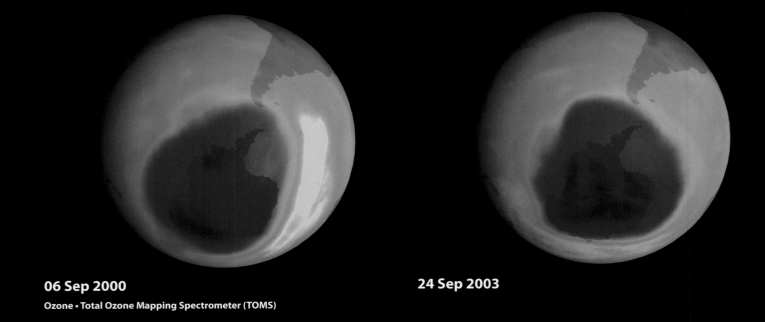

06 Sep 2000

24 Sep 2003

Ozone • Total Ozone Mapping Spectrometer (TOMS)

Figure 3.6: Every austral spring, an area of severe stratospheric ozone depletion— an "ozone hole"—forms in the atmosphere over Antarctica. The ozone holes that formed in 2000 and 2003 were the largest and second largest on record, respectively. *Source: http://www.gsfc.nasa.gov/gsfc/earth/ pictures/2003/0925ozonehole/still_hires_24Sept2003.tif and http://www.gsfc.nasa.gov/ftp/pub/ozone/ozone_still_2000_09_06.tif (NASA 2004a)*

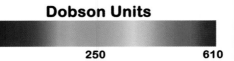

Dobson Units

90   250   610

austral spring (late August through early October). In the past few years, the Antarctic ozone hole has been about the size of North America. In 2000, the Antarctic ozone hole was the largest on record, covering 29.6 million km² (11.4 million square miles). In the austral spring of 2003, it was almost as large, covering 28.9 million km² (11.1 million square miles) (Figure 3.6).

Seasonal ozone depletion is also noticeable around the North Pole. More than 60 per cent of stratospheric ozone north of the Arctic Circle was

lost during the winter and early spring of 1999-2000 (Shah 2002).

Some ozone-depleting chemicals, such as CFCs, also contribute to global warming. Like carbon dioxide and methane, CFCs are powerful greenhouse gases that trap heat radiating from the Earth's surface and prevent it from immediately escaping into space. This causes the part of the atmosphere nearest the Earth's surface to warm, resulting in global warming. This warming in the troposphere, however, leads to colder-than-normal temperatures in the stratosphere. This, in turn, enhances the formation of certain types of stratospheric clouds that foster ozone-destroying chemical reactions in the stratosphere (Shanklin n.d.).

Fortunately, bans against the production and use of CFCs and other stratospheric ozone-destroying chemicals appear to be working to reverse the damage that has been done to the ozone layer. In the past few years, the Antarctic ozone hole has not increased significantly in size

Figure 3.7: Growth of the Antarctic ozone hole over 20 years, as observed by the satellite-borne Total Ozone Mapping Spectrometer (TOMS). Darkest blue areas represent regions of maximum ozone depletion. Atmospheric ozone concentration is measured in Dobson Units. A "normal" stratospheric ozone measurement is approximately 300 Dobson Units. Measurements of 220 Dobson Units and below represent significant ozone depletion. *Source: http://www.gsfc.nasa.gov/topstory/2004/0517aura.html*

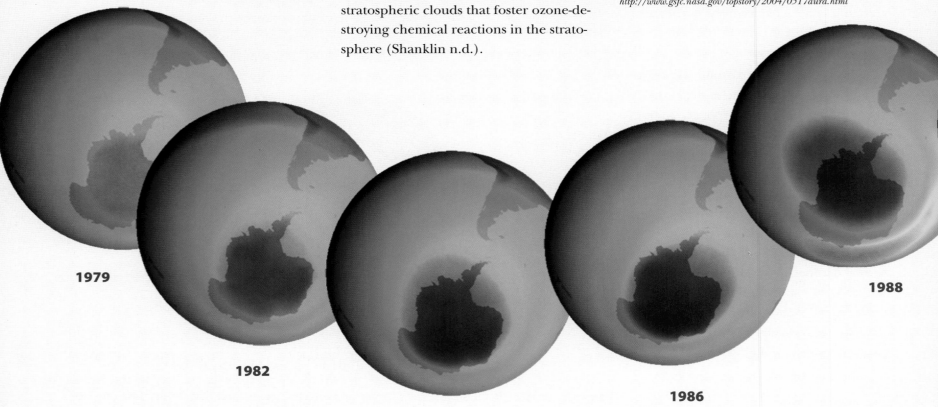

1979

1982

1984

1986

1988

Credit: John Bortniak/UNEP/NOAA

or intensity. Some researchers predict that if atmospheric concentrations of ozone-destroying chemicals drop to pre-ozone-hole levels, the Antarctic ozone hole should disappear in approximately 50 years (Figure 3.7) (WMO-UNEP 2002).

Global Warming

Atmospheric temperature and chemistry are strongly influenced by the amount and types of trace gases present in the atmosphere. Examples of human-made trace gases are chlorofluorocarbons, such as CFC-11, CFC-12, and halons. Carbon dioxide, nitrous oxide, and methane (CH_4) are naturally formed trace gases produced by the burning of fossil fuels, released by living and dead biomass,

and resulting from various metabolic processes of microorganisms in the soil, wetlands, and oceans. There is increasing evidence that the percentages of environmentally significant trace gases (greenhouse gases) are changing due to both natural and human factors, and contributing to global warming.

Global warming is recognized as one of the greatest environmental threats facing the world today. Global warming is the gradual rise of the Earth's average surface temperature caused by an enhancing of the planet's natural greenhouse effect. Radiant energy leaving the planet is naturally retained in the atmosphere thanks to the presence of certain gases such as water vapor and carbon dioxide. This heat-trapping effect is, in fact, what makes life on the Earth possible.

Global warming, by contrast, is an intensification of the Earth's greenhouse

effect. The Earth's average surface temperature, which has been relatively stable for more than 1 000 years, has risen by about 0.5 degrees Celsius in the past 100 years (Figure 3.8). The nine warmest years in the 20th century have all occurred since 1980; the 1990s were probably the warmest decade of the second millennium (IPCC 2001).

Global warming has occurred in the distant past as the result of natural influences. However, since the industrial era, the term is most often used to refer to the current warming predicted as a result of increases in the atmospheric concentrations of certain heat-trapping greenhouse gases generated by human activities (Figure 3.9). Most scientists

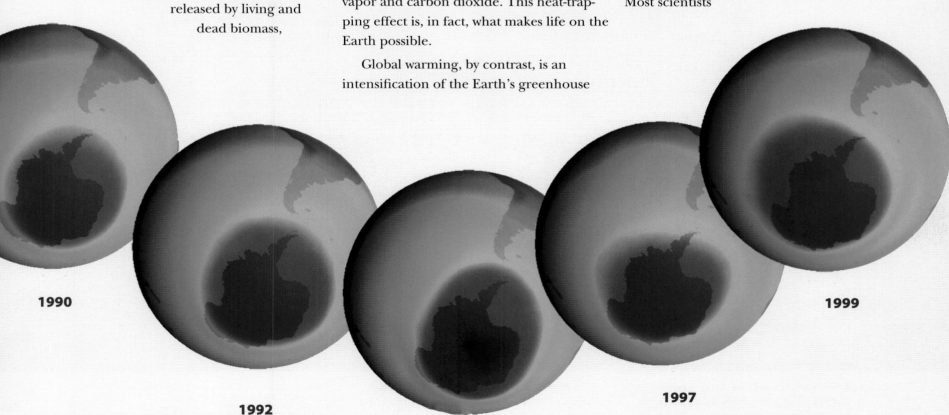

1990

1992

1994

1997

1999

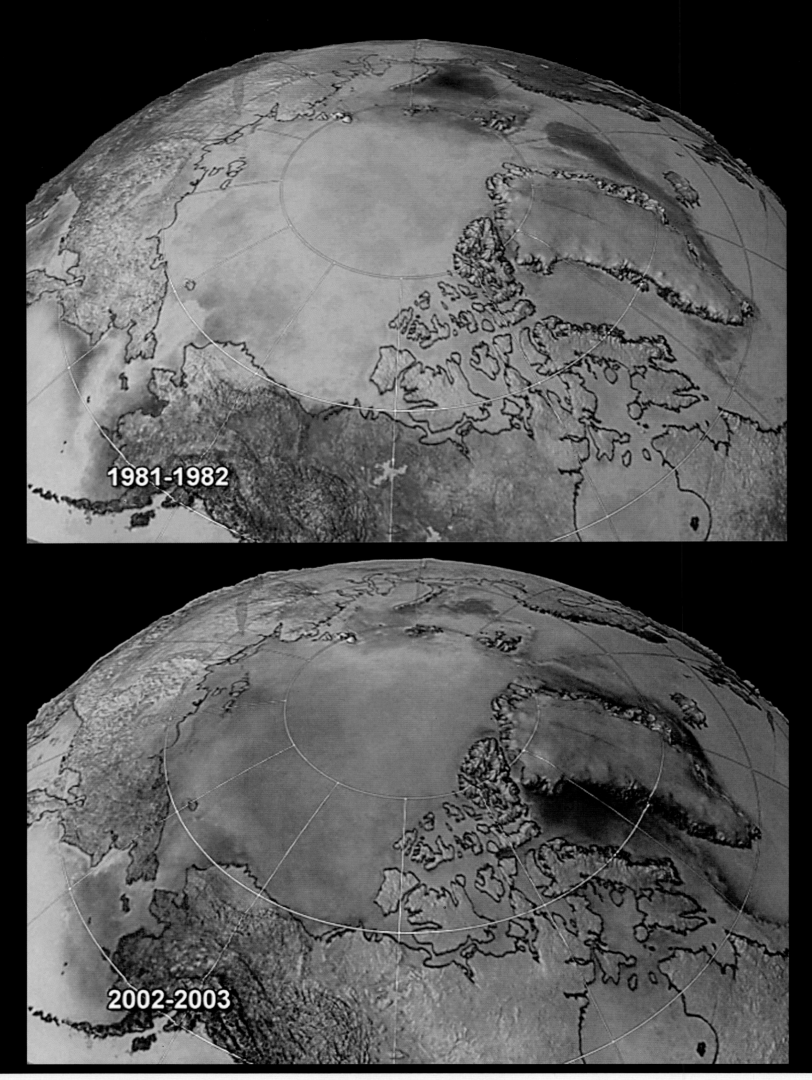

Figure 3.8: Satellite data of Arctic regions show warming is taking place there at an accelerating rate. These two maps show temperature anomalies in the Arctic in 1981 and in 2003. The anomalies range from 7°C (12.6°F) below normal to 7°C (12.6°F) above normal. Shades of orange and red show areas of warming; shades of blue shows areas of cooling; white represents little or no change (map adapted from Comiso). The data reveal that some regions are warming faster than 2.5°C (4.5°F) per decade (NASA 2003b).

Source: http://svs.gsfc.nasa.gov/vis/a000000/a002800/a002830/

Temperature Trend (°C/decade)

-2.5   0   +2.5

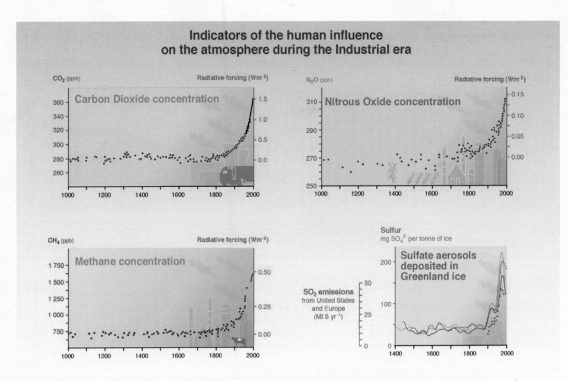

Figure 3.9: Human activities directly influence the abundance of greenhouse gases and aerosols in the atmosphere. Carbon dioxide, nitrous oxide, methane, and sulfur aerosols have all increased significantly in the past 50 years. *Source: Intergovernmental Panel on Climate Change, IPCC (2001)*

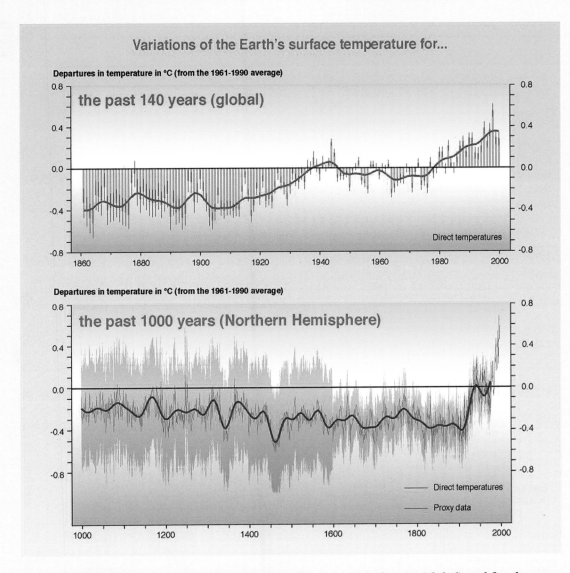

Figure 3.10: Variations in the Earth's temperature for the past 140 years (global) and for the past 1 000 years (Northern Hemisphere). *Source: IPCC (2001)*

believe that much of this global temperature increase is due to increased use of fossil fuels, which when burned, release carbon dioxide into the atmosphere where it absorbs infrared radiation that normally would pass through the atmosphere and travel out into space (Brehm 2003).

The planet is not as warm as it was approximately 1 000 years ago (Figure 3.10). Nevertheless, CO_2 currently accounts for the greatest proportion of greenhouse gas emissions. Much of the CO_2 added to the atmosphere comes from the burning of fossil fuels in vehicles, for heating, and for the production of electricity (Figure 3.11).

In addition to carbon dioxide, rising levels of methane in the atmosphere are also of concern. The relative rate of increase of methane has greatly exceeded that of carbon dioxide in the last several decades. Methane is released into the atmosphere in many ways: as a result of agriculture and ranching activities; through the decay of organic matter, including waste dumps; through deforestation; and as a by-product of the hydrocarbon economy. None of these sources are anticipated to decrease in the future. On the contrary, methane emissions are expected to increase, as each year an additional 100 million people require food and fuel as world population expands (Figure 3.12).

Most scientists believe that recent global warming is mainly due to human activities and related increases in concentrations of greenhouse gases (Figure 3.13), primarily CO_2, CH_4, nitrous oxide (N_2O), hydrofluorocarbons (HFCs), perfluorocarbons (PFCs), and sulfur hexafluoride (SF_6). These changes are driven by worldwide population and economic growth, and the underlying production and consumption of fossil fuels, as well as by the intensification of agricultural activity and changes in land use and land cover. Energy production and use, the largest sole source of CO_2 emissions and a large contributor of CH_4 and N_2O emissions, accounted for 81.7 per cent of emissions in industrialised countries in 1998 (UNFCCC 2000).

From far out in space, instruments carried aboard satellites, such as NASA's

Moderate Resolution Imaging Spectrora-diometre (MODIS) sensor, are taking the temperature of the Earth's surface. Satellite data confirm that the Earth's average surface temperature has been slowly rising for the past few decades (Figure 3.14). Satellite records are more detailed and comprehensive than previously available ground measurements, and are essential for improving climate analyses and computer modeling.

One of the more predictable effects of global warming will be a rise in sea levels (Figure 3.15). It is already under way at a pace of about a millimetre a year—a consequence of both melting land ice and the thermal expansion of the oceans (Harrison and Pearce 2001). Predictions as to how much global sea levels may rise over the next century range from half a metre (1.5 feet) (Houghton et al. 2001) to between 1 and 2 m (3 to 6 feet) (Nicholls et al. 1999). Such an increase would drown many coastal areas and atoll islands. Unless countries take action to address rising sea levels, the resulting flooding is expected to impact some 200 million people worldwide by the 2080s. In addition, around 25 per cent of the world's coastal wetlands could be lost by this time due to sea-level rise (DETR 1997).

Global warming may have some positive impacts. It could, for example, open

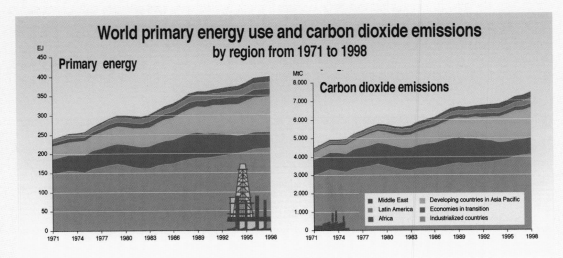

Figure 3.11: Between 1971 and 1998, energy use and carbon dioxide emissions both increased significantly, contributing to the likelihood of global warming. *Source: IPCC (2001)*

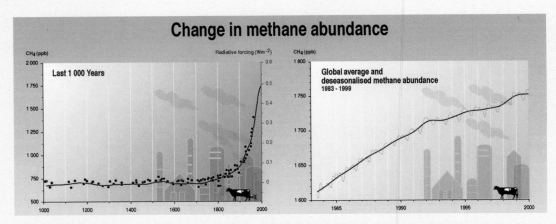

Figure 3.12: Methane is the second largest contributor to global warming and its atmospheric concentration has increased significantly over the last two decades. Methane emissions from human-related activities now represent about 70 per cent of total emissions, as opposed to less than 10 per cent 200 years ago. *Source: IPCC (2001)*

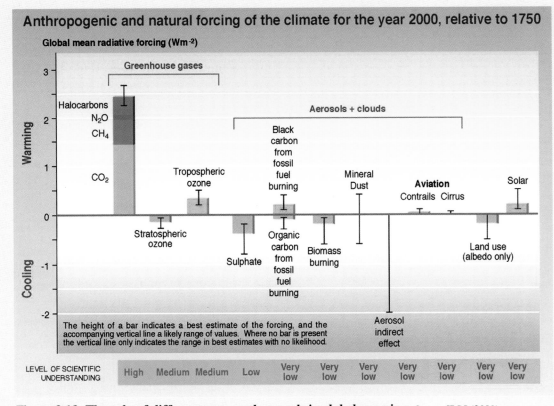

Figure 3.13: The role of different gases and aerosols in global warming. *Source: IPCC (2001)*

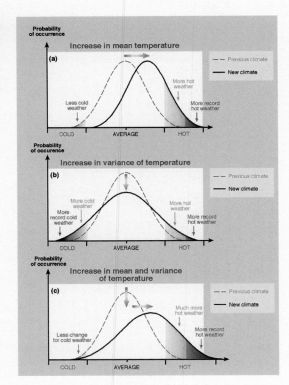

Figure 3.14: Global warming is an increase in the Earth's average surface temperature. These graphs illustrate how a shift in the mean temperature and its variance can affect weather. *Source: IPCC (2001)*

What causes the sea level to change?

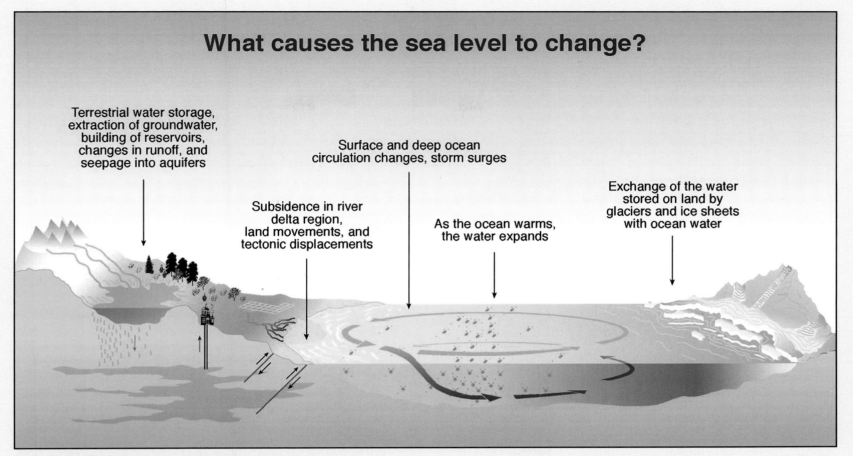

Terrestrial water storage, extraction of groundwater, building of reservoirs, changes in runoff, and seepage into aquifers

Surface and deep ocean circulation changes, storm surges

Exchange of the water stored on land by glaciers and ice sheets with ocean water

Subsidence in river delta region, land movements, and tectonic displacements

As the ocean warms, the water expands

Figure 3.15: Reasons for sea level change *Source: IPCC (2001)*

new lands for agriculture and forestry in the far north. During the past 30 years in Iceland, old farmlands have been exposed, and are being used, as the Breidamerkur-jökull Glacier has receded. All told, however, the negative impacts of unchecked global warming outweigh any positive benefits.

Climate Change

Climate is the statistical description in terms of the mean and variability of relevant measures of the atmosphere-ocean system over periods of time ranging from weeks to thousands or millions of years. Climate change is defined as a statistically significant variation in either the mean

state of the climate or in its variability, persisting for an extended period (typically decades or longer). Climate change may be due to natural internal processes or to external forcing (Figure 3.16). Volcanic gases and dust, changes in ocean circulation, fluctuations in solar output, and increased concentrations of greenhouse gases in the

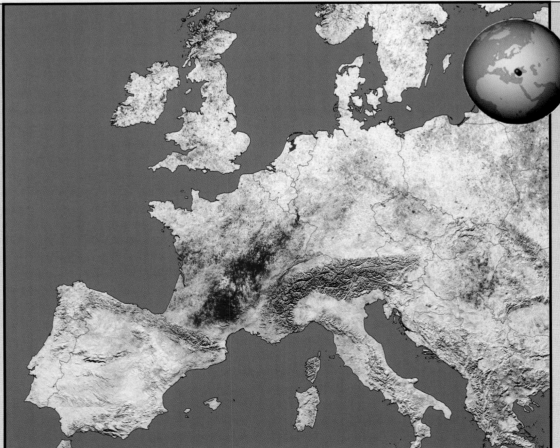

Temperature Anomaly (°C)

-10   -5   0   5   10

European Heat Wave, July 2003 *Credit: NASA— Satellite Thermometers Show Earth Has a Fever (2004)*

Case Study: European Heat Wave July 2003

This image shows the differences in daytime land surface temperatures (temperature anomalies) collected over Europe between July 2001 and July 2003 by the Moderate Resolution Imaging Spectroradiometer (MODIS) on NASA's Terra satellite. A blanket of deep red across southern and eastern France (left of image center) reveals that temperatures in this region were 10°C (18°F) hotter during 2003 than in 2001. Temperatures were similar in white areas and cooler in blue areas. Although models predict an overall increase in global average temperatures, regional differences may be pronounced, and some areas, such as mid-continental zones in North America and Asia, may actualy experience some degree of cooling (NASA 2003c).

79

atmosphere can all cause climate changes (USCCSP 2003).

Global warming, whatever its underlying causes, is expected to have adverse, possibly irreversible effects on the Earth's climate, including changes in regional temperature and rainfall patterns and more frequent extreme weather events. Climate change will affect the ecology of the planet by impacting biodiversity, causing species extinctions, altering migratory patterns, and disturbing ecosystems in countless ways. Climate change will impact human societies by affecting agriculture, water supplies, water quality, settlement patterns, and health.

Overall, climate change is likely to intensify the already increasing pressures on various sectors. Although the impact of climate change may, in some cases, be smaller than other stresses on the environment, even relatively small changes can have serious adverse effects, especially where there may be critical thresholds, where development is already marginal, or where a region is less able to implement adaptation measures (DETR 1997).

For example, climate change is likely to exacerbate already increasing pressure

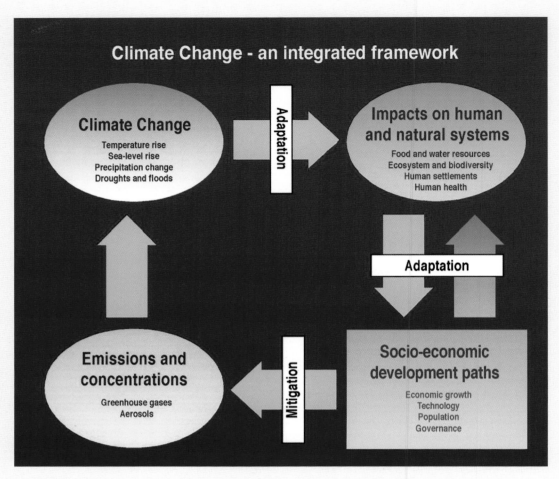

Figure 3.16: Climate change is the result of complex interactions among many factors. *Source: IPCC (2001)*

being put on water resources by a growing global population, particularly in Africa, Central America, the Indian subcontinent, and southern Europe. By the 2050s, models indicate that there may be an additional 100 million people living in countries with extreme water stress due to climate change alone (DETR 1997).

Carbon dioxide, the gas largely blamed for global warming, has reached record-high levels in the atmosphere (Hanley 2004). Carbon dioxide levels have risen by 30 per cent in the last 200 years as a result of industrial emissions, automobiles, and rapid forest burning, especially in the tropics. Much of this increase has occurred since 1960. This increase in CO_2 is thought to enhance the Earth's natural greenhouse effect and thus increase global temperatures (Figure 3.17).

The Intergovernmental Panel on Climate Change projects that, if unchecked, atmospheric carbon dioxide concentrations will range from 650 to 970 ppm by 2100. As a result, the panel estimates, average global temperature may rise by 1.4°C (2.5°F) to 5.8°C (10.4°F) between 1990 and 2100 (Hanley 2004).

Interestingly, increased levels of atmospheric CO_2 may stimulate the growth of some kinds of plants. Researchers in the Amazon have noted increased growth rates in several species of trees (Laurance et al. 2004). The forests are also becoming more dynamic, with existing trees dying and being replaced by new trees at a more rapid pace. In addition, the species composition of the forest is changing. Rising atmospheric CO_2 concentrations may explain these changes, although the effects of this and other large-scale environmental alterations remain uncertain.

Air Pollution

Air pollution is the presence of contaminants or pollutant substances in the air that interfere with human health or welfare or produce other harmful environmental effects (EEA 2004). The term "smog"—first coined in London, England, to describe a combination of smoke and fog—is now used in reference to a specific combination of airborne particles, gases, and chemicals that together affect peoples' health and their natural environment (Health Canada 2003).

Aerosols are tiny particles suspended in the air. Some occur naturally, originating from volcanoes, dust storms, forest and grassland fires, living vegetation, and sea spray. Human activities, such as the

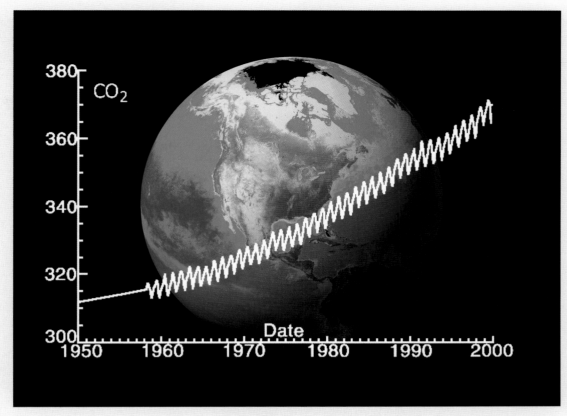

Figure 3.17: For more than 40 years, researchers at Mauna Loa Observatory, Hawaii, have tracked the steady increase of atmospheric carbon dioxide (concentration expressed in parts per million, or ppm). *Source: SIO (2004) and SeaWiFS: NASA Carbon Cycle Initiative, NASA/Goddard Space Flight Center Scientific Visualization Studio (2004). Source: NASA (2004b)*

burning of fossil fuels and the alteration of natural land cover, also generate aerosols. Averaged over the globe, human-generated aerosols currently account for about ten per cent of total atmospheric aerosols. Most of those aerosols are concentrated in the Northern Hemisphere, especially downwind of industrial sites, slash-and-burn agricultural regions, and overgrazed grasslands (Figure 3.18).

As the composition of Earth's atmosphere changes, so does its ability to absorb, reflect and retain solar energy. Greenhouse gases, including water vapor, trap heat in the atmosphere. Airborne aerosols from human and natural sources absorb or reflect solar energy based on colour, shape, size, and substance. The impact of aerosols, tropospheric ozone, and upper tropospher-

Figure 3.18: Aerosol particles larger than 1 micrometre are composed primarily of windblown dust and sea salt from sea spray. Aerosols smaller than 1 micrometre are mostly formed by condensation processes, such as the conversion of SO_2 gas released by volcanic eruptions to sulfate particles and by formation of soot and smoke during burning processes. After aerosols form, they are mixed and transported throughout the atmosphere by wind and weather; they are removed from the atmosphere primarily through cloud formation and precipitation. *Source: http://earthobservatory.nasa.gov/Library/Aerosols/ (NASA 2004c)*

ic water vapor on Earth's climate remains largely unquantified (NASA 1989).

About 25 per cent of the world's population is exposed to potentially harmful amounts of SO_2, O_3, and particulate matter in smog (Schwela 1995). Globally, some 50 per cent of cases of chronic respiratory illness are now thought to be associated with air pollution (UNEP 1999b).

Some pollutants travel long distances on the wind, causing acid deposition in the surrounding countryside and even in neighboring countries. In the 1980s, "acid rain" was identified as a major international environmental problem, moving from

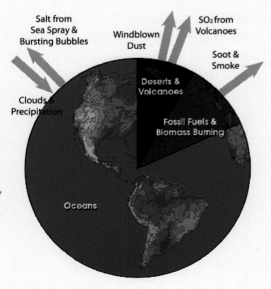

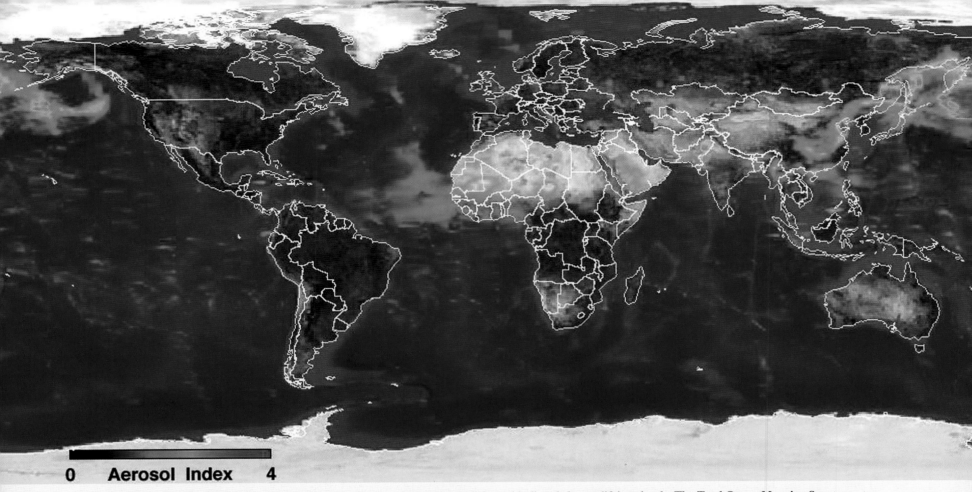

0   **Aerosol Index**   4

Fig.3.19: Aerosols affect climate both directly by reflecting and absorbing sunlight and indirectly by modifying clouds. The Total Ozone Mapping Spectrometer (TOMS) aerosol index is an indicator of smoke and dust absorption. This figure shows aerosols—the hazy green, yellow, and red patches—crossing the Atlantic and Pacific Oceans. Dust from the Sahara Desert is carried westward toward the Americas and provides nutrients for Amazon forests. Asian dust and pollution travel to the Pacific Northwest. *Source: NASA (http://www.gsfc.nasa.gov/topstory/2004/0517aura.html) (NASA 2004a)*

heavily industrialized areas of both Europe and North America into prime agricultural areas that lay downwind. Mountain regions suffered the most because their higher rainfall increased the volume of acid deposition and their often thin soils could not neutralize the acid. Lakes and streams in pristine parts of Scandinavia and Scotland became acidified, and fish populations were decimated in some areas. The most intense acid rain fallout occurred in the so-called Black Triangle region bordered by Germany, Czech Republic, and Poland (Harrison and Pearce 2001) (Figure 3.19).

Acid precipitation decreased throughout the 1980s and 1990s across large portions of North America and Europe. Many recent studies have attributed observed reversals in surface-water acidification at national and regional scales to this decline (Stoddard et al. 1999; Larssen 2004).

Decreases in acid precipitation have been achieved largely through improved flue gas treatments, fuel switching, use of low-sulfur fuels in power stations, and use of catalytic converters in automobiles. Since 1985, international treaties and heavy investment in desulphurization equipment by power station operators have cut sulfur pollution in Europe and North America by as much as 80 per cent (Harrison and Pearce 2001).

Although significant progress has been made in controlling acid-forming emissions in some countries, the global threat from acid precipitation still remains. In fact, the problem is growing rapidly in Asia, where 1990s-level SO_2 emissions could triple by 2010 if current trends continue. Curtailing the already substantial acid precipitation damage in Asia, and avoiding much more severe damage in the future, will require investments in pollution control similiar to

those made in Europe and North America over the past 20 years (Downing 1997; WRI 1998).

Nitrogen dioxide is the orange gas that is the most visible component of most air pollution. In many cities, NO_2 and other pollutants are suspended in the air to form a brownish haze commonly called smog. Nitrogen dioxide is formed when oxygen in the air combines with nitric oxide. Nitric oxide comes from automobiles, aerosols, and industrial emissions, and contributes to the formation of acid rain. In addition, this pollutant can cause a wide range of environmental damage, including eutrophication of water bodies—explosive algae growth that can deplete oxygen and kill aquatic organisms.

Case Study: Emissions in Paris 1999-2003

Paris, France, lies on a relatively flat plain. Most of the time, Paris benefits from a wet and windswept oceanic climate that encourages the dispersal of air pollution and thus cleans the air. However, under certain meteorological conditions (anticyclones and a lack of wind), pollutants can remain trapped in the atmosphere above the city, where they become concentrated, resulting in significantly higher levels of pollution. Thus, for equivalent pollutant emissions in terms of location and intensity, the levels of pollutants recorded in the atmosphere can vary by a factor of 20 according to meteorological conditions.

This explains why peaks in secondary pollutants often affect wider areas than peaks in primary pollutants. For example, when the wind blows from the city in a certain direction, the rural area surrounding the Paris region is also subject to ozone pollution. Indeed, ozone levels registered in these areas are often much higher than those in the centre of Paris itself.

In 1994, according to the Centre Interprofessionnel Technique d'Etude de la Pollution Atmosphérique (CITEPA, Inter-professional Technical Centre for Research into Air Pollution) SO_2 emissions in the Paris region corresponded to eight per cent of national emissions (mainland France and overseas territories), oxides of nitrogen (NOx) emissions to 10 per cent, Volatile Organic Compound (non-methane) (VOCNM) emissions to 12 per cent, carbon monoxide (CO) emissions to 15 per cent, and CO_2 emissions to 14 per cent. Given that 19 per cent of the population lives in the Paris region, emissions per inhabitant in this area are less than the national average for all substances (CITEPA 1994).

Existing air-quality-monitoring tools in the greater region of Paris provide a constant indication of air pollution levels at specific background and roadside locations. In addition to standard monitoring, specific modeling applications give extensive descriptions of air-quality patterns for several significant pollutants. Despite

Annual SO₂ emmissions in the Paris region

Annual NOx emmissions in the Paris region

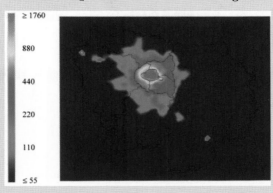

Annual CO emmissions in the Paris region

Annual VOC emmissions in the Paris region

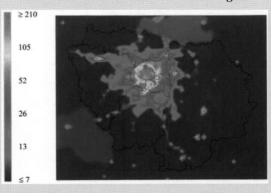

Concentrations of SO₂, NOx, CO, and VOC over Paris. *Source: http://www.airparif.asso.fr/english/polluants/default.htm*

Annual averages of SO₂ in Ile-de-France from 1999 to 2003

Regional cartography of the annual level of benzene evaluated within the framework of the European project LIFE "RESOLUTION"

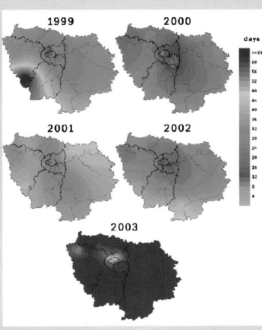

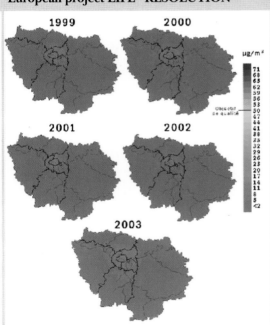

Air quality dynamics over the region of Paris, France, from 1999 to 2003. *Source: http://www.airparif.asso.fr/english/polluants/default.htm*

the involvement of the transport sector in monitoring local atmospheric emissions, there is no direct and constant traffic data feed. Recently, a project known as HEAVEN (Healthier Environment through the Abatement of Vehicle Emissions and Noise) was implemented in Paris. Its main objective was to integrate real-time traffic

information with the air quality monitoring tools. HEAVEN helped develop and demonstrate new concepts and tools to allow cities to estimate emissions from traffic in near-real time. This enhanced the identification and evaluation of the best strategies for transport demand management.

Souce: http://heaven.rec.org

Case Study: Pollution from Wild Fires 2003–2004

Whether started by people or natural events, fires add large quantities of pollutants to the atmosphere every year, primarily in the form of CO and aerosols. Satellite sensors can help researchers distinguish between wildfires and urban or industrial fires. Some can also distinguish different types of fire-generated pollutants. For instance, two sensors aboard NASA's Terra satellite—the Measurements of Pollution in the Troposphere (MOPITT) instrument and the Moderate Resolution Imaging Spectroradiometer (MODIS) instrument—gather data on CO and aerosols, respectively.

Carbon monoxide is one of the more easily mapped air pollutants. In the MOPITT-generated series of maps shown below, global seasonal variations in CO concentration are clearly visible (highest concentrations of CO appear as red). Major concentrations of CO during different seasons can be easily identified and tracked over time on such images, leading to better understanding of sources of CO pollution and its transcontinental transport (NASA 2004d). For example, in the summer image of this series, a very high concentration of CO appears over west central Africa, largely due to forest fires.

Wildfires in southern Africa are a major source of carbon monoxide pollution. Every August in southern Africa, thousands of people equipped with lighters or torches travel out onto the savanna and intentionally set the dry grasslands ablaze. Burned grasses send up tender new growth that is ideal for cattle consumption. The fires typically scorch an area the size of Montana, Wyoming, Idaho, and the Dakotas combined. Long plumes of smoke rise like hundreds of billowing smokestacks, and herds of animals are sent scurrying across open grassland.

During this fire season, a thick pall of smoke clouds the sky for many weeks. The smoke is laced with a number of pollutants, including nitrogen oxides, carbon monoxide, and hydrocarbons. Some of these substances react with the intense heat and sunlight to form ozone. Ground-level ozone contributes to respiratory diseases and can seriously damage crops. At higher levels in the troposphere, ozone molecules trap thermal radiation emanating from the Earth's surface in the same way as carbon dioxide and other greenhouse gases do. In fact, up to 20 per cent of the global warming experienced by the Earth over the past 150 years is thought to be from ozone.

In the spring of 2003, the MODIS and MOPITT instruments were used to monitor fires and fire-produced air pollutants in Siberia, especially in the Baikal region. These fires produced large amounts of fine carbon aerosols that spread out over the Pacific Ocean and remained suspended in the atmosphere for a few days. Carbon monoxide was also produced by the fires,

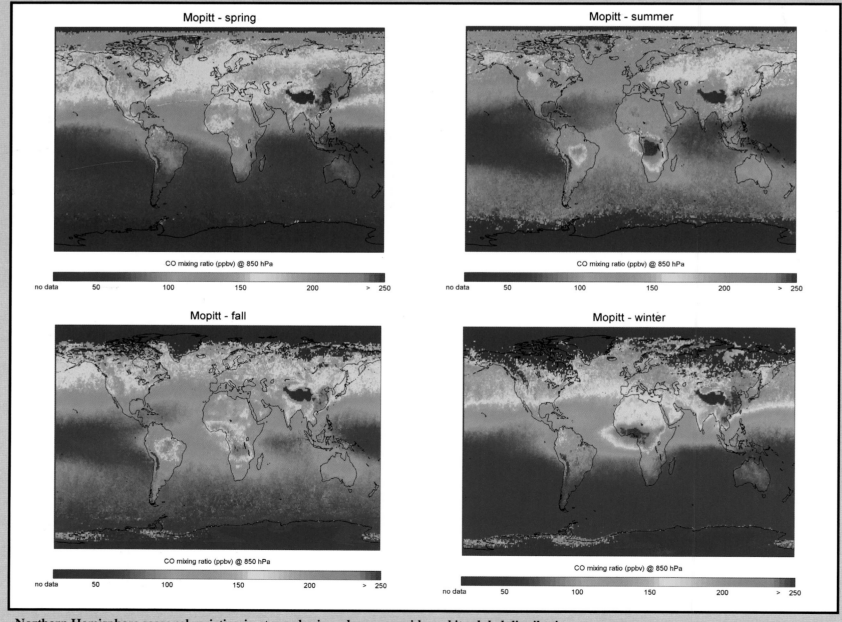

Northern Hemisphere seasonal variation in atmospheric carbon monoxide and its global distribution. *Source: http://svs.gsfc.nasa.gov/vis/a000000/a002100/a002150/ (NASA 2004d)*

but unlike the aerosols, remained airborne for a much longer period of time, allowing it to cross the Pacific Ocean and reduce air quality over North America before continuing on around the globe.

Gas and particle emissions produced as a result of fires in forests and other vegetation impact the composition of the atmosphere (WHO 2000). These gases and particles interact with those generated by fossil-fuel combustion or other technological processes, and are major causes of urban air pollution. They also create ambient pollution in rural areas. When biomass fuel is burned, the process of combustion is not complete and pollutants released include particulate matter, carbon monoxide, oxides of nitrogen, sulfur dioxide and organic compounds. Once emitted, the pollutants may undergo physical and chemical changes. Thus, vegetation fires are major contributors of toxic gaseous and particle air pollutants into the atmosphere. These fires are also sources of "greenhouse" and reactive gases. Particulate pollution affects more people globally on a continuing basis than any other type of air pollution. In 1997/98, forest fires in Southeast Asia affected at least 70 million people in Brunei Darussalam, Indonesia, Malaysia, the Philippines, Singapore, and Thailand. Thousands of people fled the fires and smoke and the increase in the number of emergency visits to hospitals demonstrated the severity of the fires and pollution they caused (WHO 2000).

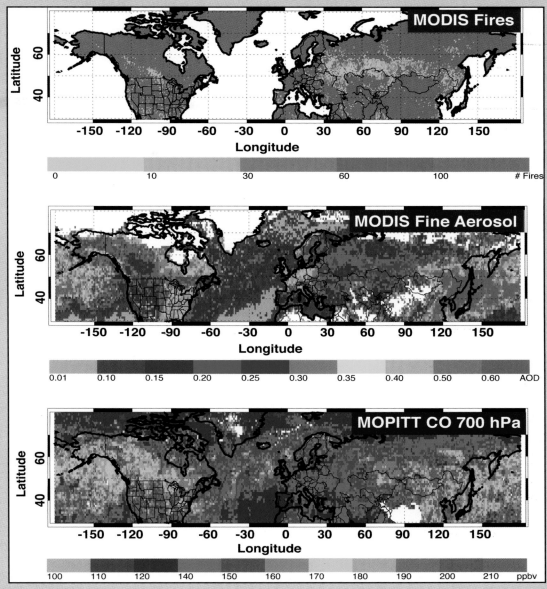

Pollution outflow from spring 2003 fires in Siberia can be seen in the top and middle image. These fires produced large amounts of fine carbon aerosol detected by MODIS instrument (bright colours) on the Terra satellite, which spread over the Pacific Ocean but lasted only a few days. They also produced carbon monoxide, which was detected by the MOPITT instrument on the Terra satellite (bottom image). This gas can last over a month, which allowed it to cross the Pacific Ocean and reduce air quality over North America before continuing on around the globe. *Credit: David Edwards, The National Center for Atmospheric Research (NCAR) (NASA 2004e)*

Case Study: African Fires 2002

Wildfires—from forest fires and brush fires to grass fires and slash-and-burn agriculture—can be sweeping and destructive conflagrations, especially in wilderness or rural areas. As biomass burns, particulates, black carbon, and gases including CO_2, CO, NOx, CH_4, and CH_3Cl are produced in great quantities. All of these pollutants can be lofted relatively high in the atmosphere due to the convective heating of a raging fire (Graedel and Crutzen 1993).

The image at right shows fire activity in Africa from 1 January 2002 to 31 December 2002. The fires are shown as tiny dots with each dot depicting the geographic region in which fire was detected. The color of a dot represents the number of days since a sizable amount of fire was detected

in that region, with red-orange representing less than 20 days, orange representing 20 to 40 days, yellow representing 40 to 60 days, and gray to black representing more than 60 days. These data were gathered by the MODIS instrument on the Terra satellite. MODIS detects fires by measuring the brightness temperature of a region in several frequency bands and looking for hot spots where this temperature is greater than the surrounding region.

Global statistics on the amount of land burned worldwide every year vary considerably. It has been estimated that from 7.5 million to 8.2 million km^2 (4.6 million to 5.1 million square miles) are burned and between 1 800 million and 10 000 million metric tonnes of dry biomass are consumed in fires annually. Global change scenarios predict an increase in total area

African Fires during 2002 *Credit: http://svs.gsfc.nasa. gov/vis/a000000/a002800/a002890/index.html (NASA 2004f)*

burned, with an increase in very large and intense fires.

Case Study: Pollution in China 2001 and 2004

During February 2004, a considerable outflow of pollution stemmed from China and Southeast Asia. The image at right shows atmospheric concentrations of carbon monoxide at an altitude of roughly 3 km (1.9 miles) over this region that were moving across the Pacific Ocean and at some points reaching the western coast of the United States.

Carbon monoxide is a good indicator of air pollution since it is produced during combustion processes, such as the burning of fossil fuels in urban and industrial areas, as well as by wildfires and biomass burning in more rural areas. Industrial emissions were mainly responsible for the high levels of carbon monoxide over China in this image, whereas emissions in Southeast Asia were due primarily to agricultural fires.

Natural processes and events can also be a source of transcontinental air pollution. In 2001, a large dust storm developed over China (see below). Prevailing winds swept particulates from this storm all the way to the eastern coast of North America.

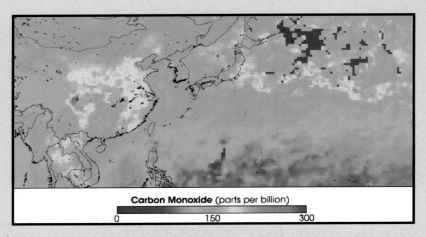

Carbon Monoxide (parts per billion)

0   150   300

This image was developed from a composite of carbon monoxide data collected over China and Southeast Asia from 1-25 February 2004, by the Measurements of Pollution in the Troposphere (MOPITT) instrument aboard NASA's Terra satellite. The colors represent the mixing ratios of carbon monoxide in the air, given in parts per billion by volume. The grey areas show where no data were collected due to persistent cloud cover. *Source: http://earthobservatory.nasa. gov/Natural (NASA 2004g)*

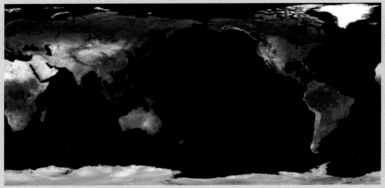

6 April 2001 normal aerosol levels are apparent on the first day of the dust storm.

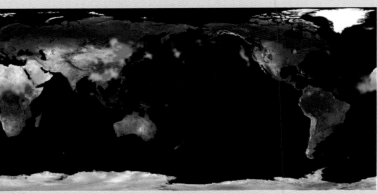

7 April 2001 - Blue represents just slightly higher-than-normal aerosols and red represents the highest concentration of aerosols.

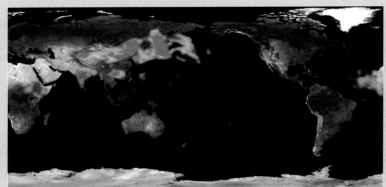

10 April 2001 - The aerosol impact from the dust storm can clearly be seen over China, Mongolia, Russia, Korea, Japan, and the Pacific Ocean.

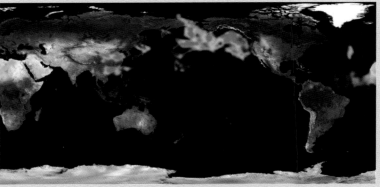

13 April 2001 - The aerosol impact from the dust storm can clearly be seen over Japan, the Pacific Ocean, Alaska, and the United States.

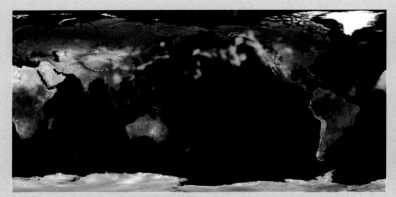

14 April 2001 - The aerosol impact from the dust storm can clearly be seen in the Pacific Ocean, and the United States.

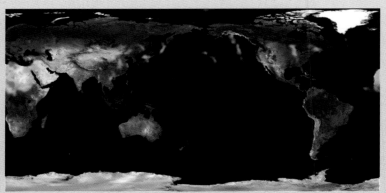

17 April 2001 - High levels of aerosols are visible over the East Coast of the United States, especially Maine.

A large dust storm developed over China on 6-7 April 2001. This series of images shows air-borne dust from the storm moving over China, Russia, Japan, the Pacific Ocean, Canada, and ultimately over the United States on 17 April 2001 (NASA 2004g). *Visualization Credit: NASA/Goddard Space Flight Center Scientific Visualization Studio Source: http://svs.gsfc.nasa.gov/vis/a000000/a002900/a002957/index.html*

3.2 Coastal Areas

Oceans cover roughly 70 per cent of the Earth's surface and make up some 90 per cent of space habitable by living things. They contain a vast, and largely unexplored, diversity of life, from the smallest of microorganisms to blue whales, the largest mammals on the Earth. Oceans are essential for the ecological functions and resources they provide, including food, medicines, and energy for millions of people worldwide (UNEP-WCMC 2003).

The world's oceans have a great effect on global climate. Water has a high capacity for retaining heat. Because so much of the Earth's surface is covered by oceans, the temperature of the atmosphere is kept fairly constant and within the range necessary to support life. Currently, the oceans also moderate climate change by absorbing a third of the carbon dioxide (CO_2) emitted into the air by human activity (Harrison and Pearce 2001). However, global warming may reduce the ocean's capacity to act as a CO_2 sink by 10 to 20 per cent over the next century (Houghton et al. 2001).

Different parts of the global ocean affect climate in different ways. The Indian Ocean/West Pacific Warm Pool, for example, is an enormous expanse of warm surface water. It extends almost half way around the Earth, stretching along the equator south of India, through the waters off Sumatra, Java, Borneo, and New Guinea, and into the central Pacific Ocean. The waters of the Warm Pool are warmer than any other area of open ocean on the Earth. Because these waters are warm to enough to drive moisture high into the atmosphere, the Warm Pool has a large effect on the climate of the lands that surround it. The slow fluctuations in size and intensity of the Warm Pool may be linked with the intensity of the El Niño phenomenon.

In addition to the ocean's climate-buffering capacity, its salty waters contain billions of tiny algae and other planktonic organisms. These life forms carry out a large part of the oxygen-generating photosynthetic processes that occur on the planet (Biomes Group 1996). For people, oceans represent one of the greatest sources of food on the Earth as well. Global fish production exceeds that of cattle, sheep, poultry, or eggs. It represents the largest source of wild or domestic protein in the world (UNEP-WCMC 2003). Marine fish catch rose from 51 million metric tonnes in 1975 to nearly 70 million metric tonnes in 1999 (UNEP 2002).

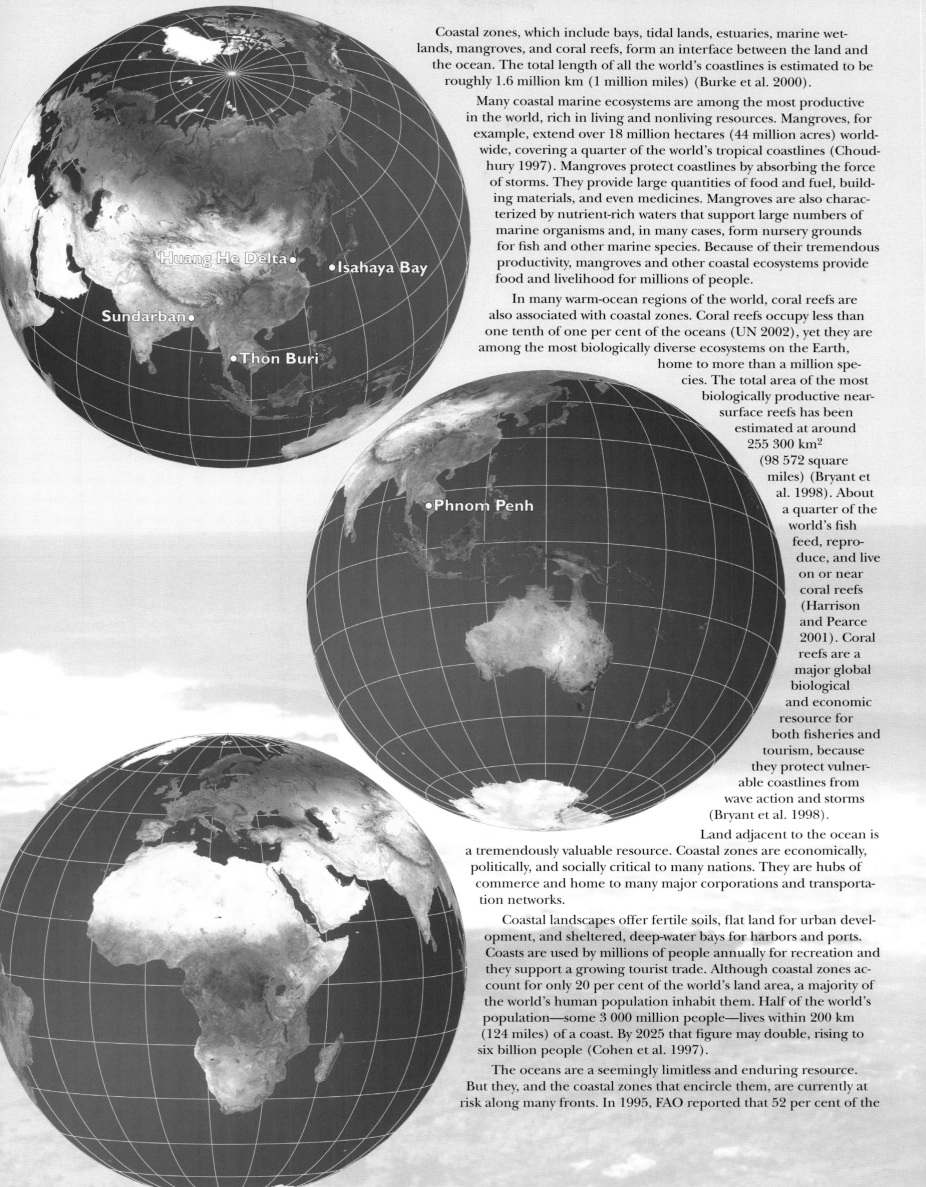

Coastal zones, which include bays, tidal lands, estuaries, marine wet-lands, mangroves, and coral reefs, form an interface between the land and the ocean. The total length of all the world's coastlines is estimated to be roughly 1.6 million km (1 million miles) (Burke et al. 2000).

Many coastal marine ecosystems are among the most productive in the world, rich in living and nonliving resources. Mangroves, for example, extend over 18 million hectares (44 million acres) world-wide, covering a quarter of the world's tropical coastlines (Choud-hury 1997). Mangroves protect coastlines by absorbing the force of storms. They provide large quantities of food and fuel, build-ing materials, and even medicines. Mangroves are also charac-terized by nutrient-rich waters that support large numbers of marine organisms and, in many cases, form nursery grounds for fish and other marine species. Because of their tremendous productivity, mangroves and other coastal ecosystems provide food and livelihood for millions of people.

In many warm-ocean regions of the world, coral reefs are also associated with coastal zones. Coral reefs occupy less than one tenth of one per cent of the oceans (UN 2002), yet they are among the most biologically diverse ecosystems on the Earth, home to more than a million spe-cies. The total area of the most biologically productive near-surface reefs has been estimated at around 255 300 km^2 (98 572 square miles) (Bryant et al. 1998). About a quarter of the world's fish feed, repro-duce, and live on or near coral reefs (Harrison and Pearce 2001). Coral reefs are a major global biological and economic resource for both fisheries and tourism, because they protect vulner-able coastlines from wave action and storms (Bryant et al. 1998).

Land adjacent to the ocean is a tremendously valuable resource. Coastal zones are economically, politically, and socially critical to many nations. They are hubs of commerce and home to many major corporations and transporta-tion networks.

Coastal landscapes offer fertile soils, flat land for urban devel-opment, and sheltered, deep-water bays for harbors and ports. Coasts are used by millions of people annually for recreation and they support a growing tourist trade. Although coastal zones ac-count for only 20 per cent of the world's land area, a majority of the world's human population inhabit them. Half of the world's population—some 3 000 million people—lives within 200 km (124 miles) of a coast. By 2025 that figure may double, rising to six billion people (Cohen et al. 1997).

The oceans are a seemingly limitless and enduring resource. But they, and the coastal zones that encircle them, are currently at risk along many fronts. In 1995, FAO reported that 52 per cent of the

Credit: Chansareek/UNEP/Topfoto

oceans' wild fish stocks were being exploited at the maximum sustainable limit, 16 per cent were already overexploited, and seven per cent were depleted. Only 23 percent will be able to sustain further expansion. (FAO 2005). In 24 countries for which sufficient data were available, trawling grounds encompass 8.8 million km² (3.4 million square miles), of which about 5.2 million km² (2 million square miles) are located on continental shelves. This represents about 57 per cent of the total continental shelf area of these countries (Burke et al. 2000).

However, not all of the decline in ocean fisheries may be attributed to fishing. Global warming may also be partly to blame (Beaugrand et al. 2003). The depletion of natural fish stocks has prompted expansion of aquaculture—the farming of fish—in many areas, a reaction to ocean degradation that does provide employment and food, but also carries with it the potential for pollution and other concerns.

Case Study: Dumping of Radioactive Waste at Sea

The Report of the United Nations Conference on Human Environment held in Stockholm in 1972 defined the principles for environmental protection, specifically for the assessment and control of marine pollution. These were forwarded to an Inter-Governmental Conference held in London later that year, where the Convention on the Prevention of Marine Pollution by Dumping of Wastes and Other Matter (also known as the London Convention of 1972) was adopted and which entered into force on 30 August 1975.

The contracting parties to the London Convention agreed to promote the effective control of all sources of pollution of the marine environment and to take all practicable steps to prevent the pollution of the ocean by dumping of waste and other matter that is liable to create hazards to human health and to harm living resources and marine life. The International Atomic Energy Agency (IAEA) was designated as the international body that should oversee matters related to the disposal of radioactive wastes in the ocean.

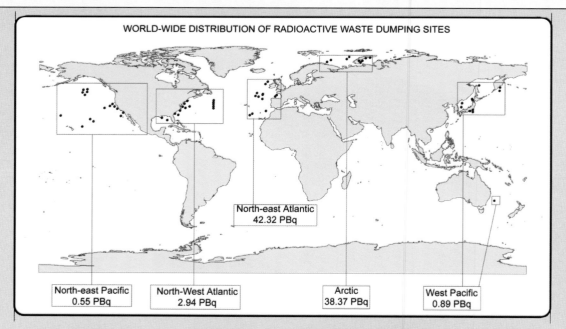

The first reported ocean disposal operation of radioactive waste was carried out by the USA in 1946 in the North-East Pacific Ocean and the latest was carried out by the Russian Federation in 1993 in the Japan Sea/East Sea. During the 48 year history of sea disposal, 14 countries have used more than 80 sites to dispose of approximately 85 PBq (1 PBq = 1015 Bq) of radioactive waste.

The figure shows the geographical distribution of disposal operations.

Source: Modified from http://www.oceansatlas.org/servlet/CDSServlet?status=ND0xNDExMyY3PWVuJjY xPSomNjU9a2%z. Figure has been modified from http://www.oceansatlas.com/unatlas/about/physicalandchemicalproperties/radiosp/index.htm.

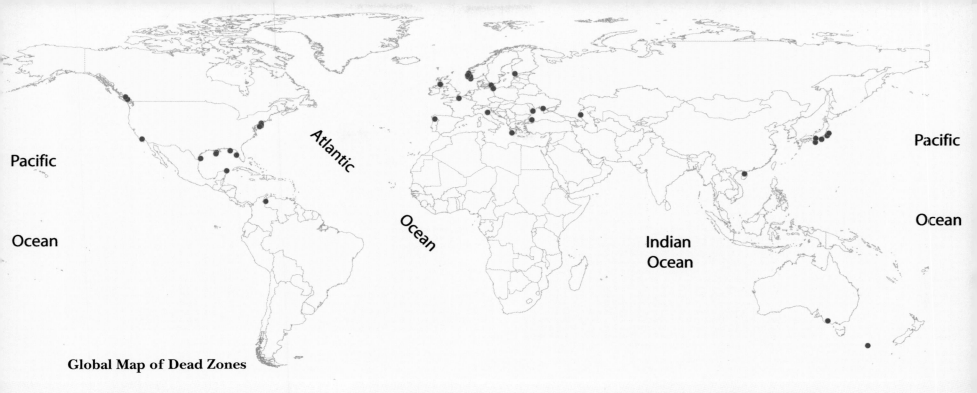

Global Map of Dead Zones

Ocean pollution is a growing and serious problem. Most of the wastes and contaminants produced by human activities eventually end up in the oceans. Chemical contamination and litter exist from the poles to the tropics and from beaches to ocean depths. Some pollutants are directly drained or dumped into the oceans, either intentionally, or accidentally as in the case of oil spills. Currently, the oceans must absorb an estimated 3 million tonnes of oil spilled annually from ships and, predominantly, from sources on land (Harrison and Pearce 2001). Other ocean pollutants first enter the atmosphere and later settle on ocean waters. Rivers that flow into oceans carry runoff from city streets, sewage, industrial wastes, pesticides and fertilizers from cropland, and silt from land-clearing and construction projects.

Because of their proximity to land, coastal waters tend to have far higher concentrations of pollutants than the open ocean (UNEP-WCMC 2003).

Polluted waters can upset the ecological balance among species in marine systems. For example, the number of poisonous algal species identified by scientists has nearly tripled since 1984, increasing fish kills, beach closures, and economic losses. Large parts of the Gulf of Mexico are now considered biological dead zones due to algal blooms (UN 2002). Dead zones are coastal areas in which bottom waters (those near the ocean floor) contain very little dissolved oxygen. This condition is called hypoxia, meaning "low oxygen" (ESA 2003). Very few organisms are able to survive in ocean dead zones (Figure 3.20). The dead zone in the Gulf of

Figure 3.20: The map above shows 58 reported ocean dead zones in 1995. The oldest and most well-studied marine dead zones are found in the Gulf of Mexico, the Black Sea, and the Baltic Sea. In 1995, the most severe case of hypoxia was in the Baltic Sea, in which about one-third, or 98 800 km² (38 000 square miles), of the body of water was reported lifeless. *Source: Modified from: http://daac.gsfc.nasa.gov/CAMPAIGN_DOCS/OCDST/dead_zones.html*

Mexico is the largest hypoxic zone in the Western Hemisphere and is also one of the largest in the world (Downing 1999; Greenhalgh 2003). In many places, the hypoxic waters of dead zones are gradually spreading out to cover larger and larger areas of ocean floor (Kempler 2000).

Marine bio-invasions have also become major global environmental and economic problems. At any one time, several thousand species are estimated to be in

Case Study: Mississippi Dead Zone 2004

Recent reports indicate that the large region of oxygen-depleted water—a dead zone—spreads across nearly 15 080 km² (5 800 square miles) of the Gulf of Mexico in what appears to be an annual event. NASA satellites monitor the health of the oceans and spot the conditions that lead to a dead zone. The photo (right) shows sediment-choked water from the Neuse River flowing out into the Gulf of Mexico near the coast (NASA 2004h).

Source: Mississippi Dead Zone, 2004
http://www1.nasa.gov/vision/earth/environment/dead_zone.html

Credit: Unknown/UNEP/NASA

Phytoplankton bloom off Denmark, 2004 (shown in light blue color). *Source: http://rapidfire.sci.gsfc.nasa.gov/gallery/?2004153-0601/Denmark.A2004153.1145.1km.jpg*

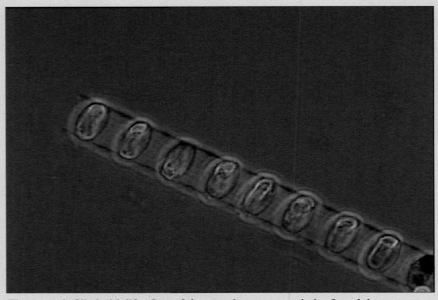

The ocean is filled with life. One of the most important varieties found there are photosynthetic phytoplankton, tiny organisms that form the base of the oceanic food web. *Source: http://www1.nasa.gov/vision/earth/environment/dead_zone.html*

Coccolithophore bloom off Brittany, France, 2004. *Source: http://rapidfire.sci.gsfc.nasa.gov/gallery/?2004167-0615/France.A2004167.1335.148.1km.jpg*

Case Study: Red Tides

Throughout the oceans, there are places where strong currents bring nutrient-rich deep waters to the surface. These upwelling nutrients nourish tiny, free-floating microscopic algae and other photosynthesizing planktonic organisms collectively known as phytoplankton. Most species of phytoplankton are not harmful. Rather, they form the base of the marine food web. Occasionally, however, phytoplankton grow and reproduce very quickly—they "bloom"—and accumulate into dense, visible patches near the water's surface. Some phytoplankton blooms are called "red tides," especially when the species involved contain red pigments and so cause the water to turn pink or red when a bloom is in progress. Phytoplankton blooms, however, have nothing to do with tides.

When phytoplankton die, they sink to the ocean floor where their remains are broken down by different kinds of bacteria. Some of these bacteria emit hydrogen sulfide gas as a by-product of decay reactions. The gas collects on the ocean floor until it bubbles up toward the surface and combines with oxygen to yield water and sulfur. The sulfur precipitates as a white solid.

Hydrogen sulfide gas by itself is toxic to fish. But its ability to bind with oxygen also depletes that essential gas from the water column. If enough oxygen is removed, deadly low-oxygen (hypoxic) conditions are created in the ocean.

the ballast tanks of the world's shipping fleet, enroute to other parts of the world. The Atlantic box jelly, believed to have been released in a ship's ballast water, has wrought ecological havoc in the Black Sea. Scientists estimate that in San Francisco Bay, a new foreign species takes hold every 14 weeks (UN 2002).

The problems are not confined to ocean waters. Many coastal ecosystems have been destroyed, and many more degraded, often as a result of human activities. Mangroves, wetlands, seagrass beds, and coral reefs are all disappearing at an alarming rate. Anywhere from 5 to 80 per cent of original mangrove area in various countries has been lost, particularly in the last 50 years (Burke et al. 2000). A major contributor to this loss is the conversion of mangroves to rice paddies and shrimp farms. With coastal regions set to double their human populations over the next 25 years, coastal ecosystems are coming under increasing threat (Harrison and Pearce 2001).

Coral reefs are particularly vulnerable to environmental change and damage from human activities. Nearly two thirds of all the world's coral reefs are deteriorating (Pomerance 1999). They are dismembered by souvenir-seeking divers, mined for building materials, and damaged by the anchors of cruise ships. Silt from dredging, deforestation, and urban sewage smothers and kills coral, or fosters the growth of suffocating and sometimes toxic algae (Harrison and Pearce 2001). Coral reefs are also subject to remote threats. Dust carried aloft by storms in Africa (Figure 3.21), and then spread across the Atlantic on prevailing winds, may have introduced bacterial infections to Caribbean reefs (USGS 2003).

Global warming is also a threat to coral reefs. Higher concentrations of carbon dioxide in the air make surface waters more acidic and reduce coral growth rates

Credit: Shoukyu/UNEP/Topfoto

(Kleypas et al. 1999). The rise of ocean temperatures by half a degree or more in recent decades has already placed many reefs at the top end of the temperature ranges they can tolerate without undergoing "bleaching" (Harrison and Pearce 2001). As corals reach the limits of their heat tolerance, they expel symbiotic algae from their bodies and become "bleached." Epidemic coral bleaching in the 1990s, which peaked with the El Niño-induced warming of 1998, is believed to have killed more coral in the last few years of the 20th century than from all human causes to date (Pearce 1999). Continued warming could cause sea levels to rise at a rate that coral reefs cannot match.

Coastal lands have been greatly impacted by human activities. Fifty-one per cent of the world's coastlines are under "moderate" or "high" threat from development activities (Bryant et al. 1995). Nineteen per cent of all lands within 100 km (62 miles) of the coast (excluding Antarctica and water bodies) are classified as "altered," having been turned into agricultural or urban areas; 10 per cent are "semi-altered," involving a mosaic of natural and altered vegetation; and 71 per cent fall within the "least modified" category. A large percentage of the coastal lands in this least

modified category, however, includes many found in uninhabited areas in northern latitudes (Burke et al. 2000).

The destruction of coastal ecosystems and the deterioration in the quality of ocean waters, together with overexploitation of resources, are seriously impacting the survival of the ecosystems and the people that depend on them (SIDA n.d.).

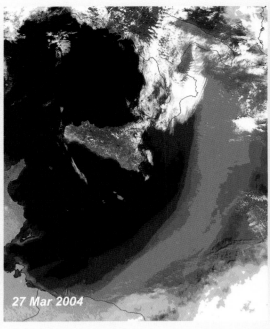

27 Mar 2004

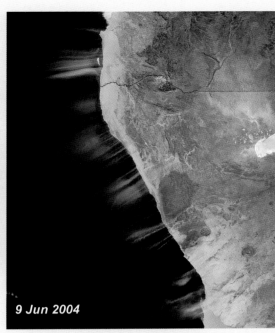

9 Jun 2004

Figure 3.21: While Saharan dust provides coral reefs with essential nutrients like iron and copper, Saharan dust also introduces bacterial infections to coral reefs. The image to the left is a Saharan dust storm spreading out over the Mediterranean in 2004. The image to the right is of large dust plumes off Namibia in 2004. *Source: http://rapidfire.sci.gsfc.nasa. gov/gallery/?2004125-0504/Libya.A2004125.0940.1km.jpg, http://rapidfire.sci.gsfc.nasa.gov/gallery/?2004161-0609/Namibia.A2004161.0930.1km.jpg*

Case Study: Africa's Kipini Wildlife and Botanical Conservancy

Christian Lambrechts

The coastal region of Kenya is famous for its natural beauty, rich culture, diverse communities, and as a recreational resource. This has ensured economic benefits for Kenya from mainly non-consumptive utilization of its natural resources through tourism. Yet this development has not taken place without considerable environmental cost. Excessive construction along the coastline, uncontrolled access to marine environments, factory fishing operations, and poor planning have led to a decline in the quantity and quality of both land and marine resources. This has been reflected in loss of biodiversity, dwindling fish stocks, and declining employment and tourism figures, although the latter are also attributable to recent security concerns.

The Kipini Wildlife and Botanical Conservancy is located along the coast of eastern Kenya. It lies within the Lamu and Tana River districts, but is connected ecologically to the Ijara District. It is a well-preserved area with high biodiversity, although few conservation projects have ever been implemented and current protection is inadequate with wildlife reserves only on paper. With increasing

Credit: Christian Lambrechts/UNEP/UNEP-GRID Nairobi

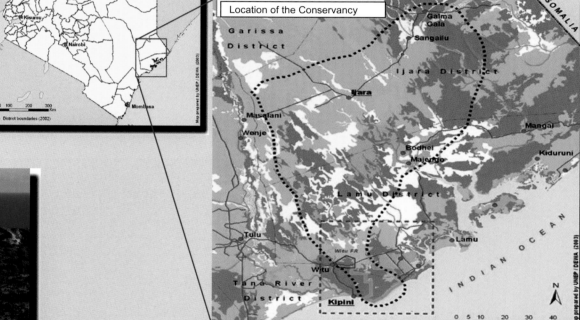

Location map

Location of the Conservancy

Credits: Christian Lambrechts/UNEP/UNEP-GRID Nairobi

population pressures on the natural resources there is a need for prompt action.

The Kipini Conservancy initially focuses on what was once known as Nairobi Ranch, an area approximately 16 000 hectares (40 000 acres) in size that is situated between the historical towns of Kipini, Witu and Lamu. The Swaleh Nguru (Sherman) family secured the land under a freehold arrangement and has maintained it even at a loss under livestock operations. Development of the ranch is necessary, as Kenya can ill afford idle land. But development is being undertaken with considerable care to ensure adequate environmental conservation. By creating an easement on this freehold land, the Swaleh Nguru family has put it in trust for future generations of Kenyans as well as visitors to the country. The conservancy will need to be vigorously managed and will involve a transition from a cattle-based ranching system to a more natural landscape populated by native species. Income generation will be based on heritage value and donor support in the short term, with increasing reliance on eco-tourism.

In the future, the Kipini Conservancy is expected to be expanded to include the range of critically endangered species. In a recent study funded by the Finnish Embassy, Ader's duiker, an extremely rare mammal once thought to be virtually extinct in Kenya, was found in the area near the Conservancy. Expansion of the Conservancy will also help to preserve highly diverse habitat corridors between the coast and the interior and will involve surrounding communities in conservation efforts.

The coastal marine ecosystem adjacent to the Conservancy is part of the Global 200 ecoregion. It supports a great diversity of animal and plant life and is known as a turtle nesting area. Several species of whales and dolphins are found in the waters offshore, as well as the globally threatened dugong (*Dugong dugon*). The part of the Conservancy that borders the Tana River delta is a stop-over and wintering grounds for many migratory bird populations. The area also provides habitat for threatened shorebirds and seabirds (UNEP/GRID–Nairobi).

Río Negro
Biological Reserve

H O N D U R A S

Gulf of
Fonseca

Estero Real
Nature Reserve

Estero Real

0   5   10
Kilometres

N

6 Jan 1987

Gulf of Fonseca

COASTAL AREAS
GULF OF FONSECA, HONDURAS

Honduras is second only to Ecuador in the production and export of cultured shrimp from Latin America. Vast areas of the delta have been converted into farms for the cultivation of shrimp.

The rapid growth of shrimp aquaculture in Honduras has caused both environmental and social

Río Negro
Biological Reserve

H O N D U R A S

Gulf of
Fonseca

Estero Real
Nature Reserve

Estero Real

0   5   40
Kilometres

N

15 Nov 1999

Shrimp aquaculture in Honduras began in the early 1970s and continued in the 1980s in the hands of both national and international enterprises.

Credit: Unknown/UNEP/Topfoto

problems. Shrimp farmers are depriving fishers, farmers and others of access to mangroves, estuaries and seasonal lagoons; destroying mangrove ecosystems, altering the hydrology of the region, destroying the habitats of other flora and fauna and precipitating declines in biodiversity; contributing to degraded water quality; and exacerbating the decline in Gulf fisheries through the indiscriminate capture of other species caught with the shrimp

post larvae that are used to stock ponds.

These two images provide a visual comparison of the increase in coverage by shrimp farms in the Gulf of Fonseca over time. It is evident from the images that between 1987 and 1999, a period of about 12 years, the total area under shrimp farming has increased tremendously.

E C U A D O R

Guayaquil

Puerto
Nuevo

Gómez Rendón

Playas

Posorja

Isla Puná

Gulf of
Guayaquil

27 Mar 1985

0   10   20
Kilometres

Gulf of Guayaquil

COASTAL AREAS
GULF OF GUAYAQUIL, ECUADOR

Guayaquil is Ecuador's largest city and primary sea
port. It is located on the Guayas River, which empties
into the huge Gulf of Guayaquil along the country's
southern coastline. Throughout the Gulf, mangroves
have been steadily converted to shrimp aquaculture
ponds for producing farmed shrimp. In a 15-year pe-

ECUADOR

Gómez Rendón

Playas

Posorja

Gulf of
Guayaquil

Isla Puná

Guayaquil

Puerto
Nuevo

23 Nov 2000

0   10   20
Kilometres

N

riod, coastal area developed for shrimp aquaculture grew by approximately 30 per cent, from 90 000 hectares (222 395 acres) in 1984 to 118 000 hectares (291 584 acres) in 2000 (CLIRSEN 2000). Roughly 70 per cent of Ecuador's shrimp farming activities are located in the Gulf of Guayaquil.

In this pair of satellite images, the loss of mangroves and growth of the aquaculture industry can be seen along the coast and in the altered dendritic patterns (branching like a tree) of coastal waterways, especially those on the large island of Puna. Mangroves provide fish breeding grounds and wildlife habitat, act as natural barriers against storm surge, and filter groundwater. Converting mangroves to aquaculture ponds has wide-reaching environmental implications.

C H I N A

Huang He (Yellow)

Xishuanghe

Dongying

Laizhou Wan

0   10   20
Kilometres

N

27 May 1979

Huang He Delta

COASTAL AREAS
HUANG HE DELTA, CHINA

The Huang He (Yellow River) is the muddiest river on Earth and is China's second longest river, running 5 475 km (3 395 miles) from eastern Tibet to the Bohai Sea. The Huang He's yellow color is caused by its tremendous load of sediment, composed primarily of mica, quartz, and feldspar particles. The sediment enters the

C H I N A

Huang He Yellow

Xishuanghe

Dongying

Laizhou Wan

0   10   20   Kilometres

N

2 May 2000

water as the river carves its way through the highly erodable loess plateau in north-central China. (Loessial soil is called huang tu, or "yellow earth," in Chinese.)

Centuries of sediment deposition and dike building along the river's course has caused it to flow above the surrounding farmland in some places, making flooding a critically dangerous problem. Where the Huang He flows into the ocean, sediments are continuously deposited in the river delta, where they gradually build up over time. Between 1979 and 2000—as these satellite images show—the delta of the Huang He expanded dramatically. Several hundred square kilometres of newly formed land were added to China's coast during this period.

IJsselmeer

Dike

Emmeloord

IJsselmeer

Enkhuizen

NETHERLANDS

Hoorn

Dronten

Markermeer

Lelystad

Flevoland

Amsterdam

21 Aug 1964

Enkhuizen

NETHERLANDS

Hoorn

Dronten

Markermeer

Lelystad

Flevoland

Harderwijk

Almere

N

0   10   20
Kilometres

23 Mar 1973

Amsterdam

• IJsselmeer

COASTAL AREAS
IJSSELMEER, NETHERLANDS

The Zuiderzee is a large body of water along Holland's northeastern coast. Between 1927 and 1932, a 30 km (19 miles) dam, known as the Afsluitdijk, was built across the Zuiderzee, separating it into the outer Waddenzee, which is open to the North Sea, and the inner IJsselmeer (Lake IJssel) where areas of reclaimed

IJsselmeer

Emmeloord

Enkhuizen

N E T H E R L A N D S

Hoorn

Dronten

Markermeer

Lelystad

Flevoland

Harderwijk

Almere

Amsterdam

10   20
|_____| Kilometres

N

30 Mar 2004

land—called polders—are used for agriculture and as villages. Dikes built since that time created additional polders that were drained using pumps and, at one time, wind mills.

These images, from 1964, 1973, and 2004, show the transformation of polders into useable farming land. The 1973 image shows a partially completed dike that, when completed allowed for the creation of the southernmost polder visible in the 1973 image. At that time, draining of the land had been completed and soil cultivation began. By 2004, this area of reclaimed land was covered with farms. The area of lighter blue water visible in the left of the 1973 and 2004 images is the Markermeer—a polder that was created but not drained. It forms a freshwater reservoir that acts as a buffer against floodwaters.

13 May 2001

J A P A N

Isahaya

Isahaya Bay

0 1 2 Kilometres

N

15 May 1993

Isahaya Bay

COASTAL AREAS
ISAHAYA BAY, JAPAN

Land reclamation began in Japan's Isahaya Bay in 1989 to separate approximately 3 000 hectares (7 413 acres) of tidal flats from the Ariake Sea and turn what was Japan's largest area of tidal lands into farmland. As these three satellites images show, the project has steadily progressed. In the 2001 image, the straight line of a 7-km (4 mile) sea wall is visible separating areas

J A P A N

Isahaya

Isahaya
Bay

0   1   2
                    Kilometres

N

23 Oct 2003

of light- and dark-colored water. Behind the sea wall, tidal flats can be seen drying as water is slowly drained away. In the 2003 image, that area has been fully reclaimed from the sea.

The Isahaya Bay Reclamation project has been fraught with controversy. Environmental groups have criticized the project for its destruction of wetland habitat. The Isahaya Bay area is known for its production of nori

(seaweed), and local farmers have complained that the reclamation project has negatively impacted the quality and abundance of the nori growing in the bay. The Isahaya project prompted the formation of the Japan Wetlands Action Network, a group of grassroots and national conservation organizations who are protesting the project and recommending that the sea wall gates be opened to restore ecological balance.

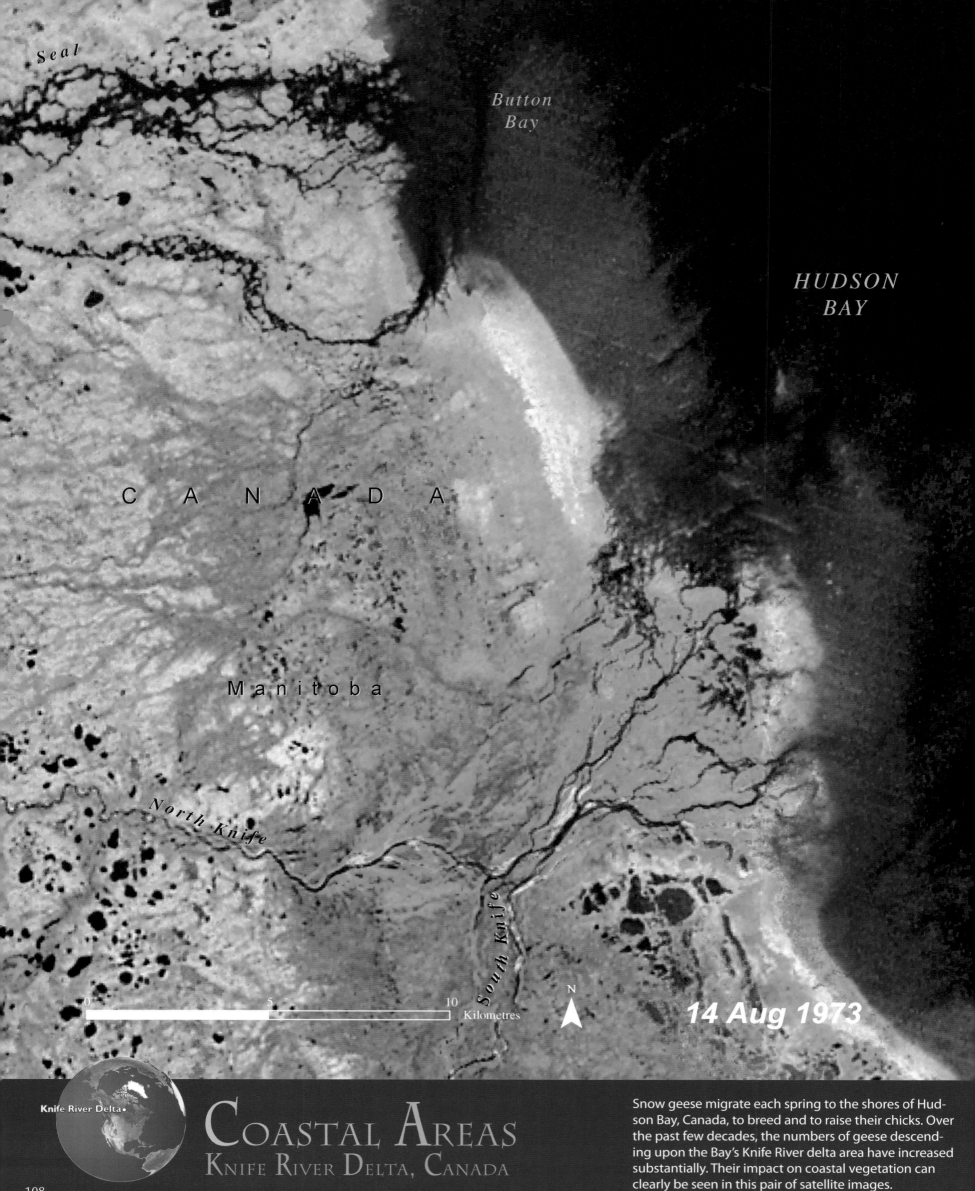

Seal

Button
Bay

HUDSON
BAY

CANADA

Manitoba

North Knife

South Knife

0   5   10
Kilometres

N

14 Aug 1973

Knife River Delta

COASTAL AREAS
KNIFE RIVER DELTA, CANADA

Snow geese migrate each spring to the shores of Hudson Bay, Canada, to breed and to raise their chicks. Over the past few decades, the numbers of geese descending upon the Bay's Knife River delta area have increased substantially. Their impact on coastal vegetation can clearly be seen in this pair of satellite images.

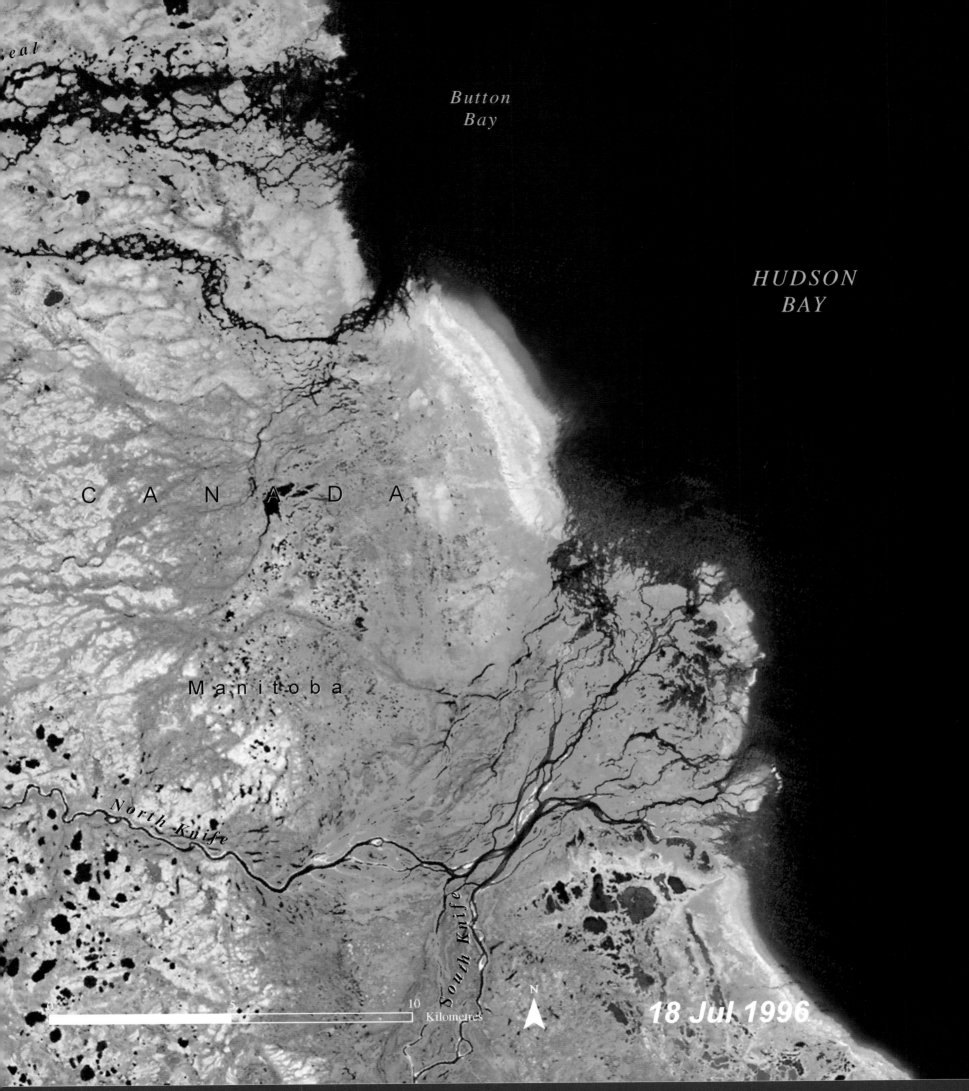

eal

Button
Bay

HUDSON
BAY

C A N A D A

M a n i t o b a

North Knife

South Knife

N

0   5   10
Kilometres

18 Jul 1996

In the image on the right, notice how the vegetation (green) has receded from the shoreline north of the delta. Snow geese have overgrazed this area and turned the shoreline into an enormous mudflat. Having denuded the shoreline of vegetation, the geese have also moved inland in search of food on the tundra, where overgrazed soil quickly becomes barren and develops a crust of salt due to evaporation. The salty layer prevents the regrowth of plants, and ultimately leads to erosion. Some researchers have suggested lifting restrictions on the hunting of snow geese in an attempt to reduce their numbers and control the overgrazing problem. Others believe such measures are "too little, too late."

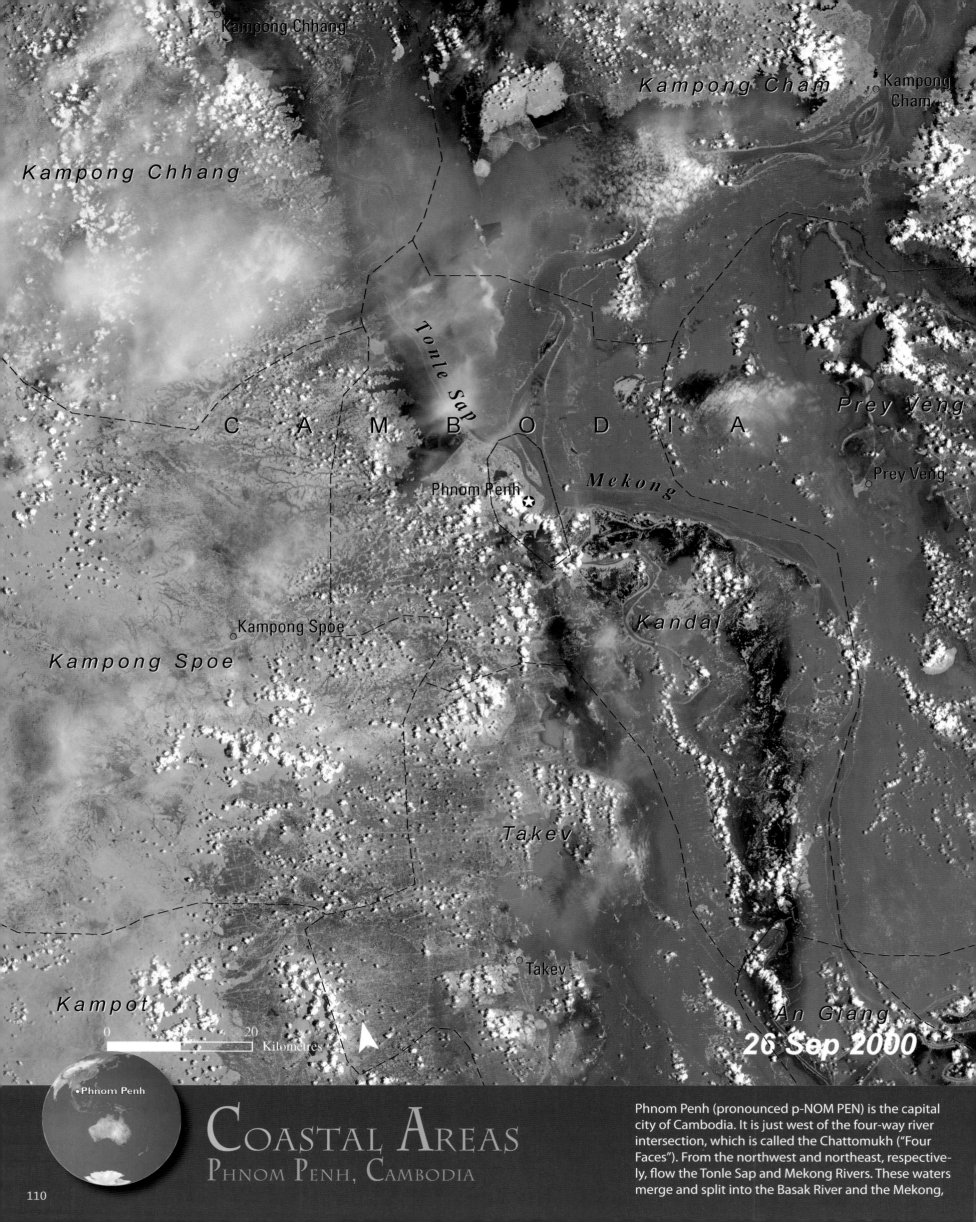

Kampong Chhang

Kampong Cham

Kampong Chham

Kampong Cham

CAMBODIA

Tonle Sap

Prey Veng

Phnom Penh

Mekong

Prey Veng

Kandal

Kampong Spoe

Kampong Spoe

Takev

Takev

Kampot

An Giang

26 Sep 2000

0   20
Kilometres

N

•Phnom Penh

COASTAL AREAS
PHNOM PENH, CAMBODIA

Phnom Penh (pronounced p-NOM PEN) is the capital city of Cambodia. It is just west of the four-way river intersection, which is called the Chattomukh ("Four Faces"). From the northwest and northeast, respectively, flow the Tonle Sap and Mekong Rivers. These waters merge and split into the Basak River and the Mekong,

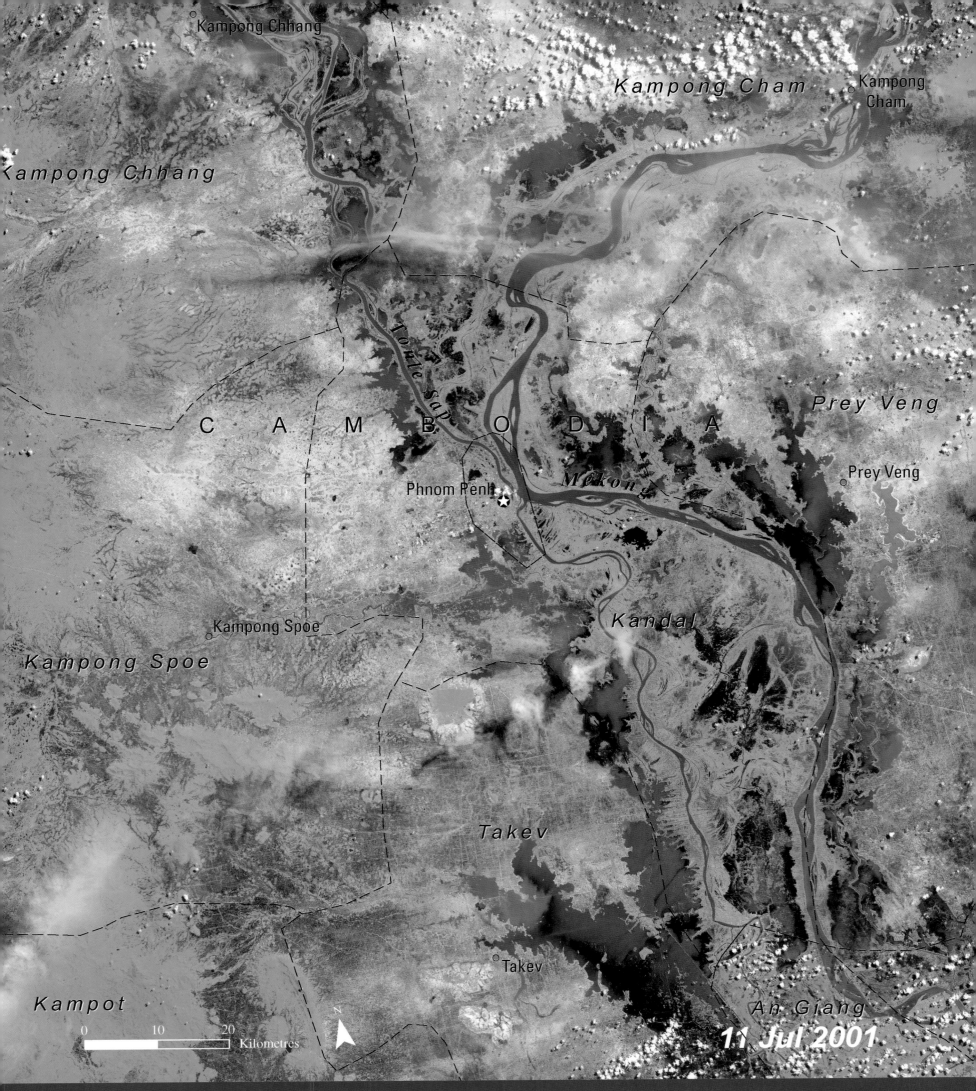

Kampong Chhang

Kampong Cham

Kampong Chhang

Kampong Cham

Tonle Sap

Prey Veng

C A M B O D I A

Mekong

Prey Veng

Phnom Penh

Kandal

Kampong Spoe

Kampong Spoe

Takev

Kampot

An Giang

Takev

0   10   20
Kilometres

N

11 Jul 2001

which flow southeast to the South China Sea. The Mekong River is the 12th longest in the world, flowing from western China to the Mekong Delta in southern Vietnam. Every autumn, monsoon rains are too great for the Mekong to carry, and it floods a large area of Cambodia. This flood even reverses the flow of the Tonle Sap River, northward to the Tonle Sap ("Great Lake") which can expand to ten times its normal size.

This pair of images show the extent of flooding associated with the two rivers. The 2000 image was taken during a period of flooding while the 2001 image was taken after the flood waters had receded. Visible also in the images, especially in the south-central area of the 2001 image, are extensive ditches and canals that are used in irrigation.

Calcutta

Kaliganj

Port Canning

INDIA

BANGLADESH

G a n g e s   D e l t a

0   20   40   Kilometres

N

9 Feb 1977 - 8 Feb 1977

Bay of Bengal

Calcutta

Shrimp Farms

Kaliganj

Port Canning

INDIA

BANGLADESH

G a n g e s   D e l t a

0   20   40   Kilometres

N

15 Nov 1999 - 28 Feb 2000

Bay of Bengal

•Sundarban

COASTAL AREAS
SUNDARBAN, INDIA/BANGLADESH

Sundarban, the largest mangrove forest of the world, is situated in the southwestern part of Bangladesh and in the West Bengal of India. Guarded by the Bay of Bengal, Sunderban is an excellent example of the coexistence of human and terrestrial plant and animal life. Despite

INDIA

BANGLADESH

Dhaka ★

INDIA

INDIA

Khulna ○

Calcutta ○

Khulna ○

Bay of Bengal

N

0   50   100
Kilometres

20 Oct 2004

high population pressure and environmental hazards such as siltation, cyclone flooding and sea level rise, the aerial extent of the mangrove forest has not changed significantly in the last 25 years. In fact, with improved management, the tiger population has increased from a mere 350 in 1993 to 500-700 in 2000 and ecotourism is progressing well. However, while sufficient data is not available, several reports suggest that forest degradation has been occur-

ring in many parts of Sundarban. The Sundarban's mangrove forests are also becoming more vulnerable due to the significant rise of shrimp farming in the region. The increase of shrimp farming has negatively affected agriculture and also contributed to the loss of mangrove forests during the past two decades.

Krung Thep

Bangkok
(Krung Thep)

Thon Buri

Chachoengsao

T H A I L A N D

Phra
Pradaeng

Chao Phraya

Samut Prakan

Samut Prakan

Bang Pakong

Chon
Buri

Shrimp farm
Credit: H. Gyde Lund/UNEP

Chon
Buri

*Gulf of
Thailand*

6 Jan 1973-16Dec 1978

Si Racha

0   5   10   Kilometres

N

COASTAL AREAS
THON BURI, THAILAND

As the city of Bangkok, Thailand, has grown, the need to provide food and an additional economic base for its burgeoning population has been a primary concern. Parts of the Thai coastline, including those near Bangkok, offer conditions favorable to aquaculture,

Krung Thep

★ Bangkok
(Krung Thep)

Thon Buri

T H A I L A N D

Chachoengsao

Phra
Pradaeng

Chao Phraya

Samut Prakan

Samut Prakan

Bang Pakong

Chon
Buri

Chon
Buri

Mangrove trees
Credit: H. Gyde Lund/UNEP

Gulf of
Thailand

Si Racha

0   5   10
Kilometres

N

8 Jan 2002

especially shrimp aquaculture. Over time, as these satellite images from 1978 and 2002 reveal, the mangroves that once lined the coast near Bangkok, as well as the rice paddies that lay further inland, have been replaced by aquaculture ponds (blue patches inland) and urban structures (light purple). The promotion and development of aquaculture has led to the current situation, where farmed shrimp and fish production now exceeds that of shrimp and fish capture by traditional methods. The development of this coastal industry has raised environmental concerns, as extensive areas of mangroves have been destroyed to make way for aquaculture ponds. The challenge of balancing the needs of people living in coastal areas versus the welfare of the coastal areas themselves is ongoing, and repeated along many coastlines worldwide.

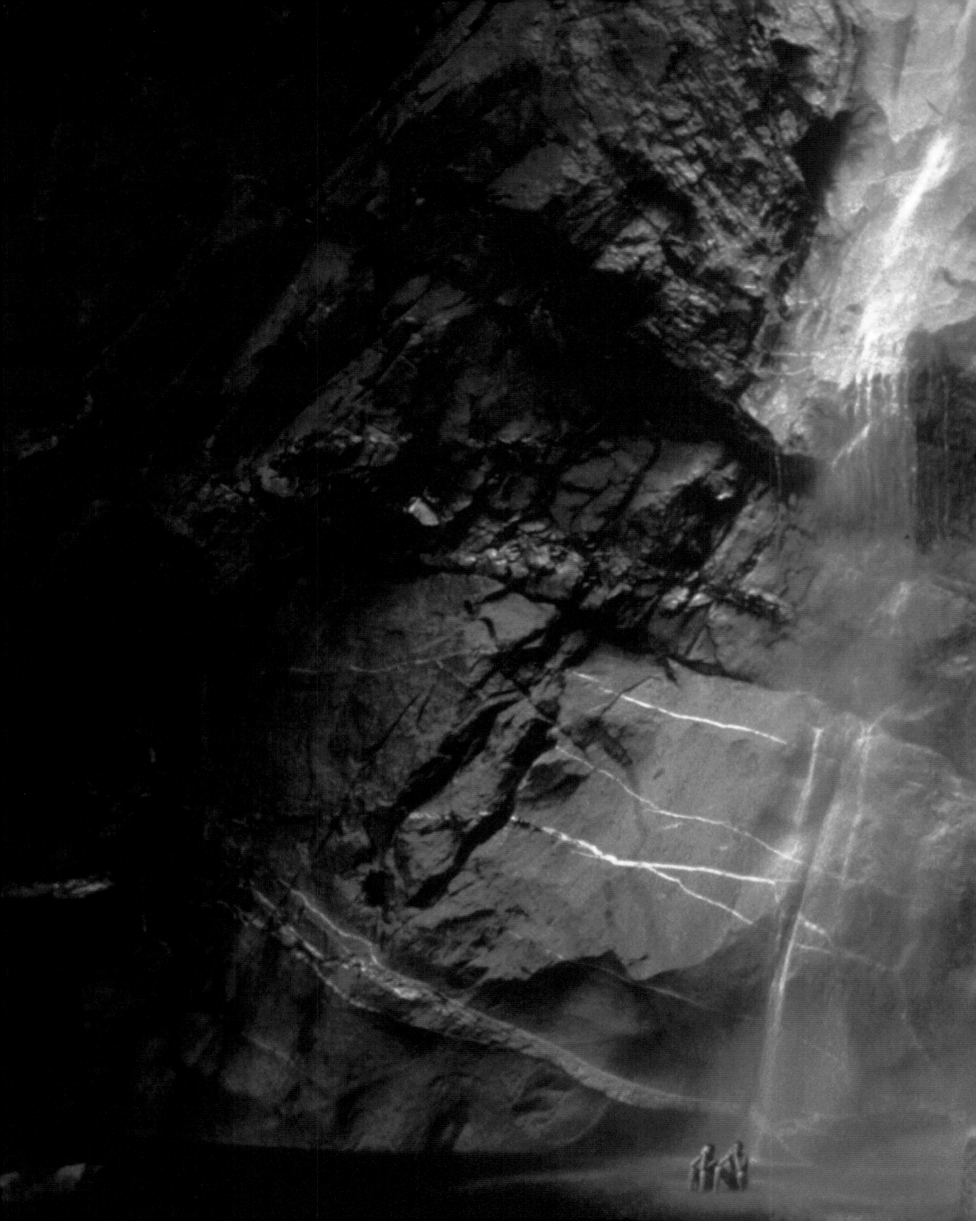

Credit: Laurent /UNEP/Topfoto

3.3 Water

Water is fundamental to almost all living things on the Earth. Human health—and survival—depends on a clean and reliable supply of fresh drinking water, as well as water for crop irrigation and sanitation (UNESCO 2000). Fresh water is water that has a very low salt content—usually less than one per cent. Only about 2.5 per cent of all water on the planet is fresh. Of that amount, only about 0.5 per cent is surface water (found in lakes, rivers, wetlands) or accessible groundwater. Rainfall is also a source of fresh water. But rainfall is unpredictable and amounts vary dramatically from place to place and season to season around the world (UNFPA 2001).

During the past century, world population has tripled. Over roughly the same period of time (1900 to 1995) water use worldwide has increased six-fold. Experts predict that by 2025, global water needs will increase even more, with 40 per cent more water needed for cities and 20 per cent more water for growing crops (Paden 2000). Yet while needs increase, the amount of available fresh water is dwindling worldwide.

Water withdrawals from rivers and underground reserves have grown by 2.5 to 3 per cent annually since 1940, significantly ahead of population growth. So much water is withdrawn from several of the world's major rivers, including the Colorado River in the United States, the Nile River in Egypt, and the Yellow River in China, that there is little to no water left by the time these rivers meet the sea (Postel et al. 1996). Demands on groundwater are equally great; water tables are falling on every continent.

Over the next two decades, it is estimated that the average supply of water per person will drop by one-third. Annually, lack of clean drinking water can be linked to roughly 250 million cases of water-related disease and between 5 and 10 million deaths worldwide. Thus, water shortages indirectly condemn millions of people to an avoidable premature death each year.

Water shortages are also impacting global grain markets, as arid countries that rely on irrigation for crop production switch from growing grain to importing it (Harrison and Pearce 2001). Irrigation accounts for 70 per cent of direct water consumption worldwide. It has been estimated that practices such as drip irrigation and inexpensive moisture monitors could cut agricultural water use by as much as 40 per cent (Wall 2001).

- Gabcikova
- Atatürk Dam
- Everglades
- Lake Chapala

The exploitation of the world's water resources has occurred at no small cost to the environment. Worldwide, all major rivers have water works that change flow regimes to some extent and therefore impact riparian ecosystems. Many endemic riparian species are disappearing (Ramsar Convention Bureau 1998), and twenty per cent of all freshwater fish are currently endangered. Few aquatic ecosystems have been as severely impacted as wetlands, however. During the 20th century, half of the Earth's wetland ecosystems—such as marshes, fens, swamps, and estuaries—disappeared (UNESCO 2003). Approximately 40 000 hectares (98 842 acres) of wetlands are destroyed each year as the result of human activities (Center for Environmental Resource Management n.d.). Drainage for agricultural production is the principal cause of wetland loss.

At the same time that global water supplies are declining, so is the quality of the water that remains. Water pollution is the presence of harmful and objectionable material—such as sewage, industrial wastes and chemicals, and run-off from land development or agriculture—in sufficient concentrations as to make water unfit for use (EEA 2004). Water pollution is a serious threat to the world's water supply. It is also a growing threat to the oceans that cover more than 70 per cent of the planet. People have long viewed these immense bodies of water as limitless dumping grounds for wastes. Over time, however, raw sewage, garbage, industrial wastes, and oil spills have begun to overwhelm the diluting capabilities of the oceans. Most coastal waters are now polluted, often severely (Revenga and Mock 2000, Revenga et al. 2000).

The 21st Century brings with it a global water crisis. Unless corrective and conservation measures are taken, it is estimated that by 2030 global demands for fresh water will exceed the supply (NSW EPA 2003, UNESCO 2000). Serious water and food security problems already exist in some developing countries and regions, and these demand urgent attention (FAO 2003).

Growing population in urban areas is exerting great pressure on water resources. Even if the world maintained the pace of the 1990s in water-supply development, this would not be enough to ensure that everyone had access to safe drinking water by the year 2025. The impacts of climate change—including changes in temperature, precipitation and sea levels—are expected to have varying consequences for the availability of freshwater around the world. Current indications are that if climate change occurs gradually, the impacts by 2025 may be minor, with some countries experiencing positive impacts while most experience negative ones. Climate change impacts are projected to become increasingly strong during the decades following 2025.

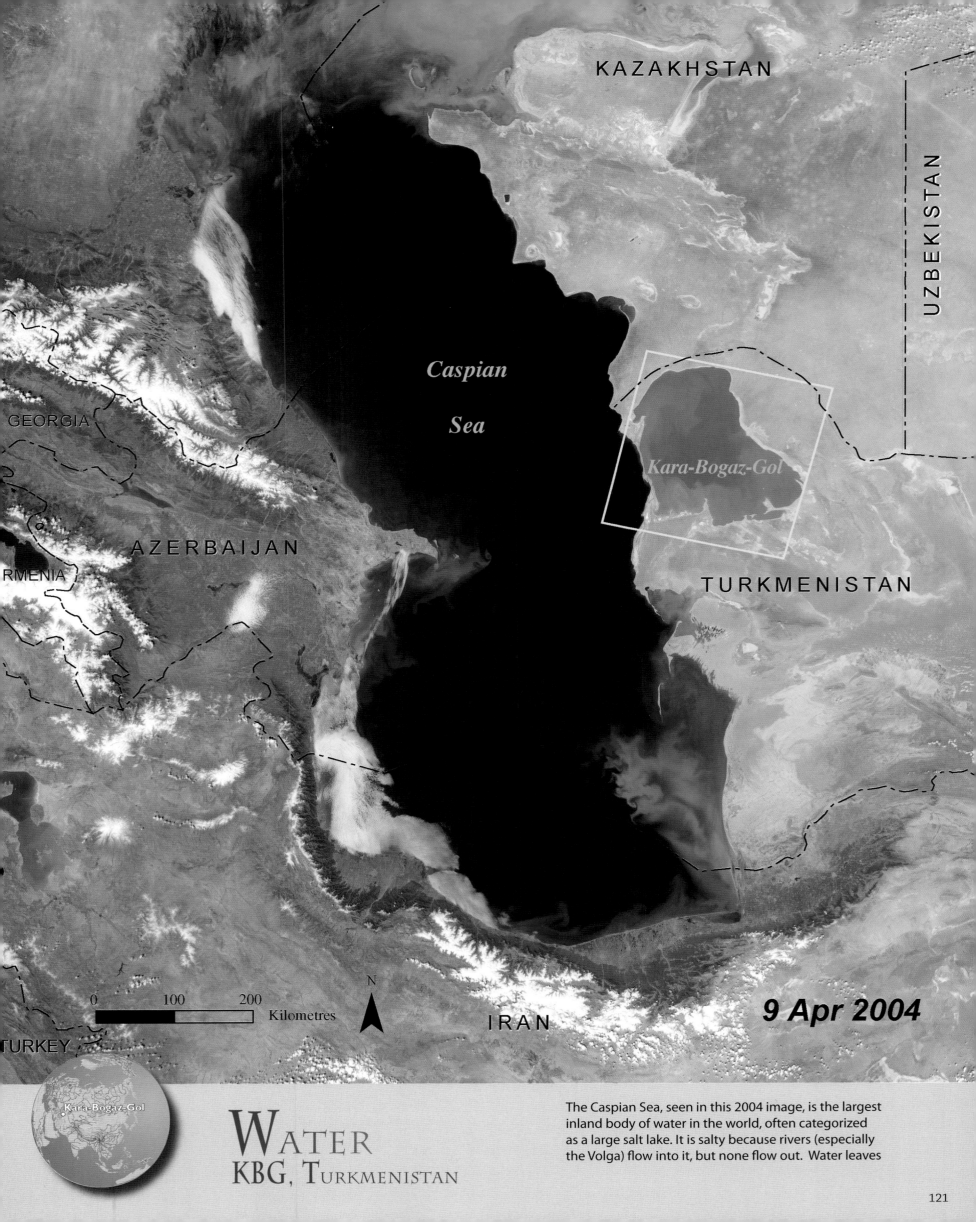

KAZAKHSTAN

UZBEKISTAN

Caspian

Sea

Kara-Bogaz-Gol

GEORGIA

AZERBAIJAN

TURKMENISTAN

RMENIA

IRAN

9 Apr 2004

0   100   200   Kilometres

N

TURKEY

Kara-Bogaz-Gol

WATER
KBG, TURKMENISTAN

The Caspian Sea, seen in this 2004 image, is the largest
inland body of water in the world, often categorized
as a large salt lake. It is salty because rivers (especially
the Volga) flow into it, but none flow out. Water leaves

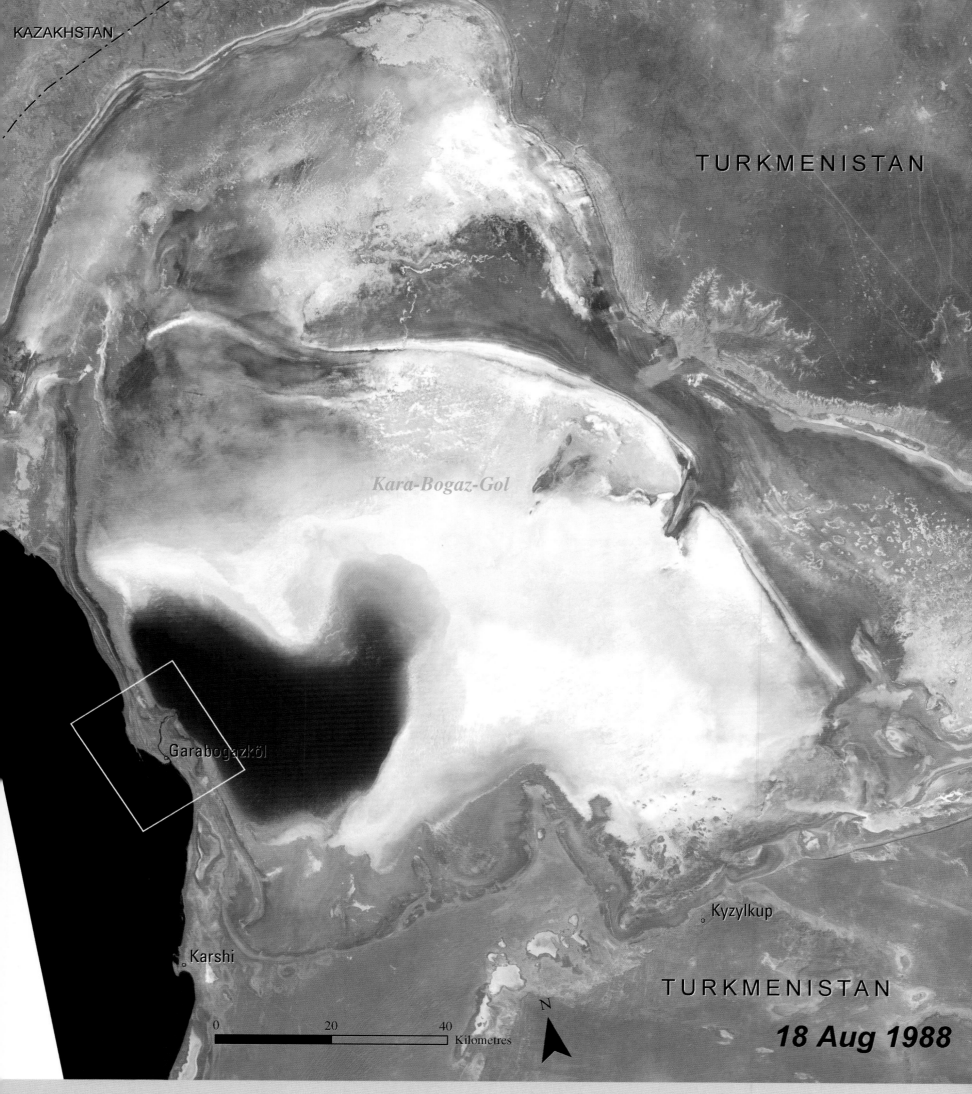

KAZAKHSTAN

TURKMENISTAN

Kara-Bogaz-Gol

Garabogazköl

Kyzylkup

Karshi

TURKMENISTAN

0   20   40 Kilometres

N

18 Aug 1988

through evaporation, and the dissolved salts remain. Changes in water levels are common in the sea, resulting both from changing climatic factors and water diversion by humans. The 2004 image highlights the area of change—the Kara-Bogaz-Gol (KBG). KBG is a large, shallow lagoon of the Caspian Sea, normally about 18 200 km² (7 000 square miles) and a

few metres deep. The Caspian's changing water level has been a concern since the 1970s. The KBG's water flows in from the Caspian Sea, and its fluctuations have affected the KBG dramatically.

In the 1980s, a dam blocked the KBG's inflow, resulting in the formation of a "salt bowl" that caused widespread problems of blowing salt,

KAZAKHSTAN

TURKMENISTAN

Kara-Bogaz-Gol

Garabogazköl

In 1980, in response to the rapidly dropping sea level, a dam was constructed to prevent water from flowing into the shallow and restricted Kara-Bogaz-Gol basin, resulting in the drying up of the bay. The dam was partially opened a few years later, and completely opened in 1992 when Caspian water levels started to rise quickly. *Credit: NASA Johnson Space Center*

Kyzylkup

Karshi

TURKMENISTAN

0   20   40
Kilometres

N

16 Jun 2000

reportedly poisoning soil, and causing health problems for people living hundreds of kilometers downwind to the east. While the dam was in place, not only did the KBG's water level rapidly drop by 2 m (7 ft) or more, but the lagoon's shallow bottom also rose 0.5 m (2 ft), due to the accumulation of salts.

The dam was partially opened in 1985, and completely opened in 1992 when Caspian Sea water levels started to rise quickly. Today, sea levels are more than 2.6 m (9 ft) higher than the 1978 levels and water flows freely into the salty waters of the Kara-Bogaz-Gol.

Aktyubinsk

Kulandy

Kokaral

Bugun'

KAZAKHSTAN

Zhanatal

Kzyl-Orda

Aral Sea

Komsomol'sk

UZBEKISTAN
Karakalpakstan

1986

Kokaral

Bugun'

Zhanata

KAZAKHSTAN

Kzyl-Orda

Aral Sea

Komsomol'sk

UZBEKISTAN

1973

Karakalpakstan

0   20   40
Kilometres

N

Aral Sea

Water
Aral Sea, Kazakhstan

The name " Aral Sea " comes from the word "aral" mean-ing island. The sea's name reflects the fact that it is a vast basin that lies as an island among waterless des-erts. The Aral Sea was once the world's fourth-largest inland sea. Its problems began in the 1960s and 1970s

Kokaral

Bugun'

Zhanatal

KAZAKHSTAN

Kzyl-Orda

Aral
Sea

1999

Komsomol'sk

UZBEKISTAN

Karakalpakstan

0   20   40
Kilometres   N

22 Sep 2004

with the diversion of the main rivers that feed it to provide for cotton cultivation in arid Soviet Central Asia. The surface of the Aral Sea once measured 66 100 km^2 (25 521 square miles). By 1987, about 60 per cent of the Aral Sea's volume had been lost, its depth had declined by 14 m (45 feet), and its salt concentration had doubled, killing the commercial fishing trade. Wind storms became toxic, carrying fine grains of clay and salts deposited on ex-

posed sea floor. Life expectancies in the districts near the sea are significantly lower than in surrounding areas.

The sea is now a quarter of the size it was 50 years ago and has broken into two parts, the North Aral Sea and the South Aral Sea. Re-engineering along the Syr Darya River delta in the north will retain water in the North Aral Sea, thereby drying the South Aral Sea completely, perhaps within 15 years.

Malatya

Adiyaman

Kâhta

Adiyaman

T U R K

Ataturk Dam

Euphrates

Bozova

Urfa

Sanliurfa

Harran

Biercik

Nizip

Gaziantep

0   10   20
Kilometres

N

Halab

S Y R I A

2 Sep 1976

The power-generating station at the Atatürk Dam already provides 8.9 billion kilowatt hours of electricity—roughly 22 per cent of the electricity the country is expected to need by 2010.
Credit: Unknown/UNEP/USDA-FAS

Atatürk Dam

WATER
ATATÜRK DAM, TURKEY

Built in 1990, the Atatürk Dam on the Euphrates River in southeastern Turkey is the centrepiece of the Southeastern Anatolia Project. The Atatürk Dam is the largest in a series of 22 dams and 19 hydroelectric stations built on the Euphrates and Tigris Rivers in order to provide irrigation water and electricity to this arid region

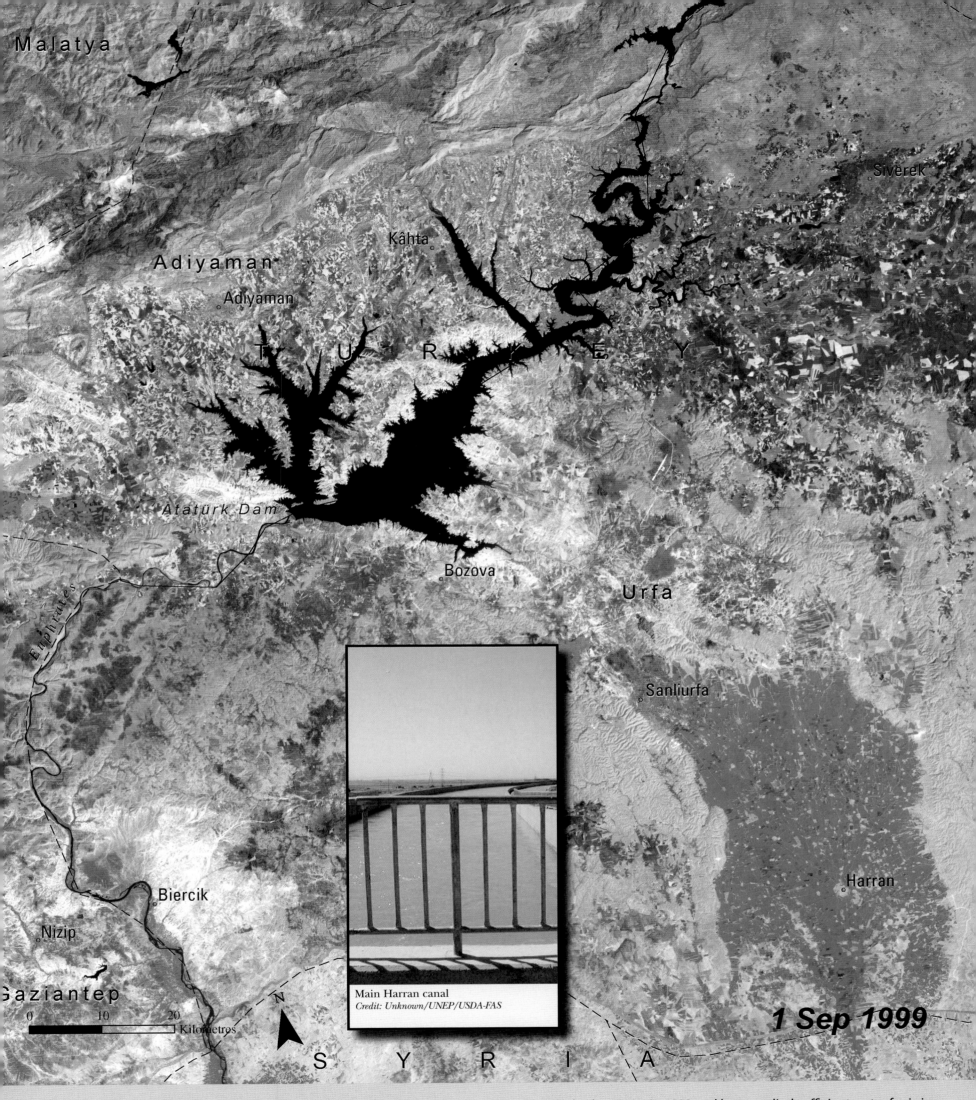

Malatya

Adiyaman

Kâhta

Siverek

T U R K E Y

Adiyaman

Atatürk Dam

Bozova

Urfa

Euphrates

Sanliurfa

Main Harran canal
Credit: Unknown/UNEP/USDA-FAS

Biercik

Nizip

Harran

Gaziantep

0   10   20
Kilometres

N

1 Sep 1999

S Y R I A

of the country. When the project's entire system of reservoirs, power generation stations, and irrigation channels is operational (projected to occur in 2010), the irrigation of approximately 1.7 million hectares (4.2 million acres) of land will be possible..

In these two Landsat images, acquired in 1976 and 1999, respectively, the transformation of the region around the dam is strikingly apparent. The dam's reservoir reached capacity in 1992 and has supplied sufficient water for irrigation to turn a once-arid landscape into a green one. This is especially obvious in the lower right-hand corner of the 1999 image, where irrigated fields completely surround the town of Harran. The development of the Harran region could not have occurred without the Atatürk Dam project, especially since the town is many kilometres from the river.

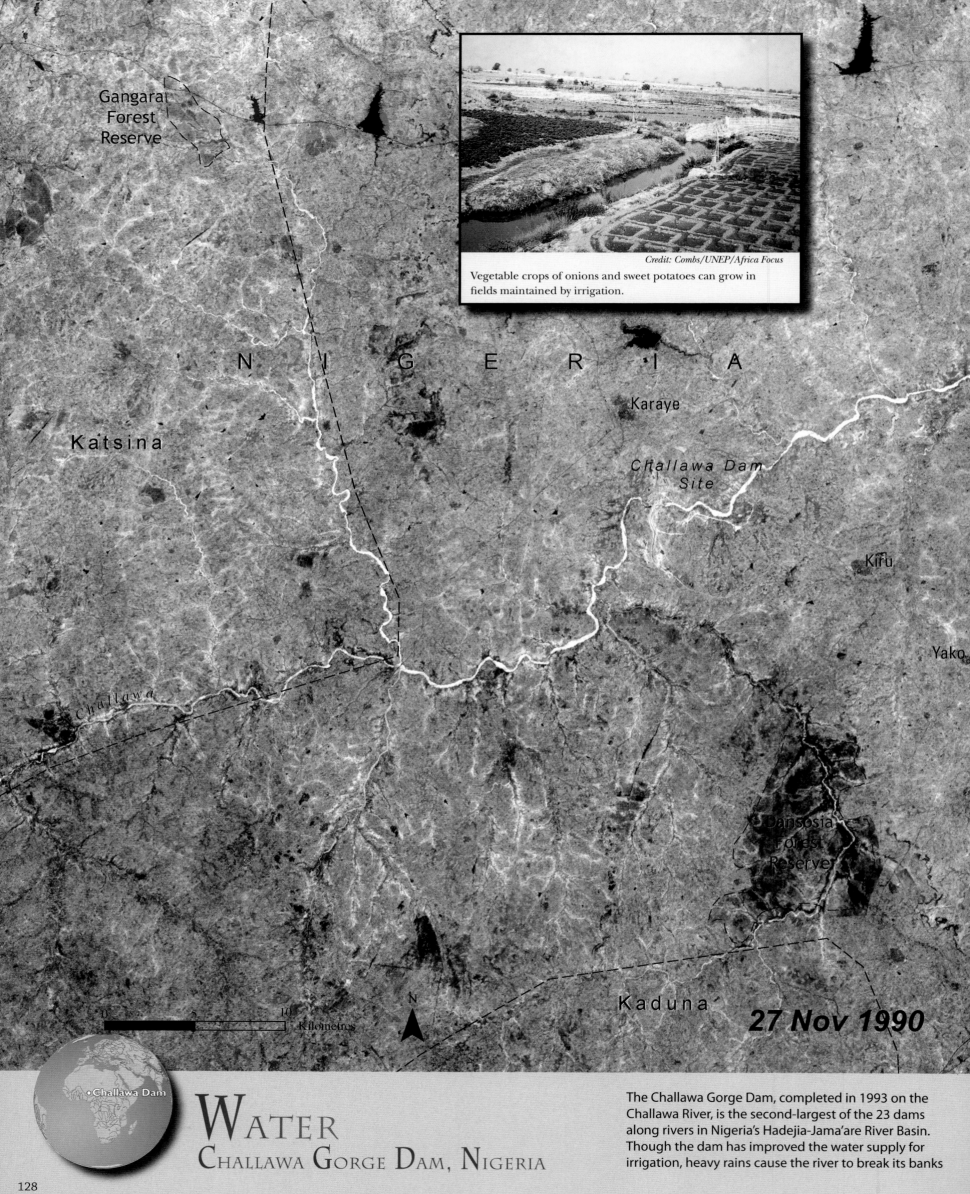

Gangara Forest Reserve

Credit: Combs/UNEP/Africa Focus

Vegetable crops of onions and sweet potatoes can grow in fields maintained by irrigation.

N I G E R I A

Karaye

Katsina

Challawa Dam Site

Kiru

Yako

Challawa

Dansosia Forest Reserve

0   10
Kilometres

N

Kaduna

27 Nov 1990

Challawa Dam

WATER
CHALLAWA GORGE DAM, NIGERIA

The Challawa Gorge Dam, completed in 1993 on the Challawa River, is the second-largest of the 23 dams along rivers in Nigeria's Hadejia-Jama'are River Basin. Though the dam has improved the water supply for irrigation, heavy rains cause the river to break its banks

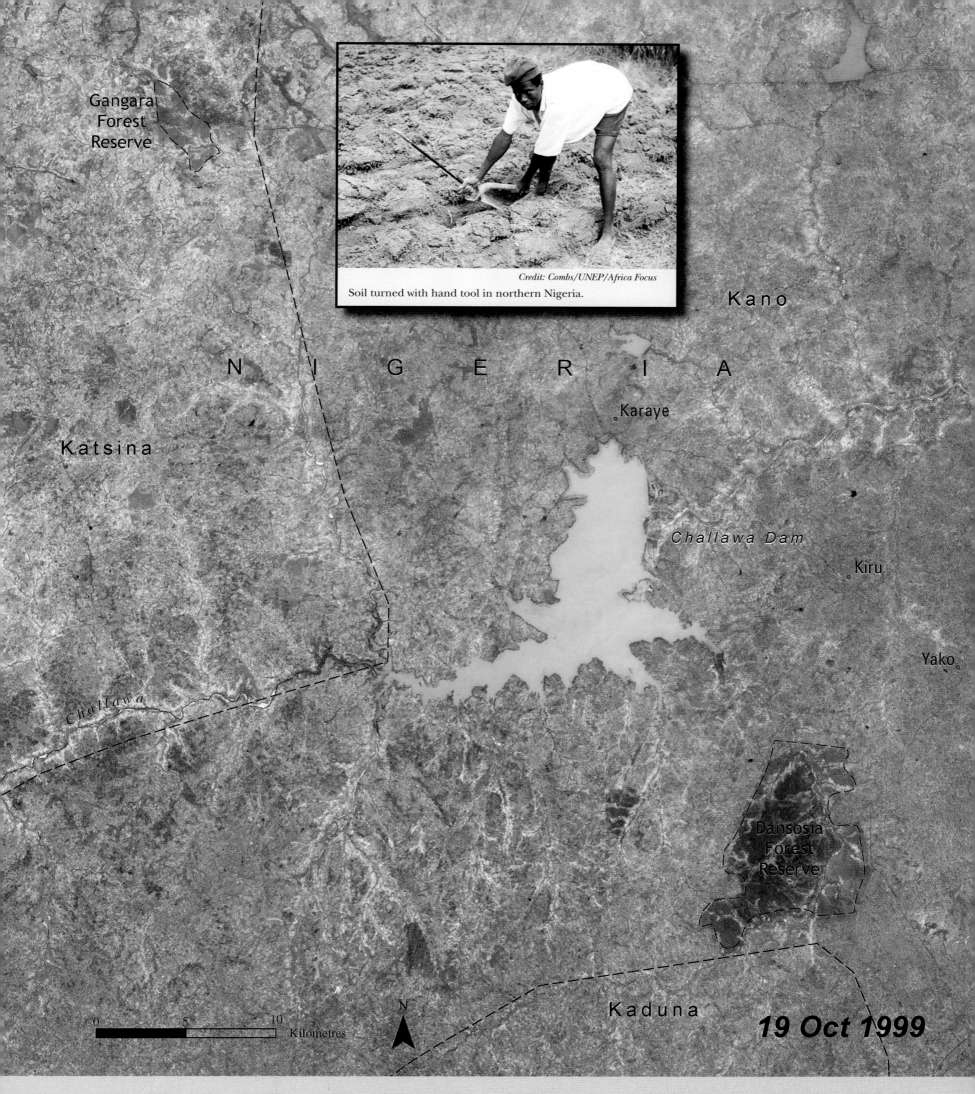

Gangara
Forest
Reserve

Credit: Combs/UNEP/Africa Focus
Soil turned with hand tool in northern Nigeria.

Kano

N I G E R I A

Karaye

Katsina

Challawa Dam

Kiru

Yako

Challawa

Dansosia
Forest
Reserve

0   5   10
Kilometres

N

Kaduna

19 Oct 1999

upstream from the dam; farmers are driven out as the rising water floods their farms and adjoining lands. Areas downstream from the dam, on the other hand, do not receive enough water to maintain the wetlands that border the river. Under these conditions, the soil dries out and overgrazing occurs, which in turn leads to wind erosion of the topsoil.

This satellite image pair gives a comparison of the area before and after construction of the dam. The 1999 image shows the degree to which flooding upstream from the dam impacts the landscape, and how the lack of water downstream negatively affects riverine wetlands and cropland. The colour of the water in the flooded area is also indicative of high-sediment content.

WEST BANK

Dead Sea

WEST BANK

Dead Sea

'Arad

Al Mazra'ah

Al Karak

ISRAEL

JORDAN

As Safi

8 Apr 1987

'Arad

Al Kar

ISRAEL

JORDAN

As Safi

0   5   10
Kilometres

N

1 Jan 1973

Dead Sea

WATER
DEAD SEA, JORDAN

For decades, heavy demands have been placed on the
land-locked Dead Sea to meet the needs of growing
populations in the countries that border it. Both Israel
and Jordan draw water from rivers that flow into the
Dead Sea, reducing the amount of water that would

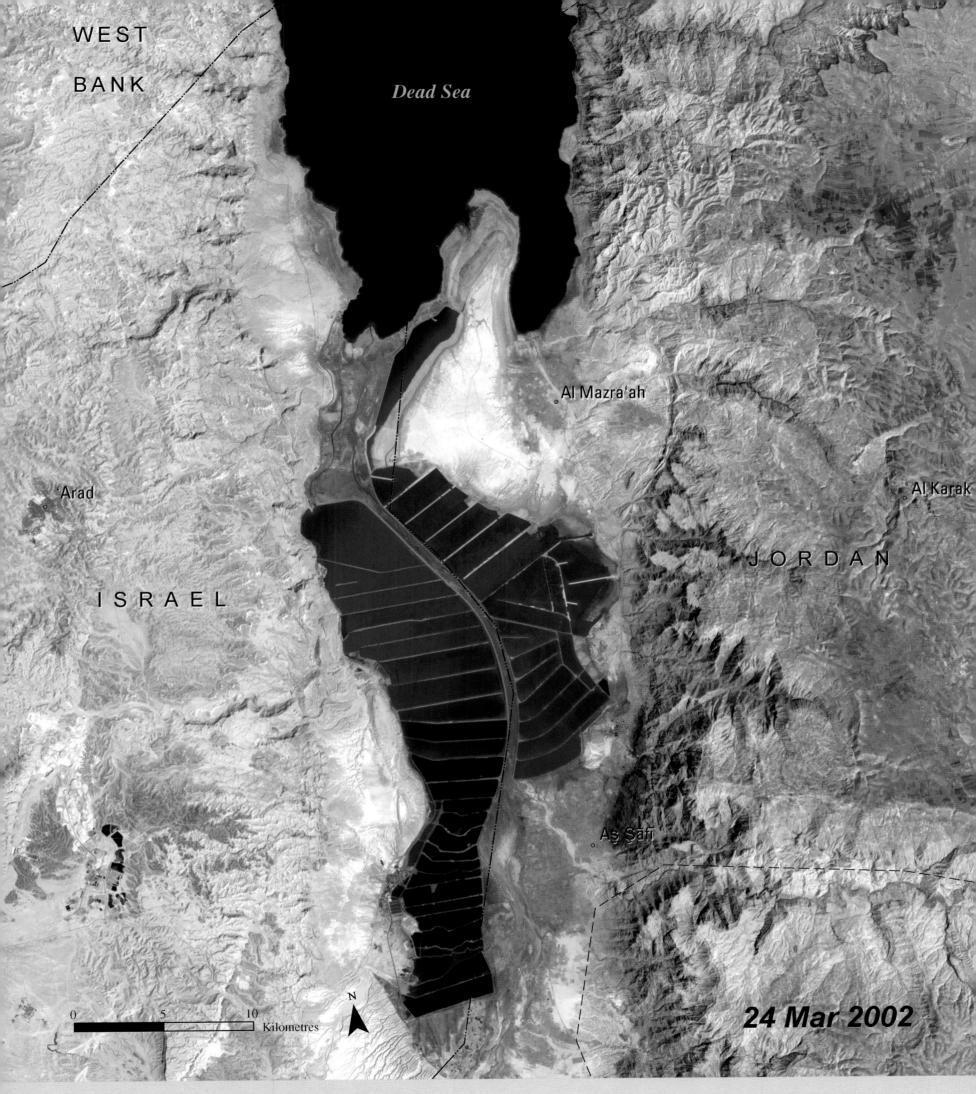

WEST BANK

Dead Sea

Al Mazra'ah

'Arad

Al Karak

ISRAEL

J O R D A N

Aş Şāfī

N

0   5   10
Kilometres

24 Mar 2002

naturally replenish it. The amount of area devoted to evaporation ponds for producing salt has greatly expanded over the past three decades. The creation of salt works tends to accelerate evaporation, further contributing to the reduction in water level. Currently, it is estimated that the water level of the Dead Sea is dropping at a rate of about one metre (3 feet) per year.

These two images, from 1973 and 2002, reveal dramatic changes in the Dead Sea over a period of about 30 years. Declining water levels, coupled with impoundments and land reclamation projects, have greatly increased the amount of exposed arid land along the coastline. The near-complete closing off of the southern part of the Sea by dry land (2002 image) reveals the severity of water level decline.

North New River Canal

Hillsboro Canal

Loxahatchee
National Wildlife
Refu[ge]

Aerial view of urban developement in the Everglades.
Credit: Unknown/UNEP/SFWMD

UNITED STATES

Miami Canal

Big Cypress
National Preserve

Big Cypress
Indian Res.

Florida

Fort Lauderdale

Big Cypress
National Preserve

Tamiami Canal

Miami

Goulds

Biscayne
Bay

Biscayne
National Park

Homestead

N

Gulf of
Mexico

Everglades
National Park

N

0   10   20
Kilometres

22 Mar 1973

•Everglades

Water

Everglades, United States

These two Landsat images of southern Florida in the United States reveal some of the changes that have occurred in this region over the past 30 years. One of the most obvious is the growth of the Fort Lauderdale-Miami urban area. Urban expansion has led to the con-

North New River Canal

Hillsboro Canal

Loxahatchee
National Wildlife
Refuge

UNITED STATES

Miami Canal

Big Cypress
National Preserve

Big Cypress
Indian Res.

Fort L

The Florida panther.
Credit: Unknown/UNEP/SFWMD

Florida

Big Cypress
National Preserve

Miami

Tamiami Canal

*Biscayne
Bay*

Gould's

Biscayne
National Park

Homestead

N

*Gulf of
Mexico*

Everglades
National Park

N

0   10   20
                    Kilometres

9 Jan 2002

version of what were once farmlands to cityscapes. The city of Miami has also expanded greatly to the southwest. The advance of urban areas westward across the peninsula threatens the continued existence of the vast wetlands area known as the Everglades. The Everglades ecosystem naturally filters groundwater and helps to recharge the Biscayne Aquifer. It is also home to a remarkable collection of plants and animals for which southern Florida is

famous. As urban areas encroach upon the Everglades, water resources and wildlife habitat are placed at serious risk. Protecting the Everglades to maintain its essential water filtering capacity and remarkable biodiversity is part of the mission of the Federal "Smart Growth" Task Force, which is working to better manage urban sprawl and its negative consequences.

SLOVAKIA

Dunajská Streda

Šamorín

Gabcíkovo

photographs view

Danube

Mosonmagyaróvár

Former Danube river bottom exposed as water is redirected into the diversion canal.

Credit: Peter Bardo-Déak/UNEP/WWF-DCP

0   5   10
Kilometres

1 Nov 1973

HUNGARY

WATER
GABCIKOVA, SLOVAKIA

• Gabcikova

The Gabcikova-Nagymoros hydraulic project on the Danube River was started in order to generate electric power, create an inland waterway, help manage water supplies, and aid in the region's economic development. The river was to be dammed and its water diverted into a canal. Four decades after it was initiated the

A dam on the Danube River
Credit: Peter Bardo-Déak/UNEP/WWF-DCP

21 Oct 2000

Cunovo Dam began operation in Slovakia in October 1992. The dam diverted 80 to 90 per cent of Danube River water down a diversion canal to support a hydroelectric power station.

This pair of images from 1973 and 2000 reveal the striking changes the massive re-channeling of river water has brought to the region. The dam altered the hydraulic regime of the Danube River valley from a natural water-way to a controlled patchwork of channels and islands. The diversion of water by the dam brought an end to the natural, beneficial flooding that added moisture and nutrients to the soil. It also reduced the ability of wetlands and marshes to filter surface water and trap sediments. Consequently, water quality and soil nutrients levels in the region have declined. Generation of electricity has come with significant environmental cost.

Butha-Buthe

Lesotho National Park

L E S O T H O

Leribe

Thaba Tseka

Sengu

9 Mar 1989

0   5   10
Kilometres

N

Lesotho Highlands

WATER
LESOTHO HIGHLANDS WATER PROJECT

The Lesotho Highlands Water Project (LHWP) is one of the largest infrastructure projects ever undertaken on the African continent. The project is designed to divert water from Lesotho's Maloti Mountains to South Africa's urban and industrial Gauteng Province. While South Africa is set to benefit from an increased supply of

Butha-Buthe

Lesotho National Park

L E S O T H O

Leribe

Katse Dam

Senqu

Thaba Tseka

0 5 10 Kilometres

N

2 Mar 2001

much-needed water, Lesotho would gain through the generation of hydro-electric power and profits from the sale of water. An 82-km (51-mile) water transfer-and-delivery system is already in place for delivering water to South Africa. On completion of the full project, a total of four dams will be placed in key locations. However, many questions remain unanswered about the social and environmental impacts the completed dams will have. The first dam in the multi-dam scheme, called Katse, located on the Orange River, closed its gates in 1995, creating an enormous reservoir along with serious social and environmental concerns.

These two images provide a comparison of the area before and after completion of the Katse dam. The effects and extent of the Katse Dam can clearly be seen in the 2001 image.

Akerme

Balqash Köli
(Lake Balkhash)

K A Z A K H S T A N

Alma-Ata

Chimpek

Dzhambul

Burylbaytal

0   5   10
Kilometres

N

21 Sep 1975 - 31 Aug 1979

Lake Balkhash

Water
Lake Balkhash, Kazakhstan

Located in Kazakhstan, Central Asia, Lake Balkhash is
replenished from the Ili River catchment area, most of
which is located in northwestern China. The lake is a
very important resource for the surrounding population. Water from the lake and its tributary rivers is used

Akerme

Balqash Köli
(Lake Balkhash)

K A Z A K H S T A N

Alma-Ata

Chimpek

Dzhambul

Burylbaytal

0   5   10
Kilometres

N

2 Aug 2001

for irrigation and both municipal and industrial purposes, including supplying the water needs of the Balkhash Copper Melting Plant. Lake fish are also an important food source. However, artificially low water prices have encouraged excessive use and waste of lake water. The United Nations has warned that Lake Balkhash, which is the second largest lake in Central Asia after the Aral Sea, could dry up if current trends are not reversed.

These two satellite images reveal an alarming drop in the lake's water levels in just over twenty years. Smaller, neighbouring lakes, to the southeast of Balkhash, have become detached from the main water body; they have dramatically decreased in size and appear to be drying up.

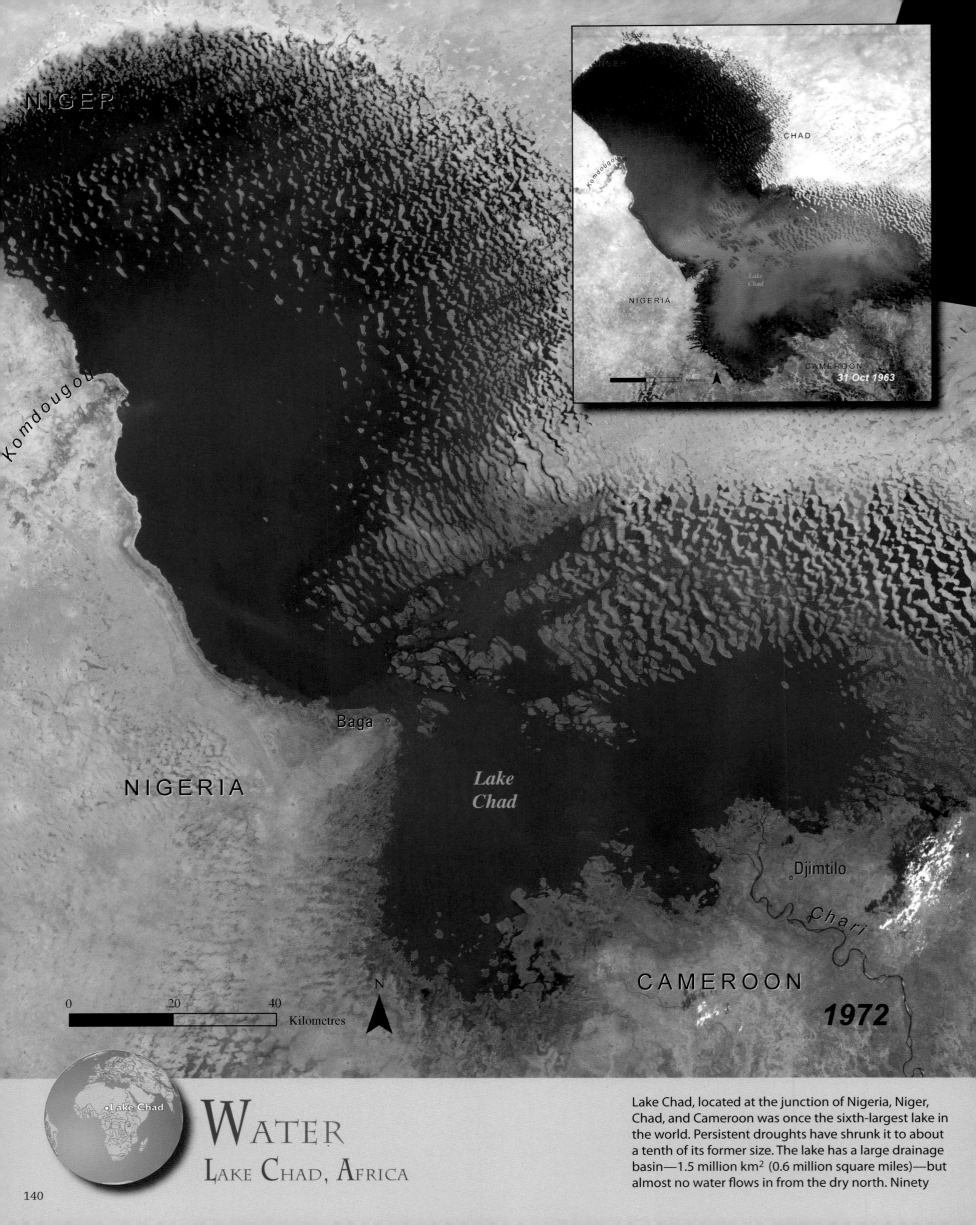

NIGER

Komdougou

Baga °

NIGERIA

Lake Chad

Djimtilo °

Chari

CAMEROON

1972

0   20   40
Kilometres

N

31 Oct 1963

NIGER

CHAD

Komdougou

Lake Chad

NIGERIA

CAMEROON

•Lake Chad

W ATER
LAKE CHAD, AFRICA

Lake Chad, located at the junction of Nigeria, Niger,
Chad, and Cameroon was once the sixth-largest lake in
the world. Persistent droughts have shrunk it to about
a tenth of its former size. The lake has a large drainage
basin—1.5 million km² (0.6 million square miles)—but
almost no water flows in from the dry north. Ninety

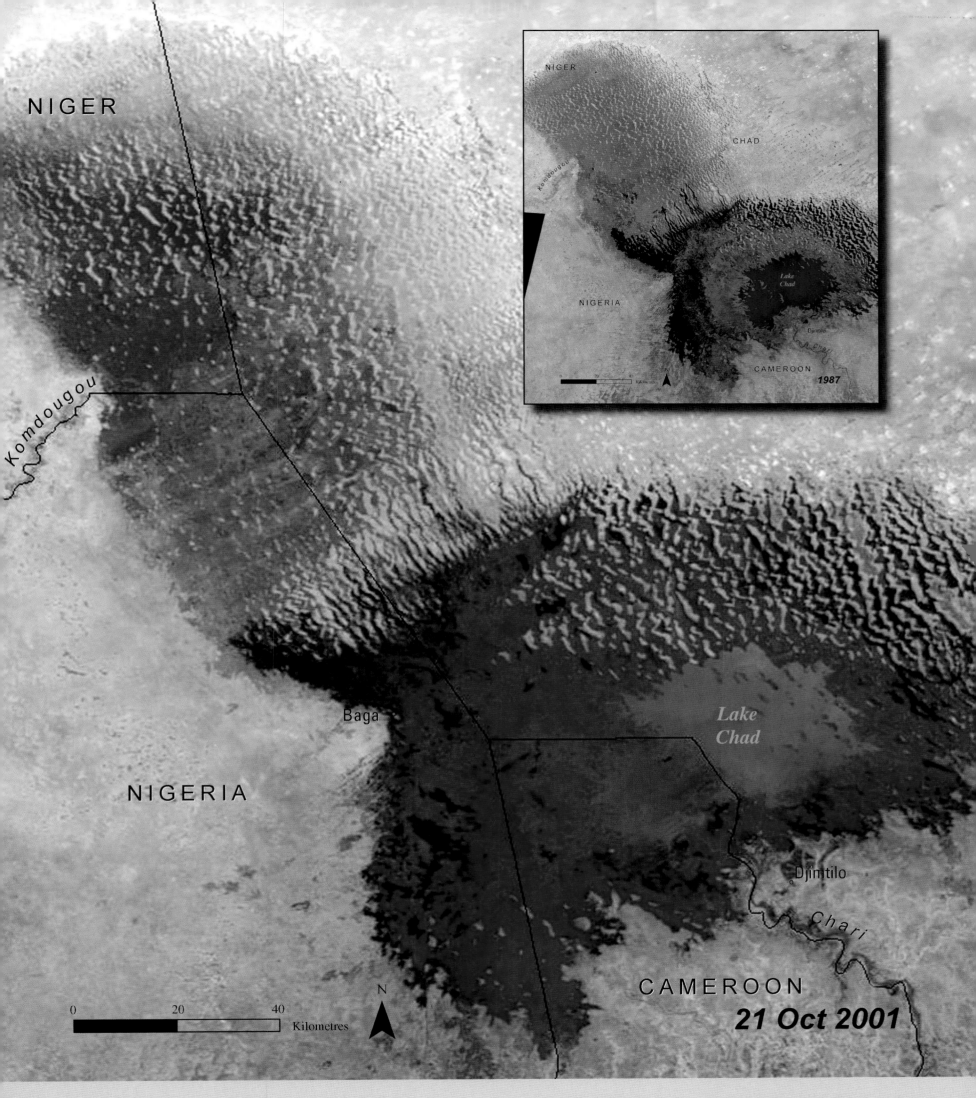

NIGER

Komdougou

Baga

NIGERIA

Djimtilo

Chari

Lake
Chad

CAMEROON
21 Oct 2001

0   20   40   Kilometres

N

Inset map (1987):

NIGER

CHAD

Komdougou

Baga

NIGERIA

*Lake
Chad*

Djimtilo

Chari

CAMEROON
1987

per cent of lake's water flows in from the Chari River. The lakebed is flat and shallow; even before the drought, the lake was no more than 5-8 m (16-26 ft) deep. Considered a deep wetland, Lake Chad was once the second largest wetland in Africa, highly productive, and supporting a diversity of wildlife.

The lake is very responsive to changes in rainfall. When rains fail, the lake drops rapidly because annual inflow is 20-85 per cent of the lake's volume.

Human diversion from the lake and from the Chari River may be significant at times of low flow, but rainfall is still the determining factor in lake level.

This image set displays a continued decline in lake surface area from 22 902 km² (8 843 square miles) in 1963 to a meager 304 km² (117 square miles) in 2001.

Guadalajara

Zapotlanejo

Jalisco

Poncitlan

Ocotlan

M E X I C O

Chapala

Laguna de Chapala

Tizapan el Alto

Sahuayo

Michoacan

0   5   10
Kilometres

N

13 Apr 1983

Lake Chapala

WATER
LAKE CHAPALA, MEXICO

Mexico's Lake Chapala, lying in the heart of an extremely arid region, is the country's largest natural lake. The lake is one of the most important wetlands in the region and home to more than 70 endemic species. Since the 1950s, Lake Chapala has undergone many changes as a result of water abstraction for agricultural

Guadalajara

Zapotlanejo

Jalisco

Poncitlan

Ocotlan

M E X I C O

Chapala

Laguna de Chapala

Tizapan el Alto

Sahuayo

M i c h o a c a n

0   5   10
Kilometres

N

29 Mar 2001

use both inside and outside the region and for a rapidly growing population. The level of the lake has declined, and there have been noticeable decreases in surrounding wetland areas as well as changes in the hydrological system connecting various springs.

Some of these changes are visible in this pair of satellite images, including alterations in the contours of the shoreline, obvious extensions of land near various townships around the lake, and the appearance of remarkably large areas of reclaimed land at the lake's eastern end. Like all arid areas, the land around Lake Chapala is prone to salinization. If the lake continues to shrink, researchers predict both a decrease in water availability and an increase in the salt content of the region's soil.

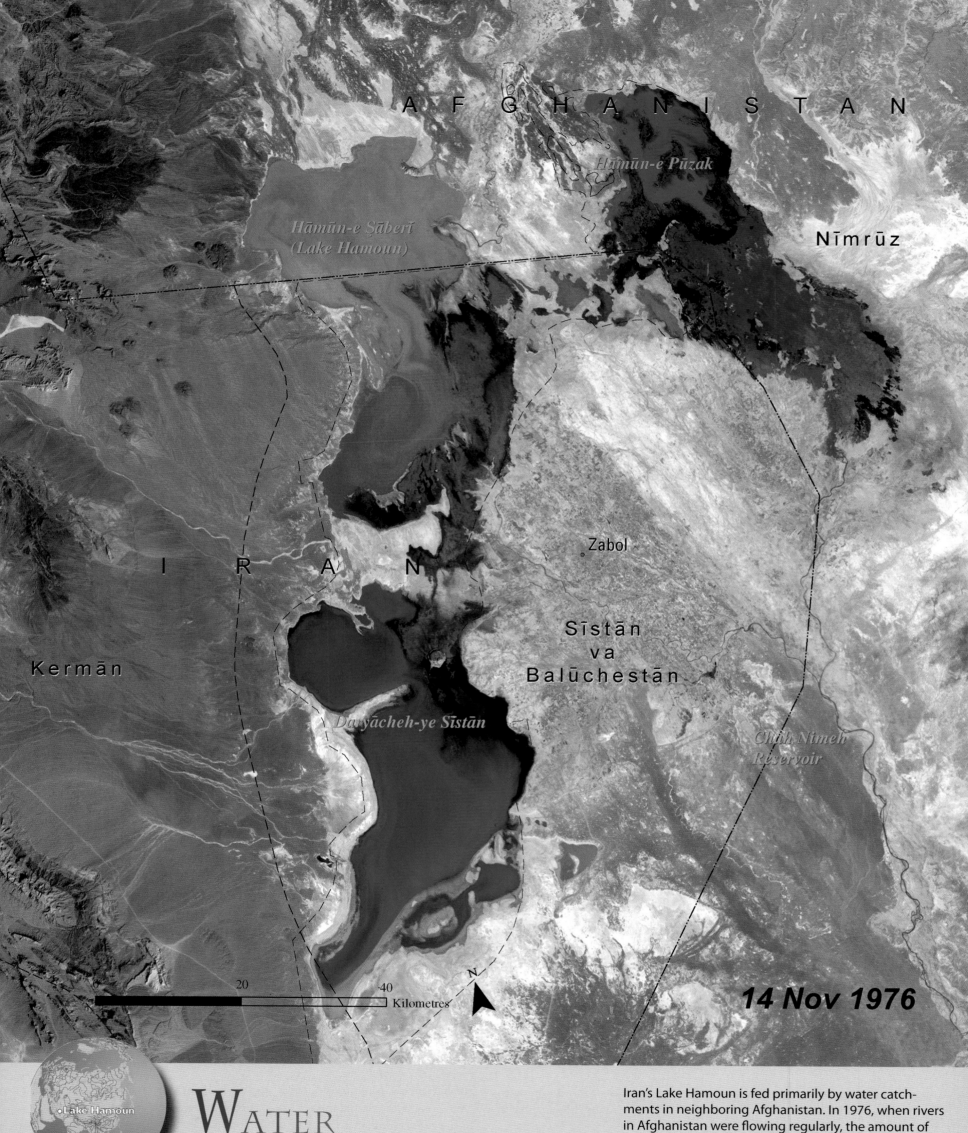

AFGHANISTAN

Hāmūn-e Pūzak

Nīmrūz

*Hāmūn-e Ṣāberī
(Lake Hamoun)*

I R A N

Zabol

Sīstān
va
Balūchestān

Kermān

Daryācheh-ye Sīstān

*Chāh Nimeh
Reservoir*

N

0   20   40
▬▬▬▬▬▬▬▬▬▬▬
Kilometres

14 Nov 1976

Lake Hamoun

WATER
LAKE HAMOUN, IRAN

Iran's Lake Hamoun is fed primarily by water catchments in neighboring Afghanistan. In 1976, when rivers in Afghanistan were flowing regularly, the amount of water in the lake was relatively high. Between 1999 and

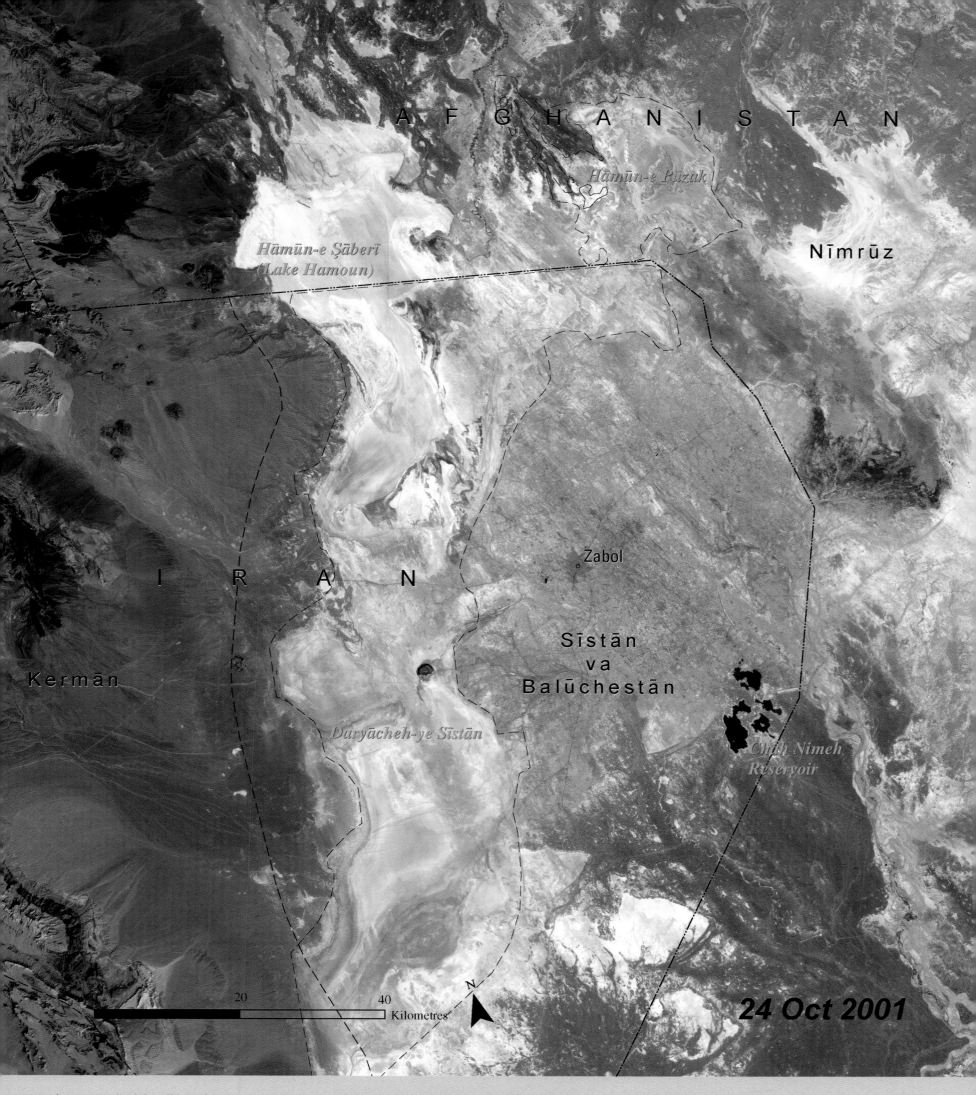

AFGHANISTAN

Hāmūn-e Pūzak

Nīmrūz

Hāmūn-e Şāberī
(Lake Hamoun)

I R A N

Zabol

Sīstān
va
Balūchestān

Kermān

Daryācheh-ye Sīstān

Chāh Nimeh
Reservoir

20   40
Kilometres   N

24 Oct 2001

2001, however, the lake all but dried up and disappeared, as can be seen in the 2001 satellite image above.

The "dry phase" of Lake Hamoun is a striking example of how competition for scarce water resources can transform a landscape. When droughts occur in Afghanistan, or the water in watersheds that support Lake Hamoun is drawn down by other natural or human-induced reasons, the end result is a dry lakebed in Iran. In addition, when the lake is dry, seasonal winds blow fine sands off the exposed lakebed. The sand is swirled into huge dunes that may cover a hundred or more fishing villages along the former lakeshore. Wildlife around the lake is negatively impacted and fisheries are brought to a halt. Changes in water policies and substantial rains in the region saw a return of much of the water in Lake Hamoun by 2003.

Menengai
FR

Bahati
FR

Nakuru FR

Nakuru

Lake
Nakuru
Water
Catchment

Lake
Nakuru

K e n y a

Eastern Mau
FR

Excised
Forest
(2001)

Nakuru
National
Park

Mau
Narok
FR

Ebu

Flamingos on Lake Nakuru *Credit : Gray Tappan/UNEP*

31 Jan 1973

0   6   12
Kilometres

N

Lake Nakuru

WATER
LAKE NAKURU, KENYA

Lake Nakuru is located in the Eastern Rift Valley in
southwest Kenya. Lake Nakuru National Park is the
second most visited protected area in Kenya. It hosts
the world's largest concentration of flamingos, as well
as many of the animal species that make Kenya a highly
valued tourism destination, including lions, leopards,

146

Menengai
FR

Nakuru FR

Nakuru

Vegetation around Lake Nakuru
Credit : Gray Tappan/UNEP

Lake
Nakuru
Water
Catchment

Lake
Nakuru

K e n y a

Nakuru
National
Park

Eastern Mau
FR

Excised
Forest
(2001)

Mau
Narok
FR

Eburru FR

0   6   12
Kilometres

N

27 Jan 2000

rhinoceros, and water buffalo. In its total area of 188 km² (73 square miles), there are over 450 bird species and 56 mammal species. Recognized as a wetland of international importance, Lake Nakuru was declared a Ramsar Site in 1990.

The threat of land cover degradation in the catchments of the lake is likely to increase flow fluctuation and decrease water quality. These images show

the land cover degradation in the lake's catchment between 1973 and 2000.

In 2001, the Government of Kenya announced its intention to excise 353 km² (136 square miles) of forest in the Eastern Mau Forest Reserve (area with white boundary in the 2000 image). As a result, most of the forest cover in the upper catchment of the main rivers that feed Lake Nakuru will disappear.

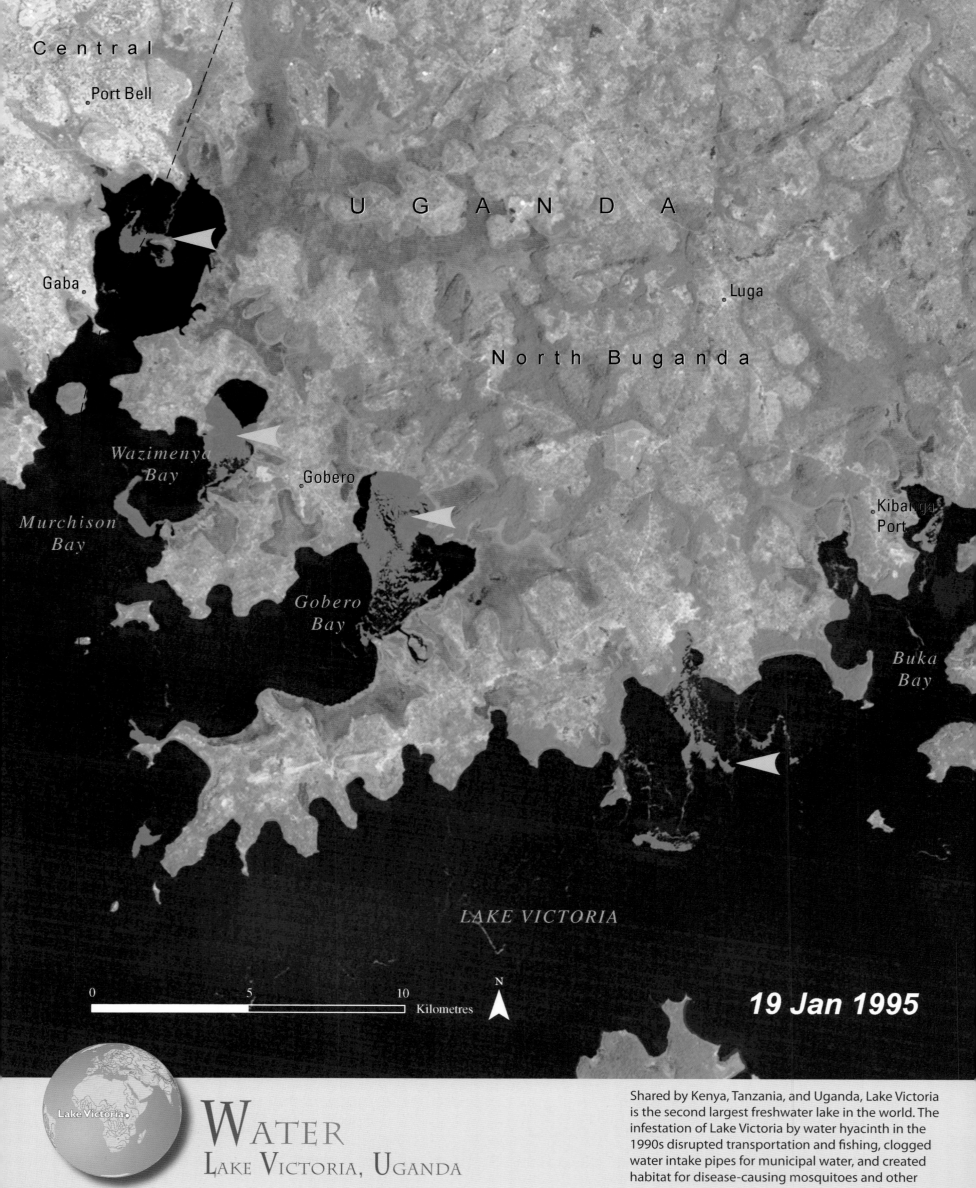

Central

Port Bell

U G A N D A

Gaba

Luga

North Buganda

Wazimenya
Bay

Gobero

Kibanga
Port

Murchison
Bay

Gobero
Bay

Buka
Bay

LAKE VICTORIA

0   5   10
Kilometres

N

19 Jan 1995

Shared by Kenya, Tanzania, and Uganda, Lake Victoria
is the second largest freshwater lake in the world. The
infestation of Lake Victoria by water hyacinth in the
1990s disrupted transportation and fishing, clogged
water intake pipes for municipal water, and created
habitat for disease-causing mosquitoes and other

WATER
LAKE VICTORIA, UGANDA

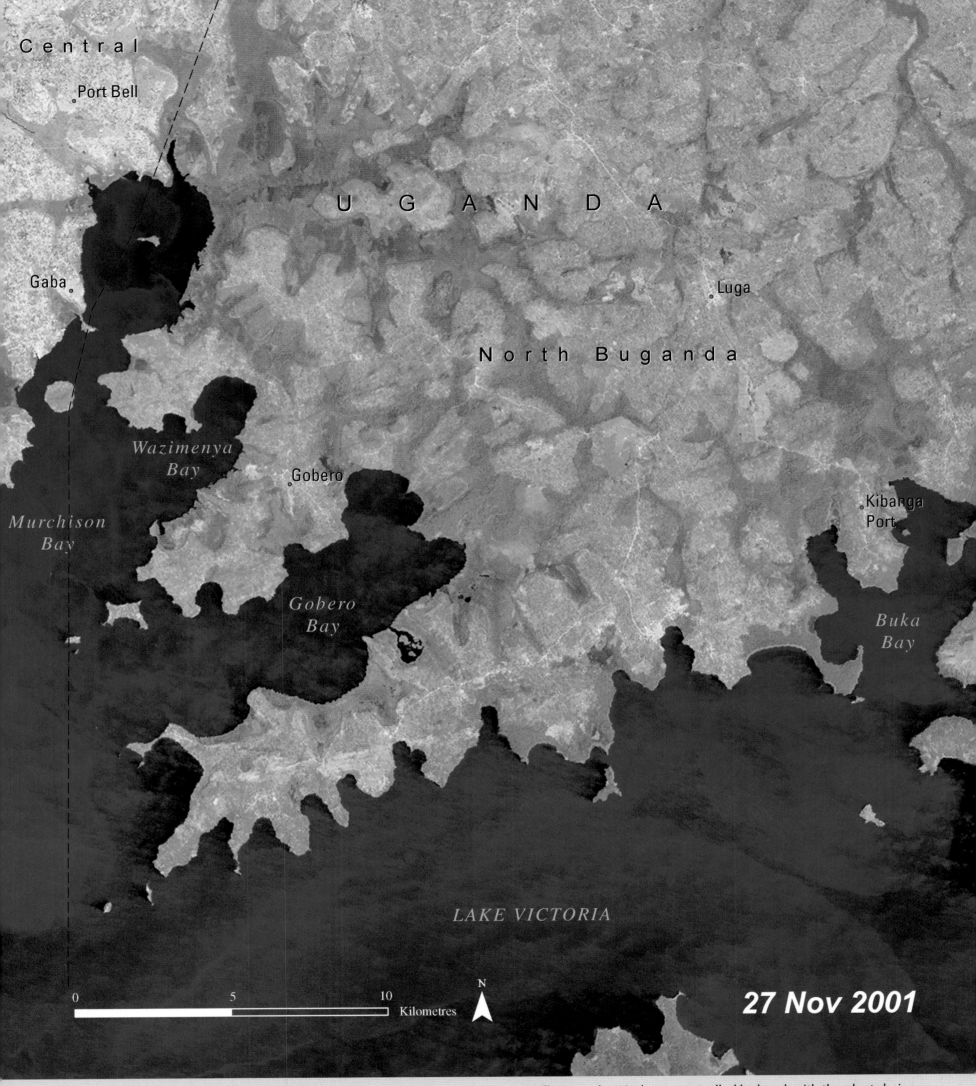

Central

Port Bell

U G A N D A

Gaba

Luga

North Buganda

Wazimenya
Bay

Gobero

Murchison
Bay

Kibanga
Port

Gobero
Bay

Buka
Bay

LAKE VICTORIA

0   5   10
Kilometres

N

27 Nov 2001

insects. This led to the initiation of the Lake Victoria Environmental Management Project in 1994. The focus of the Project was to combat hyacinth infestations on the lake, particularly the region bordered by Uganda, which was one of the most severely affected areas.

The 1995 image shows several water-hyacinth-choked bays: Murchison Bay near Gaba; large parts of Gobero and Wazimenya Bays; an area outside Buka Bay; and near Kibanga Port (yellow arrows).

Initially, water hyacinth was controlled by hand, with the plants being manually removed from the lake. But re-growth quickly occurred. A more recent control measure has been the careful introduction of natural insect predators of water hyacinth. As the 2001 image shows, this approach seems to have been successful, as the floating weeds have disappeared from all the locations noted above.

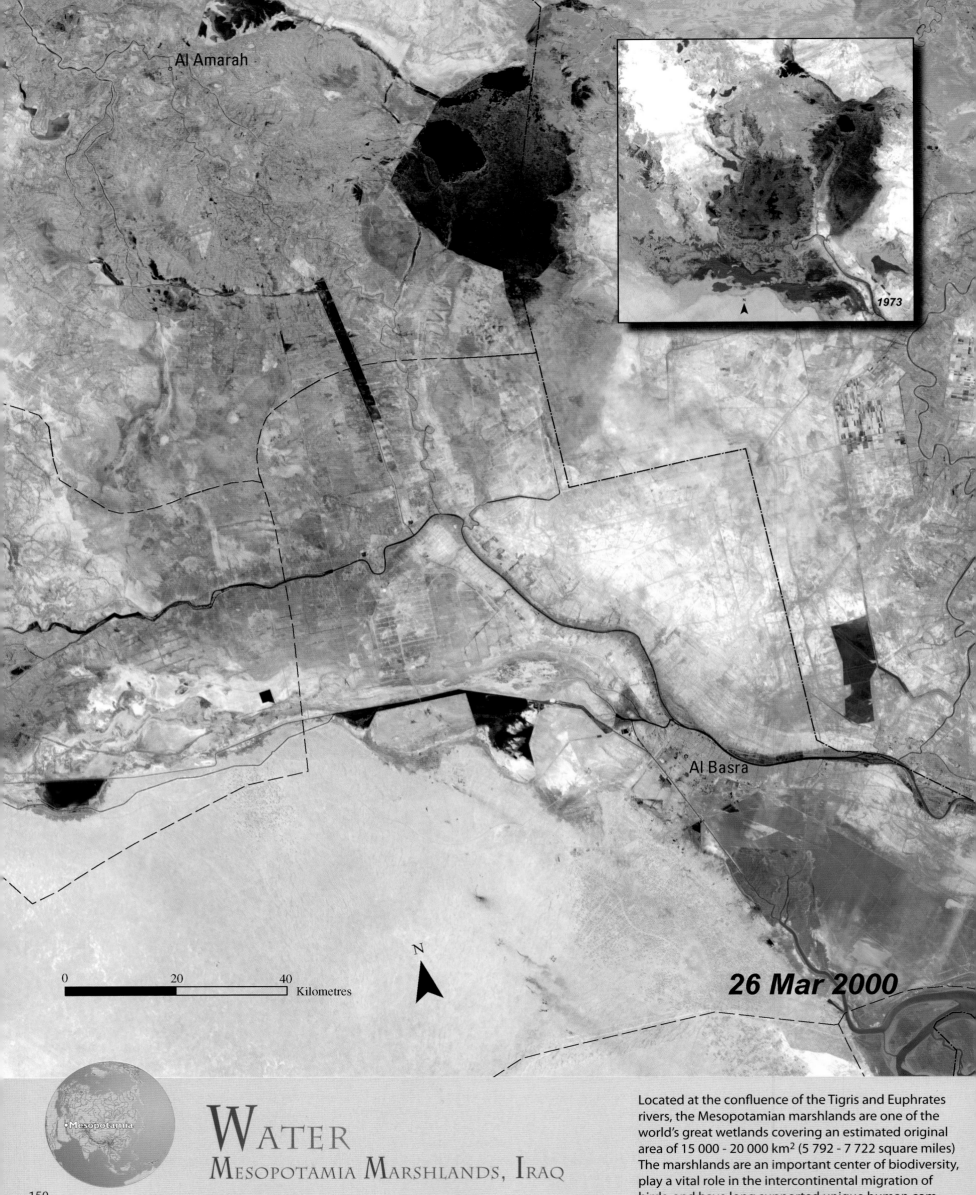

Al Amarah

1973

N

Al Basra

0   20   40
Kilometres

N

26 Mar 2000

•Mesopotamia

Water
Mesopotamia Marshlands, Iraq

Located at the confluence of the Tigris and Euphrates rivers, the Mesopotamian marshlands are one of the world's great wetlands covering an estimated original area of 15 000 - 20 000 km² (5 792 - 7 722 square miles) The marshlands are an important center of biodiversity, play a vital role in the intercontinental migration of birds, and have long supported unique human com-

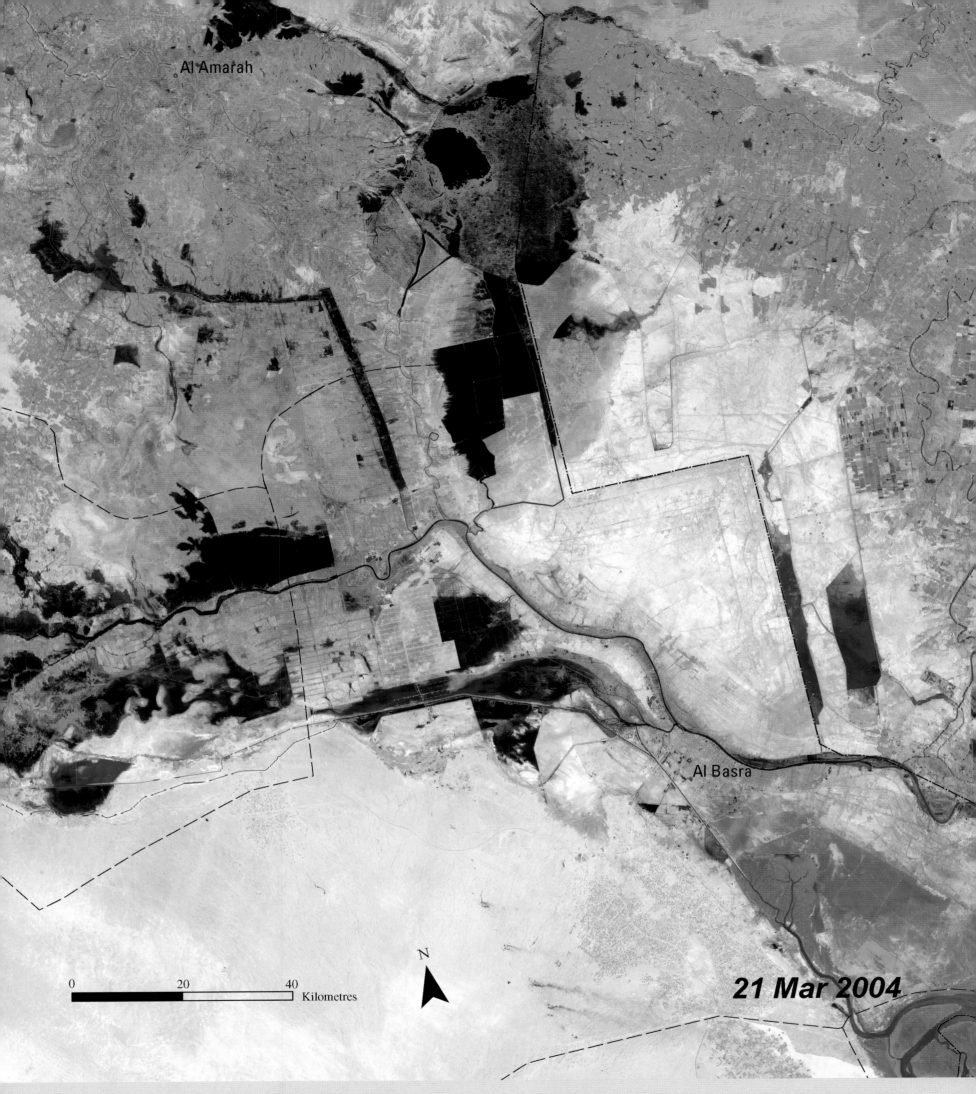

Al Amarah

Al Basra

21 Mar 2004

0   20   40
|████████|████████| Kilometres

N ▲

munities. Upstream damming as well as drainage activities in the marshlands themselves have significantly reduced the quantity of water entering the marshes. Together these factors have led to the collapse of the ecosystem. Restoration of the marshlands, mainly through reflooding by breaching of dykes and drainage canals has begun. As a result of these activities, vegetation and wildlife have returned to some parts of the marshes.

This set of images provides a synoptic illustration of the changes. While the 1973 image (inset left) shows the extent of the original marshlands, the 2000 image reveals the area after being drained, with most of the wetlands having disappeared. On the other hand, the 2004 image illustrates recovery in progress with major portions in the central and western sections having been restored to some extent.

Chang Jiang (Yangtze)

Badong

Zigui

C H I N A

H u b e i

Longping

Three Gorges Dam Site Before Construction

0   2   4
Kilometres   N

17 Apr 1987

0   10   20
Kilometres   N

Three Gorges Dam

WATER
THREE GORGES DAM, CHINA

The Three Gorges Dam on the Chang Jiang (Yangtze) River in China is one of the largest single construction projects ever attempted on the planet. The dam was constructed to supply approximately one-ninth of China's electricity—as much power as could be gener-

Chang Jiang (Yangtze)

Badong

Zigui (submerged)

C H I N A

Hubei

Longping

Three Gorges Dam

0 2 4
Kilometres
N

0 10 20
Kilometres
N

9 May 2004

ated by at least fifteen nuclear power plants. It is a relatively environmentally clean option compared to coal burning or nuclear power plants. It is also hoped that the dam will help control flooding on a river where seasonal floods during the past century has caused death of over one million people. However, the Three Gorges Dam project has also had negative environmental and social impacts as a result of the massive construction efforts and the sub-

mergence of land along the river above the dam. The former village of Zigui (top center of image) has already been submerged.

The 1987 image shows the nature of the river and surrounding landscape before work on the dam was begun. In the May 2004 image, the enormous Three Gorges Dam is clearly visible, as is the reservoir of impounded river water that has been created behind it.

153

3.4 Forests

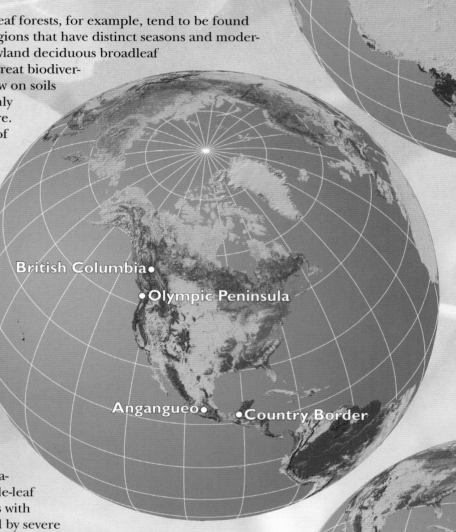

Much of the Earth's above ground biomass and biodiversity is held within its forests. Forests are sources of food, fuel, construction materials, fibers, and biological diversity. Forests are also important in water and air filtration, carbon sequestration, soil stabilization, and tourism (Williams 1994, 1990). It has been estimated that roughly three-fourths of a hectare of forestland is now needed to supply each person on the planet with shelter and fuel (Lund and Iremonger 1998).

Around the globe, there are a number of different types of forests. They are classified according to a scheme that is largely determined by climate and landform and that distinguishes evergreen and deciduous forests as well as broadleaf and needle-leaf forests.

Deciduous broadleaf forests, for example, tend to be found in temperate zone regions that have distinct seasons and moderate precipitation. Lowland deciduous broadleaf forests—a source of great biodiversity—also tend to grow on soils considered to be highly suitable for agriculture. Consequently, many of the lowland forests in the Northern Hemisphere have been cut down and the land they once occupied converted to cropland. Lowland forests are currently facing the same fate in South America and Africa.

Evergreen broadleaf forests— often called "rain forests"— tend to occur in tropical climates with high levels of precipitation. Deciduous needle-leaf forests occupy regions with climates characterized by severe temperature variations such as eastern Siberia, a region with extremely cold winters and the Sahel, a region with extremely hot, dry summers. Evergreen needle-leaf forests occur throughout polar, temperate, and subtropical regions. Mixed forests occupy transition zones. Swamp forests and mangroves exist in areas where soils are saturated, or even inundated with fresh water or ocean water, respectively.

Like any type of ecosystem, forests undergo changes, both natural and human-induced. Attempts to determine how forests have changed over time, however, are often complicated by the fact that there is no single definition as to what constitutes a "forest." Land parcels may be classified according to their cur-

Several different terms are often associated with forests and the changes they may undergo. *Deforestation* is the removal of tree cover and the conversion of the resulting deforested land to some other use. *Reforestation* refers to re-establishing tree cover on previously forested land. *Afforestation* is the creation of forest cover where it did not previously exist or where it existed only in the distant past. *Forest degradation* refers to changing a forest to such a degree that the species that inhabit that forest can no longer compete for survival; degradation typically leads to long term changes in species composition.

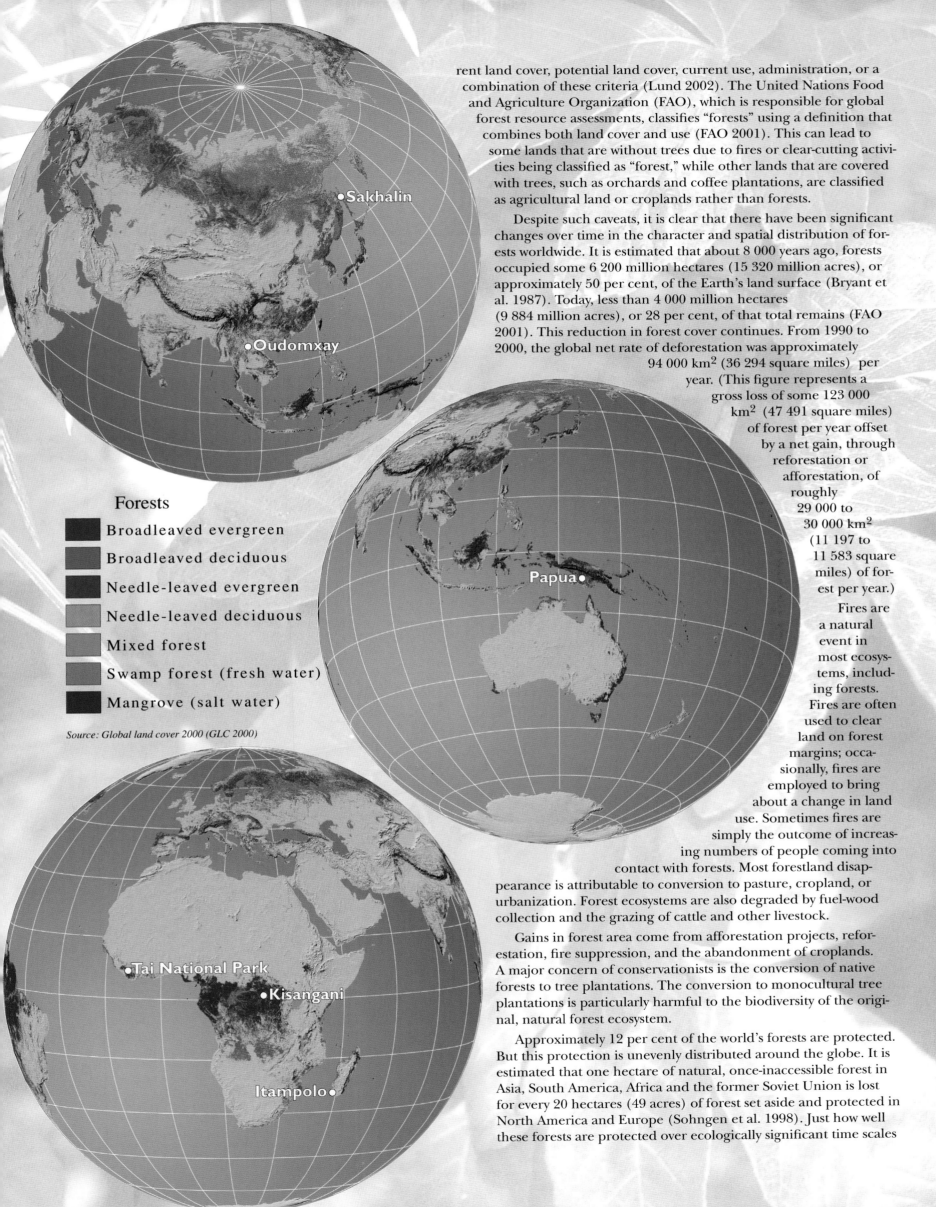

Forests

- Broadleaved evergreen
- Broadleaved deciduous
- Needle-leaved evergreen
- Needle-leaved deciduous
- Mixed forest
- Swamp forest (fresh water)
- Mangrove (salt water)

Source: Global land cover 2000 (GLC 2000)

rent land cover, potential land cover, current use, administration, or a combination of these criteria (Lund 2002). The United Nations Food and Agriculture Organization (FAO), which is responsible for global forest resource assessments, classifies "forests" using a definition that combines both land cover and use (FAO 2001). This can lead to some lands that are without trees due to fires or clear-cutting activities being classified as "forest," while other lands that are covered with trees, such as orchards and coffee plantations, are classified as agricultural land or croplands rather than forests.

Despite such caveats, it is clear that there have been significant changes over time in the character and spatial distribution of forests worldwide. It is estimated that about 8 000 years ago, forests occupied some 6 200 million hectares (15 320 million acres), or approximately 50 per cent, of the Earth's land surface (Bryant et al. 1987). Today, less than 4 000 million hectares (9 884 million acres), or 28 per cent, of that total remains (FAO 2001). This reduction in forest cover continues. From 1990 to 2000, the global net rate of deforestation was approximately 94 000 km^2 (36 294 square miles) per year. (This figure represents a gross loss of some 123 000 km^2 (47 491 square miles) of forest per year offset by a net gain, through reforestation or afforestation, of roughly 29 000 to 30 000 km^2 (11 197 to 11 583 square miles) of forest per year.)

Fires are a natural event in most ecosystems, including forests. Fires are often used to clear land on forest margins; occasionally, fires are employed to bring about a change in land use. Sometimes fires are simply the outcome of increasing numbers of people coming into contact with forests. Most forestland disappearance is attributable to conversion to pasture, cropland, or urbanization. Forest ecosystems are also degraded by fuel-wood collection and the grazing of cattle and other livestock.

Gains in forest area come from afforestation projects, reforestation, fire suppression, and the abandonment of croplands. A major concern of conservationists is the conversion of native forests to tree plantations. The conversion to monocultural tree plantations is particularly harmful to the biodiversity of the original, natural forest ecosystem.

Approximately 12 per cent of the world's forests are protected. But this protection is unevenly distributed around the globe. It is estimated that one hectare of natural, once-inaccessible forest in Asia, South America, Africa and the former Soviet Union is lost for every 20 hectares (49 acres) of forest set aside and protected in North America and Europe (Sohngen et al. 1998). Just how well these forests are protected over ecologically significant time scales

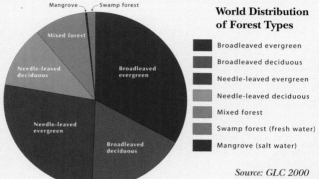

World Distribution of Forest Types

- Broadleaved evergreen
- Broadleaved deciduous
- Needle-leaved evergreen
- Needle-leaved deciduous
- Mixed forest
- Swamp forest (fresh water)
- Mangrove (salt water)

Source: GLC 2000

is also a continuing issue. The historical trend is for people to continue to convert more and more of any forested landscape for other uses. Often, these converted lands end up as low-grade pasture, with the result that forest ecosystems are destroyed for what amounts to only transient economic or social benefit. Unfortunately, this trend is likely to continue until either the human population stabilizes or demands for forest goods and services become sustainable—or both.

Whatever the future holds, many of the images in this section show the profound effect that humans have had, and continue to have, on the Earth's forests.

Forest Fire

Wildfires can have both positive and negative effects on the environment. In some ecosystems, fires play an ecologically significant role in maintaining biogeochemical cycles. The biological diversity of plant and animal life in the world's forests, prairies, and wetlands partly depends on the effects of fire. Some plants, for example, can-

not reproduce without fire; intense heat is needed to open cones or rupture seed coats so that seed dispersal and germination can take place. Fires naturally shape many types of ecosystems including the boreal forests of Canada, Alaska, and Russia and the chaparral in southern California.

In addition to "natural" fires, people have used fire for thousands of years to clear land. Whether caused by nature or by people, wildfires are a significant force for environmental change, one that can radically alter a landscape in a very short time (see box below).

The Earth's burgeoning human population, coupled with intensified economic development, has led to the serious degradation of many of the world's forests. Degraded forests are often highly

Case Study: Rodeo-Chediski Fires 2002

On the afternoon of 18 June 2002, a fire broke out near the Rodeo Fairgrounds on the Fort Apache Reservation in Arizona. By mid-morning on 20 June, the so-called Rodeo fire had expanded to 12 000 hectares (30 000 acres). Meanwhile, a second blaze began burning near Chediski Peak about 24 km (15 miles) from the Rodeo fire, where a lost hiker had started a signal fire. Two days later, on 24 June, the two fires merged to encompass more than 94 000 hectares (235 000 acres). Over the subsequent two weeks, the fire burned an additional 80 000 hectares (200 000 acres), making it the largest, most severe wildfire in Arizona. Before the blaze was brought under control, over 30 000 people were evacuated and 400 homes were destroyed. *Source: The Wilderness Society 2002*

Rodeo-Chediski Burned Area

24 June 2002

Rodeo Fire

Chediski Fire

20 June 2002

The 20 June 2002 satellite image pictured left shows the Rodeo and Chediski fires burning seperately. The 24 June 2002 image above shows the burned areas of the the Rodeo-Chediski fire. Areas where fires were burning when the images were captured appear bright red; already burned areas have a darker red coloration. *Source: USGS 2002, NASA 2002*

Credits: Unknown/UNEP/USDA Forestry Service

susceptible to fire. Logging and large-scale land clearing, for example, disturb the microclimate of forest ecosystems, increasing their fuel load and thus the potential intensity of resulting forest fires. Changes in the Earth's climate, characterized by extended droughts, higher global surface temperatures, and more violent thunderstorms, hurricanes and other types of severe weather, can further intensify the risk of forest fires (Remote Sensing Services GmbH 1998). Concomitantly, fires themselves may play an important role in climate change by emitting both greenhouse gases and smoke particles (aerosols) into the atmosphere.

In 2000, more than 350 million hectares (865 million acres) of forest were burned worldwide—an area equal to the size of India (Northoff 2003). However, when fire sweeps through a forest, many of the trees may survive. In some cases, the fire may burn only the low-growing vegetation.

Amazon Forests and Fire

Before widespread human settlement began to encroach on the borders of South America's Amazon forests, there was no such thing as an Amazon fire season. Now, fire may pose the biggest threat to the survival of the Amazon forest ecosystem.

Slash-and-burn agriculture converts forest to farmland, but that obvious destruction is only the beginning. Intentionally set fires often expand out of control and burn through the understory in areas of surrounding forest, killing, but not completely burning small trees, vines and shrubs. The dead and dying trees collapse, spilling firewood and kindling to the ground and creating openings in the forest canopy. Logging has a similar effect. The intense tropical sun, previously deflected by the green canopy, heats the forest floor, pushing fire danger even higher. In this damaged, fragmented landscape, the onset of the natural dry season brings with it an ominous threat of fire. That threat grows even greater when El Niño-driven droughts occur several times per decade.

Case Study: Pará, Brazil
2004

Once extremely rare even during dry seasons, fire is now a common occurrence in the Amazon rain forest as people use it as a land management tool. Although land management fires may not be immediately hazardous, it is not uncommon for them to escape control and become considerably more destructive. Forest fires can also impact weather, climate, human health, and natural resources.

These two satellite images show an area of the rain forest in the state of Pará, Brazil, at the point where the Tapajós River (angling up from the bottom of each image) joins the Amazon River (running across each image). In this region, forested land is being cleared for logging, ranching, and farming. Cleared areas are visible along the river banks, extending into the forest. Deforested areas appear light green, while undisturbed forest is dark green.

18 Nov 2004

The image dated 18 November 2004 reveals the locations of a number of fires (red dots) burning in different parts of the forest. Roughly three weeks later, fires are still burning and the entire scene has a hazy appearance, the result of smoke suspended in the air over the forest. Fires burned off and on in this region for more than a month.
Source: NOAA 2004

7 Dec 2004

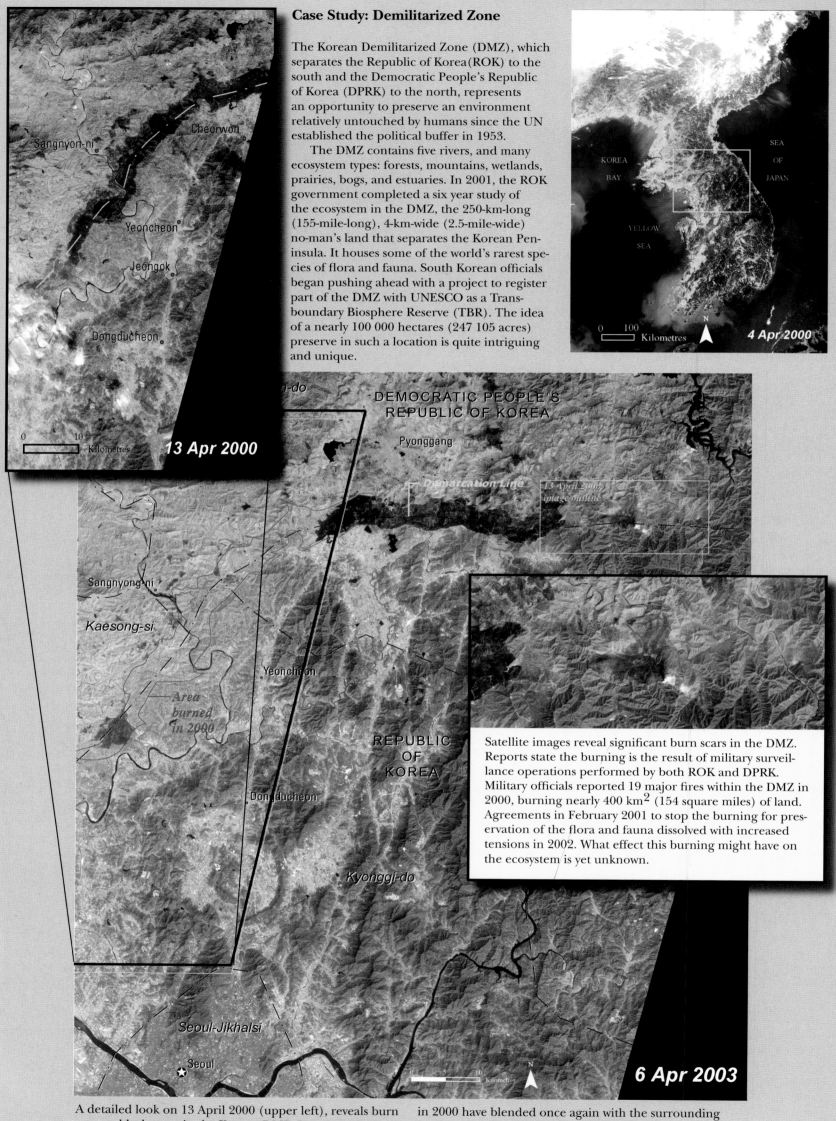

Case Study: Demilitarized Zone

The Korean Demilitarized Zone (DMZ), which separates the Republic of Korea(ROK) to the south and the Democratic People's Republic of Korea (DPRK) to the north, represents an opportunity to preserve an environment relatively untouched by humans since the UN established the political buffer in 1953.

The DMZ contains five rivers, and many ecosystem types: forests, mountains, wetlands, prairies, bogs, and estuaries. In 2001, the ROK government completed a six year study of the ecosystem in the DMZ, the 250-km-long (155-mile-long), 4-km-wide (2.5-mile-wide) no-man's land that separates the Korean Peninsula. It houses some of the world's rarest species of flora and fauna. South Korean officials began pushing ahead with a project to register part of the DMZ with UNESCO as a Transboundary Biosphere Reserve (TBR). The idea of a nearly 100 000 hectares (247 105 acres) preserve in such a location is quite intriguing and unique.

13 Apr 2000

4 Apr 2000

DEMOCRATIC PEOPLE'S
REPUBLIC OF KOREA

Pyonggang

Demarcation Line

13 April 2000 image outline

Sangnyong-ni

Kaesong-si

Yeoncheon

Area burned in 2000

REPUBLIC
OF
KOREA

Don ducheon

Satellite images reveal significant burn scars in the DMZ. Reports state the burning is the result of military surveillance operations performed by both ROK and DPRK. Military officials reported 19 major fires within the DMZ in 2000, burning nearly 400 km^2 (154 square miles) of land. Agreements in February 2001 to stop the burning for preservation of the flora and fauna dissolved with increased tensions in 2002. What effect this burning might have on the ecosystem is yet unknown.

Kyonggi-do

Seoul-Jikhalsi

Seoul

6 Apr 2003

A detailed look on 13 April 2000 (upper left), reveals burn scars as black areas in the Korean DMZ. It appears that the area has been burned on both sides of the border, probably for military surveillance purposes. Areas burned in 2000 have blended once again with the surrounding landscape, but new burns actively clear vegetation to the north and east.

Core Zone

Angangueo

Mariposa
Monarca

M E X I C O

14 Mar 1986

Bosensheve
National
Park

Credit: Beth Allen/UNEP/ Journey North

The fir forest is continually under threat of degradation due to industrial logging, charcoal production, domestic use, and agricultural
expansion. To this end, two presidential decrees (1980 and 1986) have
been issued in an attempt at preserving this important habitat.

0   5   10
Kilometres

14 Mar 1986

Core Zone

Anangueo

Mariposa
Monarca

M E X I C O

31 Mar 2001

Bosensheve
National
Park

Credit: Beth Allen/UNEP/Journey North

Thousands of Monarch butterflies congregate in the butterfly reserve.

31 Mar 2001

envelope that protects the butterflies from freezing on cold nights during the five month overwintering season. Adiabatic rainfall together
with fog condensation on the fir and pine boughs provides the moisture that prevents the butterflies from desiccating as the dry season
advances. A comparison of the 1986 image to the 2001 image reveals

that parts of the forests have been degraded severely. The two close-
up images serve to illustrate the most affected areas. In these images,
the unaffected forest is green in colour while the degraded area is tan.
It is estimated that between 1984 and 1999, 38 per cent of the forests
protected by two presidential decrees were degraded.

Arkhangel'skaya oblast'

Arkhangel'skaya oblast'

Latyuga

Latyuga

R U S S I A

R U S S I A

Bol'shaya Pyssa

Bol'shaya Pyssa

Udorskiy
Nature
Sanctuary

Udorskiy
Nature
Sanctuary

Patrakovo

Patrakovo

Mezen'

Mezen'

Respublika Komi

Respublika Komi

0   5   10
Kilometres

0   5   10
Kilometres

N

N

19 Jul 1973

20 Jul 1987

Credit: Fred Wohlert/UNEP/Topham

Mining activities and petroleum refineries are major sources of
air pollution in northern boreal forests, including those of the
Arkhangelsk region.

Credit: David P. Shorthouse/UNEP/Forestry Images

The Arkhangelsk region has suffered from severe forest degradation.

Mezenskaya Pizhma

Arkhangel'skaya oblast'

Latyuga

R U S S I A

Pysskiy Nature Sanctuary

Pyssa

Bol'shaya Pyssa

Sodzimskiy
Nature
Sanctuary

Udorskiy
Nature
Sanctuary

Patrakovo

Respublika Komi

Mezen

Melentyevo

Chernutyevo

UGA1450077

0   5   10
Kilometres

N

29 Jun 2000

The Arkhangelsk region was once cloaked with dense boreal forests.
In comparing these three satellite images, however, the widespread forest
cover disturbance is obvious. In some places, large sections of the forest
have been clear-felled and the trees completely removed. Other places
show a block pattern, where sections of relatively undisturbed forest are left

number of areas, networks of minor access roads delineate the forest cover.
The region is also home to the Plesetsk Space Center and has been impacted
by fire and pollution from falling rocket stages. Overall, forest cover in the
Arkhangelsk has been heavily disturbed—even within areas designated as
nature sanctuaries.

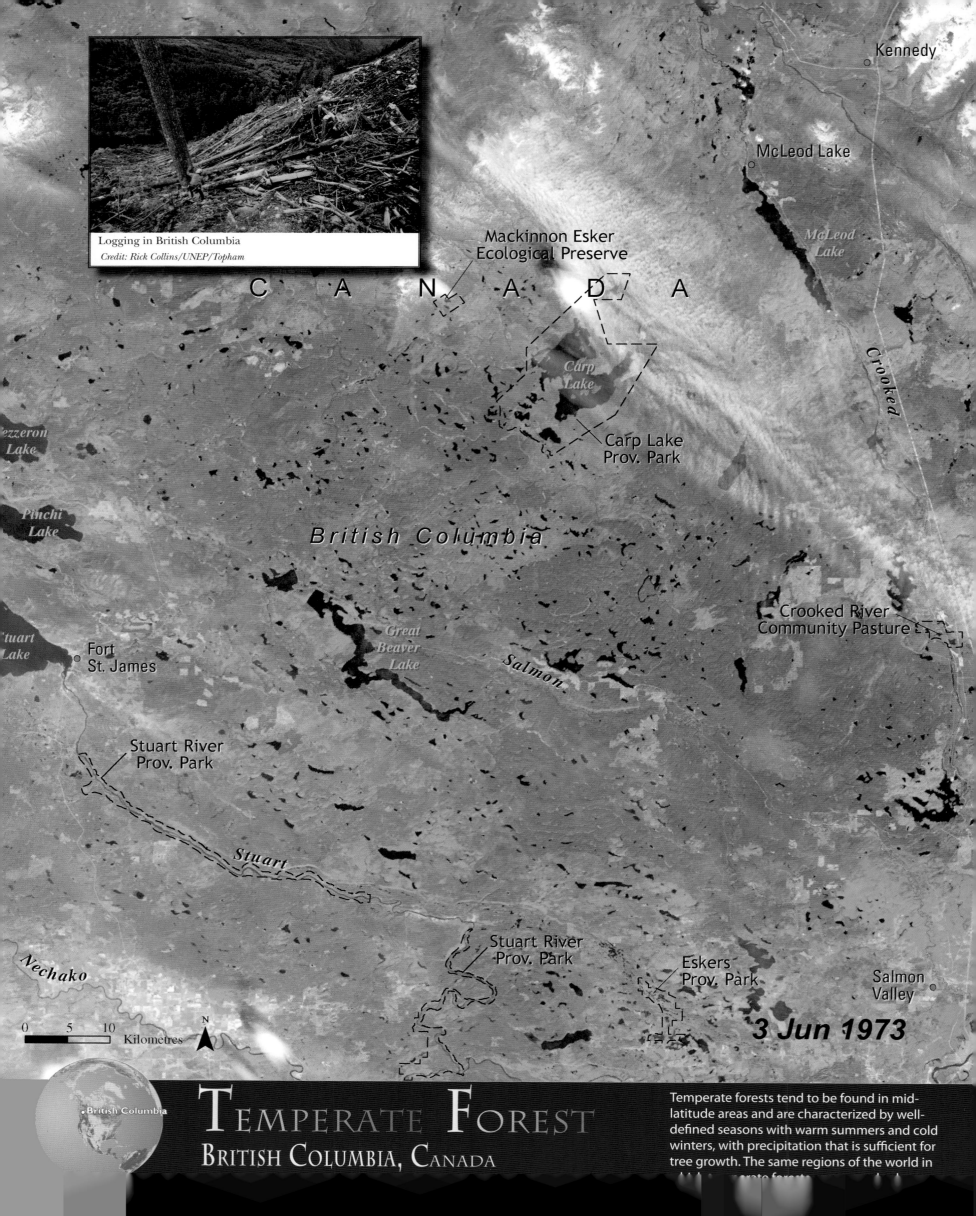

Logging in British Columbia
Credit: Rick Collins/UNEP/Topham

Kennedy

McLeod Lake

McLeod Lake

Mackinnon Esker
Ecological Preserve

C A N A D A

Crooked

Carp Lake

Carp Lake
Prov. Park

ezzeron Lake

Pinchi Lake

British Columbia

Crooked River
Community Pasture

tuart Lake

Fort
St. James

Great Beaver Lake

Salmon

Stuart River
Prov. Park

Stuart

Stuart River
Prov. Park

Eskers
Prov. Park

Salmon
Valley

Nechako

3 Jun 1973

0   5   10   Kilometres   N

British Columbia

Temperate Forest
British Columbia, Canada

Temperate forests tend to be found in mid-
latitude areas and are characterized by well-
defined seasons with warm summers and cold
winters, with precipitation that is sufficient for
tree growth. The same regions of the world in

Kennedy

McLeod Lake

McLeod Lake

Mackinnon Esker
Ecological Preserve

C A N A D A

Crooked

Carp Lake

Carp Lake
Prov. Park

Tezzeron Lake

British Columbia

Pinchi Lake

Crooked River
Community Pasture

Stuart Lake

Fort
St. James

Great Beaver Lake

Salmon

Stuart River
Prov. Park

Stuart

Stuart River
Prov. Park

Eskers
Prov. Park

Salmon
Valley

Nechako

12 Sep 1999

0   5   10   Kilometres

N

to large numbers of people. As a result, temperate forests constitute one of the most altered biomes on the planet. Only scattered remnants of the Earth's original temperate forests remain today, some of which still contain stands of trees that are in high demand for their valuable wood. The interior of British Columbia is a perfect example. Logging is a major industry in British Columbia, carried out almost exclusively in virgin forest, which is very rich in

endemic biodiversity. This pair of satellite images of the Fraser River Valley in British Columbia shows the impact of logging and other human activities during a period of about 25 years, from 1973-1999. The heavy exploitation of the forest is evidenced by the "patchwork quilt" appearance that is typical of logged-over areas.

El Triunfo

El Destino

Rio San Pedro

M E X I C O

Tikal National Park

G U A T E M A L A

Laguna
del
Tigre

Gracia de Dios

Recreo

Rio San Pedro

Progreso

Sierra del Lancondón
National Park

14-15 Feb 1974

Sierra del Lancondón
National Park

0   Kilometres

Rio Usumacinta

TROPICAL FOREST

COUNTRY BORDER GUATEMALA/MEXICO

The border between Guatemala and
Mexico runs through Mexico's Chiapas
Forest and Guatemala's El Peten. In this
pair of images, the border is easy to see,

El Triunfo

El Destino

Rio San Pedro

M E X I C O

Tikal National Park

Laguna del Tigre

G U A T E M A L A

Gracia de Dios

Recreo

Rio San Pedro

Progreso

Sierra del Lancondón National Park

27 Mar - 3 Apr 2000

Sierra del Lancondón National Park

0 5 10
Kilometres

N

Rio Usumacinta

even without the black lines that have been overlaid on the images to show the outlines of the two countries.

The region crossed by this border was once biologically very diverse. On the Guatemalan side, it still is, as most of the El Peten remains as closed canopy forest because of lower population densities and the protected status of the Sierra de Lacondon and Laguna del Tigre National Parks. Across the border in Chiapas, however, a larger and increasing population has an obvious effect on the landscape. Between 1974 and 2000, much of the forest on the Mexican side of the border has been converted to cropland or pasture.

BRAZIL

Itaipú Dam

Ciudad del Este   Foz do Iguaçu

Iguaçu
National   Iguaçu (Iguazú)
Park

Puerto Iguazú

Iguazú National Park

Iguazú Falls

PARAGUAY

ARGENTINA

Urugua-í Provincia

Paraná

Credit: John Townshend/UNEP
Sparsely populated during the 1970s, this region has undergone major development and large areas of forest have been converted to agricultural lands.

0   10   20
Kilometres   N

23 Feb 1973

•Iguazu

SUBTROPICAL FOREST
IGUAZÚ, SOUTH AMERICA

Iguazú National Park, located in Argentina near its borders with Brazil and Paraguay, contains remnants of the highly endangered Paranaense Rain Forest. Isolated from other rain forests by natural barriers, the Paranaense developed

Itaipú Dam

Itaipú Dam, one of the world's largest, was built from 1973 to 1982 and is a major source of electricity for both Brazil and Paraguay.

Ciudad del Este

Foz do Iguaçu

Iguaçu National Park

Iguaçu (Iguazú)

Puerto Iguazú

Iguazú National Park

Iguazú Falls

One of the many falls within Igauzú National Park

PARAGUAY

Over 90 per cent of the Paranaense Forest has been converted into agricultural fields, in which mainly soybeans and corn are grown.

Paraná

ARGENTINA

Uruguá-í Provincial Pa

N

Kilometres

12 May 2003

thousands of species of mammals, birds, reptiles, and amphibians unique to the area. The famous Iguazú Falls are located within the boundaries of the National Park and are shared by Argentina and Brazil.

Between1973 and 2003, dramatic changes to the landscape occurred in this region. In 1973 the forested area spread across the borders of the three

nations. By 2003, however, large areas of the forest in Paraguay and Brazil, and smaller amounts in Argentina, had been converted to other forms of land cover, creating a mosaic of differently colored land use areas. Note the variation in land cover patterns among the different countries—reflections of different land use polices and practices.

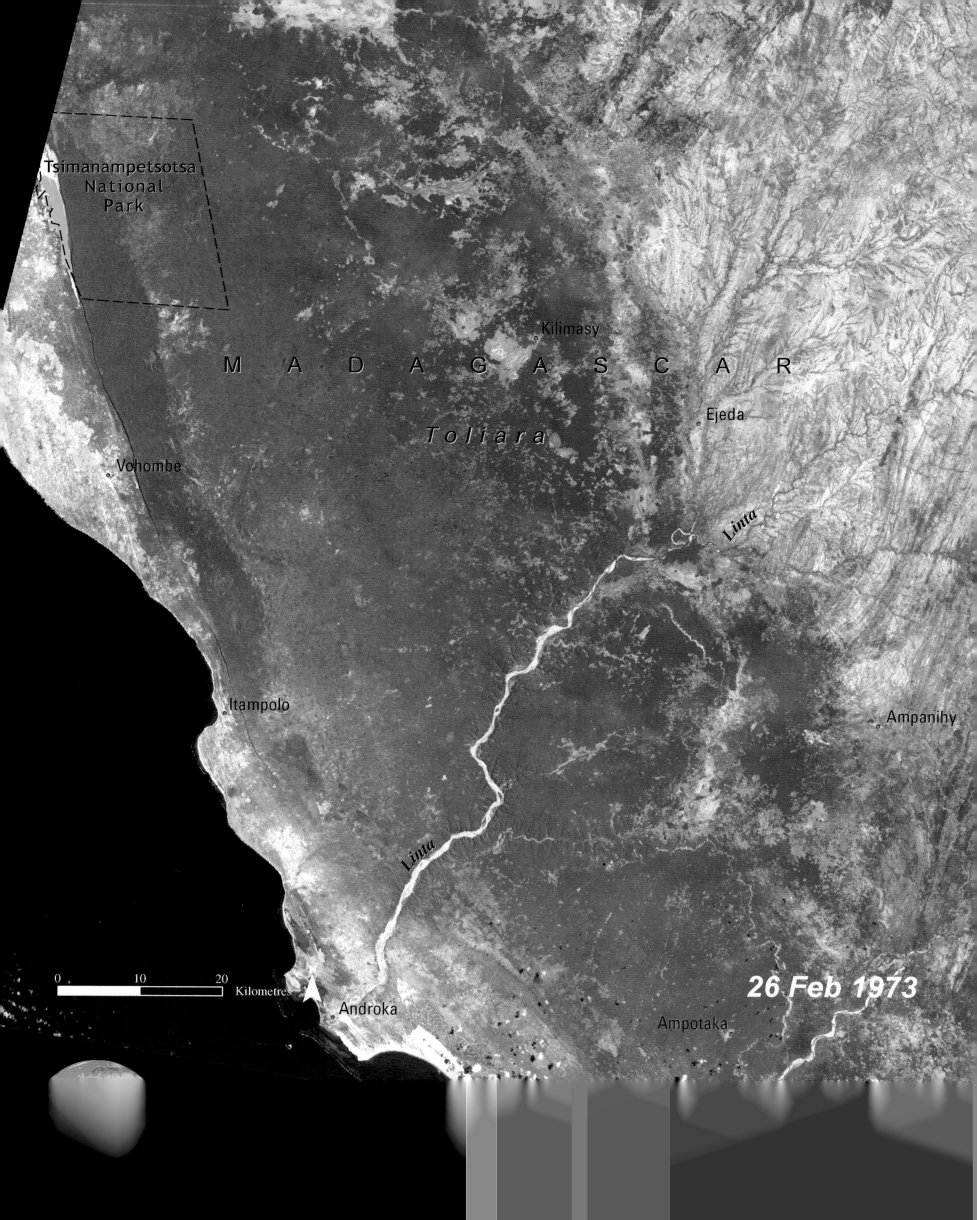

Tsimanampetsotsa
National
Park

Kilimasy

M A D A G A S C A R

T o l i a r a

Ejeda

Vohombe

Linta

Linta

Itampolo

Ampanihy

0   10   20
|————————————————————| Kilometres

26 Feb 1973

Androka

Ampotaka

Tsimanampetsotsa
National
Park

Kilimasy

M A D A G A S C A R

Toliara

Ejeda

Vohombe

Linta

Itampolo

Ampanihy

Linta

0   10   20   Kilometres

6 May 2001

Androka

Ampotaka

almost completely forested. But the practice of burning the forest to clear land for dry rice cultivation has over time denuded most of the landscape, particularly in the central highlands (tan colour in the 2001 image). Coffee production, grazing, gathering fuelwood, logging, cattle ranching, mining and other activities also have contributed

to deforestation and land degradation. This set of satellite images shows a narrow coastal plain near the Linta River of southwestern Madagascar. Between 1973 and 2001, the forests in this area have all but disappeared. Remarkably, numerous endemic species still remain in scattered forest remnants.

Mugwambuli

Tshopo

Congo

Kisangani

Congo

DEMOCRATIC REPUBLIC
OF THE CONGO

Boat on the Congo River
Credit: Lumbuenamo Raymond/UNEP

15 Jan 1975

0   5   10
Kilometres

N

Kisangani

TROPICAL FOREST
KISANGANI, D.R. OF THE CONGO

Kisangani, in the Democratic Republic of the
Congo, is located along the Congo River in
the northwestern part of the country. It is a
city of roughly a half million people.

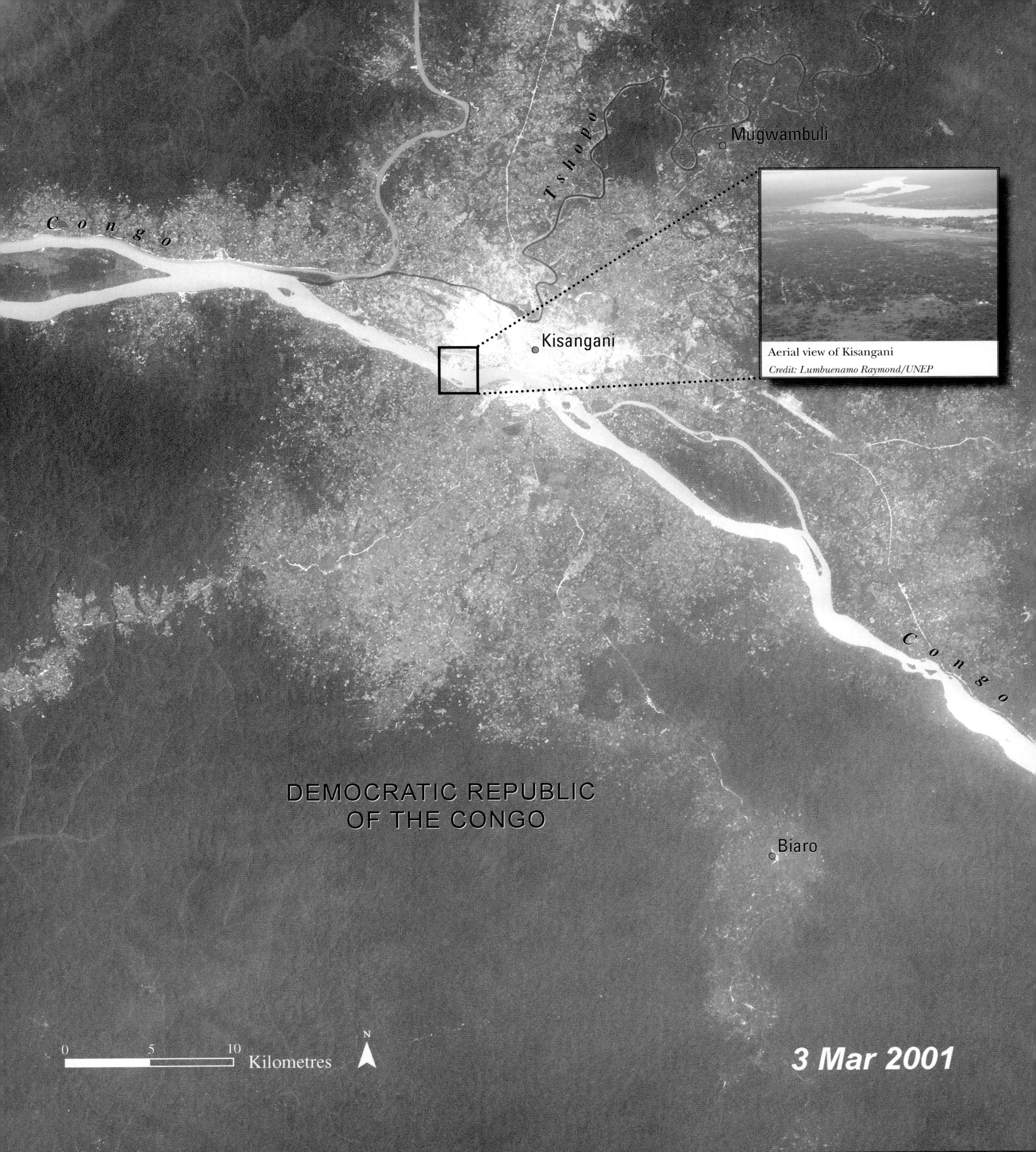
Congo
Tshopo
Mugwambuli
Kisangani
Aerial view of Kisangani
Credit: Lumbuenamo Raymond/UNEP
Congo
DEMOCRATIC REPUBLIC
OF THE CONGO
Biaro
0
5
10
Kilometres
N
3 Mar 2001

around Kisangani is a rich green colour, indicative of dense forest cover. However, directly around the city is a light green zone—evidence of deforestation and conversion of the land to other uses. In the second image, taken in 2001, the cleared area around the city has grown and become consolidated; it has also spread along the rivers and the roads.

Much of the deforestation is attributed to the influx of refugees into the country. Even the denser parts of the forest, once thought to be impenetrable, show signs of deforestation.

RUSSIA

Girvasskiy
Nature
Sanctuary

Urho Kekkosen
Kansallispuisto
National Park

F I N L A N D

lappi

Tulppio

Drainage of peat lands enables
trees to grow.
Credit: H. Gyde Lund/UNEP

Värriön Luonnonpuisto
State Nature Reserve

20 Jul 1987

Tuntsan Erämaa - Alue
Wilderness Area

0   2   4
|_____|_____| Kilometres

N

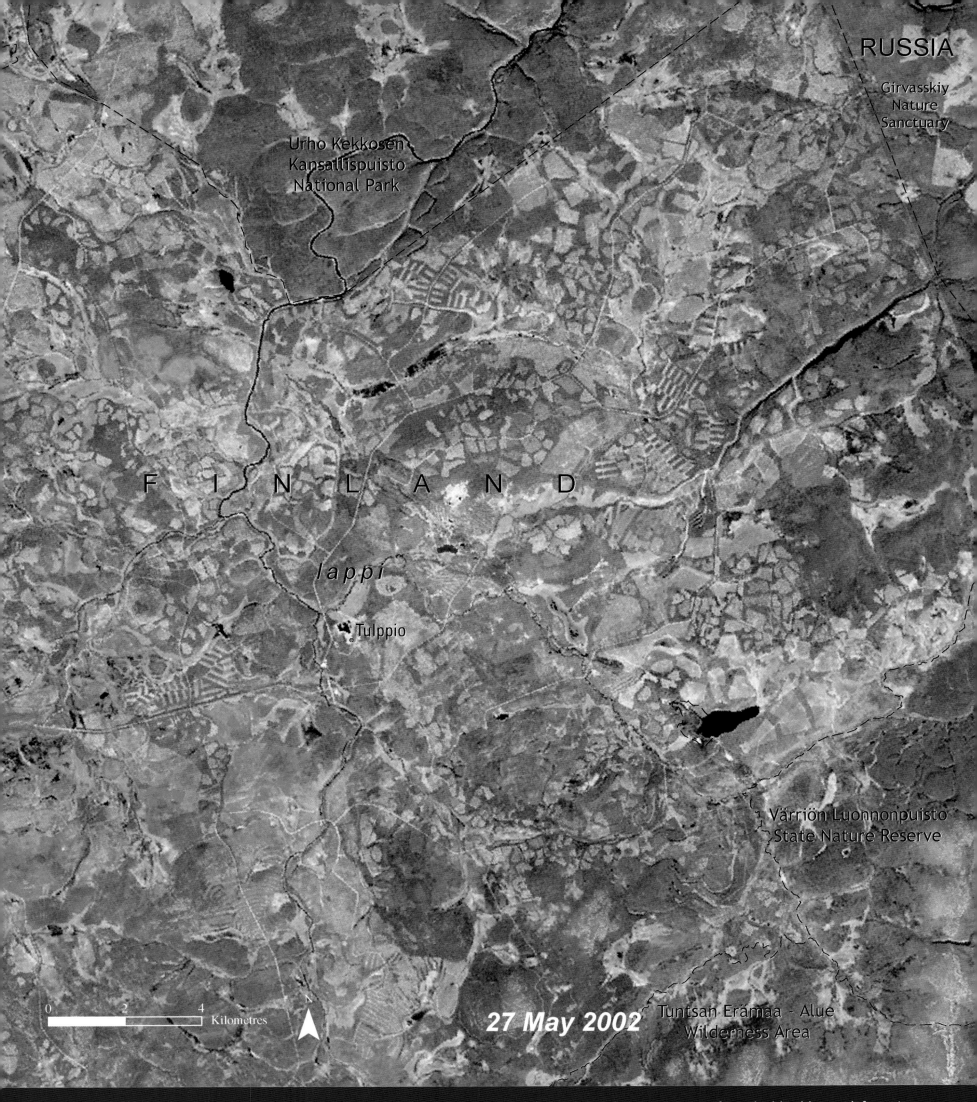

RUSSIA

Girvasskiy
Nature
Sanctuary

Urho Kekkosen
Kansallispuisto
National Park

F I N L A N D

lappi

Tulppio

Värriön Luonnonpuisto
State Nature Reserve

0   2   4
Kilometres

27 May 2002

Tuntsan Erämaa - Alue
Wilderness Area

exports. Feeding this massive paper industry is the Finnish forest industry – one of the most intensive in the world. As a result, Finland's forests—including its remaining old-growth fragments—are being exploited by clearcutting, forest thinning, road construction, and ditching of soils. The result is the severe and extensive fragmentation of natural habitat. While much of Finland's productive forest (around 62 per cent) is in the hands of

private landowners, the vast majority of its valuable old-growth forest is owned and logged by the state. These two images show a result of this logging in the northeastern areas of the country. In the 1987 image, the area has a near homogeneous forest cover (green); on the other hand, the 2002 image shows only a few patches, mainly in the protected areas with continuous forest cover. The patches of tan signify clearcut areas.

Beaver

Olympic N.F.

Olympic N.F.

Lake Crescent

Olympic N.F.

Bogachiel

Hoh

O l y m p i c

Olympic

National

△
Mt. Olympus

Park

O l y m p i c

Olympic N.F.

U N I T E D   S T A T E S

Kalaloch

Clearwater

Queets

W a s h i n g t o n

Quinault

Quinault Lake

Cushman Lake

Quinault Indian
Reservation

Quinault

Olympic N.F.

Pacific Beach

Skokomish Indian
Reservation

Humptulips

Shelton

0   5   10
Kilometres

N

12 Sep 1974

Olympic Nat. Forest

Temperate Forest
Olympic Peninsula, United States

On the slopes and the surrounding areas of Mt.
Olympus in the Olympic Peninsula of the Pacific
Northwest, one of the last remnants of temperate forests in the United States is quickly disappearing. Between 1971 and 2002, nearly half a
million hectares (1.1 million acres), or almost 29

Beaver

Olympic N.F.

Olympic N.F.

Lake Crescent

Olympic N.F.

Olympic N.F.

Bogachiel

Ioh

Olympic
National

△ **Mt. Olympus**

Park

M o u n t a i n s

Olympic N.F.

U N I T E D   S T A T E S

W a s h i n g t o n

Kalaloch

Clearwater

Queets

Quinault Indian
Reservation

*Quinault
Lake*

Quinault

*Cushman
Lake*

Quinault

Olympic N.F.

Pacific Beach

Skokomish Indian
Reservation

Humptulips

Shelton

0   5   10
Kilometres

N

30 Jun 2000

An alpine meadow located in Olympic National Park.

Photo Credit: US National Park Service

per cent of the forest covering the Peninsula, was clear-cut. That is an
area equal in size to the Olympic National Park and its five adjacent wilderness areas.

The 1974 image shows the characteristic patchwork of purple and pink areas where clear-cutting has taken place. Light green patches signify regrowth.

the Peninsula were the most severely impacted during this period of time: 48
per cent of the forests on Native lands were clear-felled. In the 2000 image,
clear-cutting is obviously still continuing, as is development to the north,
west, and south of the national park. There is evidence of good regrowth of
trees in forest reserve areas in preparation for the next clear-felling cycle.

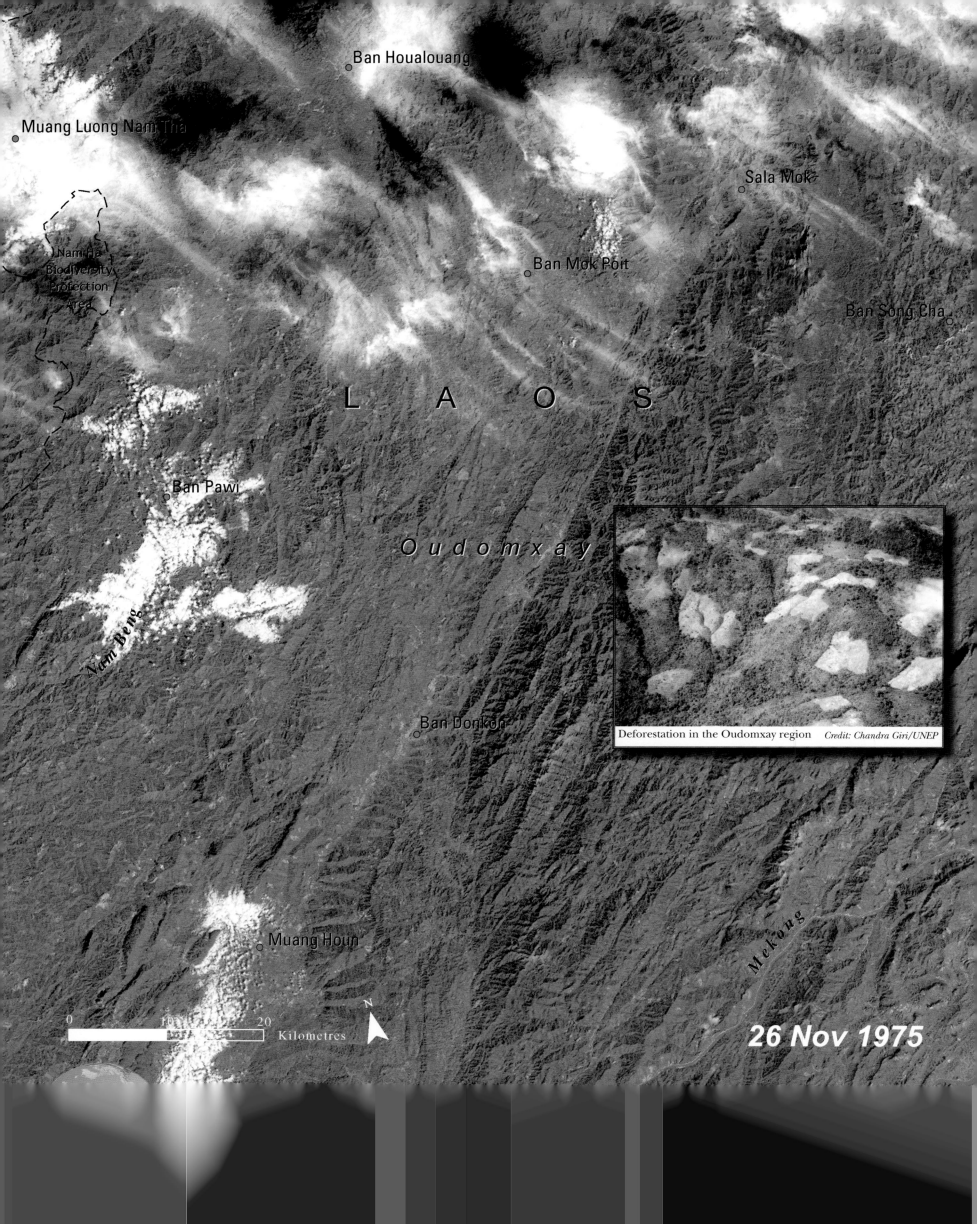

Ban Houalouang

Muang Luong Nam Tha

Sala Mok

Nam Ha
Biodiversity
Protection
Area

Ban Mok Poit

Ban Song Cha

L A O S

Ban Pawi

O u d o m x a y

Nam Beng

Ban Donkon

Deforestation in the Oudomxay region   *Credit: Chandra Giri/UNEP*

Mekong

Muang Houn

N

0   10   20   Kilometres

26 Nov 1975

Ban Houaloyang

Muang Luong Nam Tha

Sala Mok

Nam Ha
Biodiversity
Protection
Area

Ban Mok Poit

Ban Song Cha

L A O S

Ban Pawi

O u d o m x a y

Nam Beng

Ban Donkon

Recently harvested lumber next to a road
Credit: Chandra Giri/UNEP

Muang Houn

Mekong

N

0   10   20
Kilometres

2 Nov 2000

burning and shifting cultivation—a land-use pattern in which patches of forest are cleared, burned and cultivated, and then abandoned to allow regrowth and recovery before being exploited again—are the primary causes of deforestation throughout the country. A comparison

of shifting cultivation activities—and the resulting environmental damage—has increased over time. In 1975 there were only scattered areas under cultivation (white patches) in the forest. By 2000, the number and size of cultivated areas and disturbed forest cover had

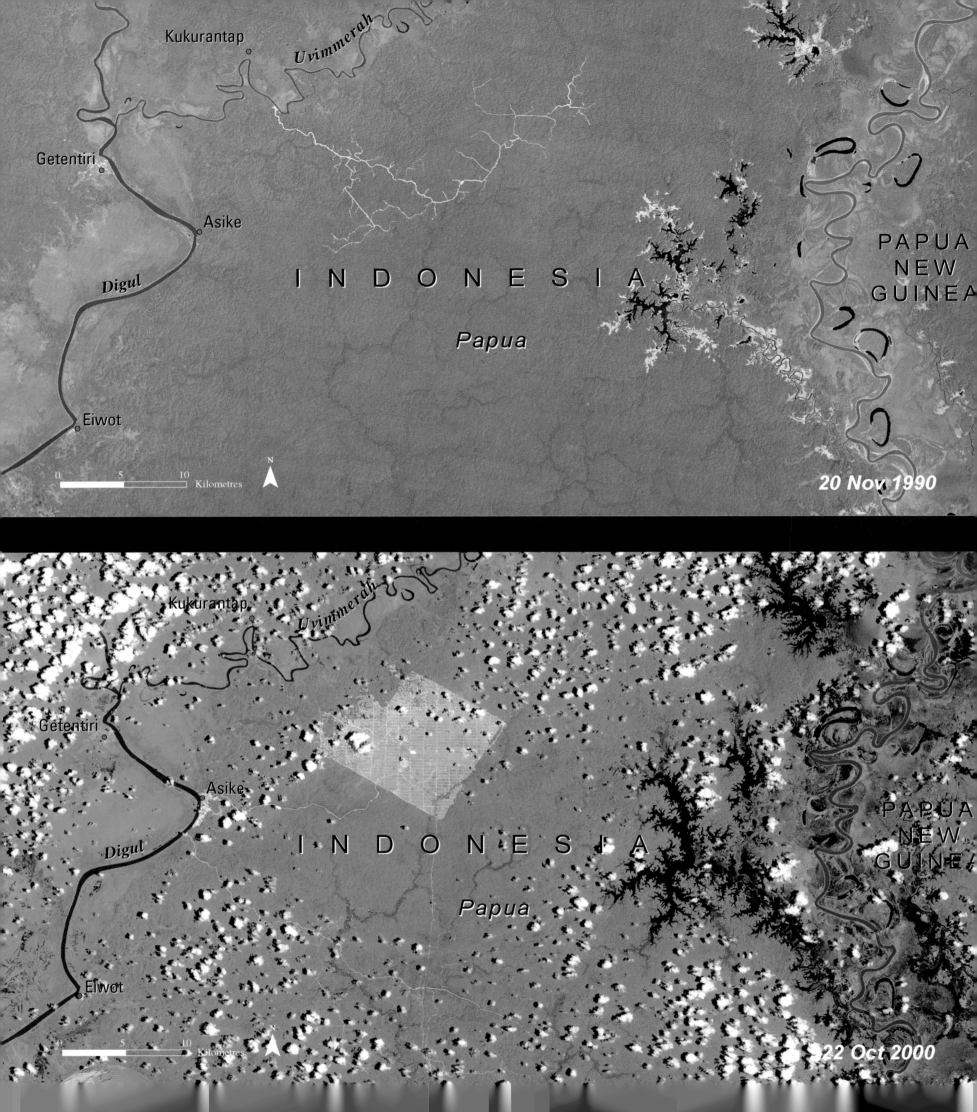

Kukurantap

Uvimmerah

— Expansion since 2000

Getentiri

Asike

Digul

I N D O N E S I A

Papua

Eiwot

Oil palm fruit in Malaysia
Credit: H. Gyde Lund/UNEP

0   5   10
▬▬▬▬▬▬▬▬▬ Kilometres

N

28 Oct 2002

satellite images reveal how a combination of transmigration, logging interests, and palm oil plantation development have transformed an area that was previously tropical lowland rain and swamp forest.

While the 1990 image shows the first signs of development in this region, with the jagged access road network forming the only break in the forest cover, the October 2000 image reveals an area being prepared for palm oil plantations, and evidence of an influx of plantation workers. The 2002 image clearly shows a checkered pattern of plantations in the primary development area, and the extension of the road network to the north, south, and southwest.

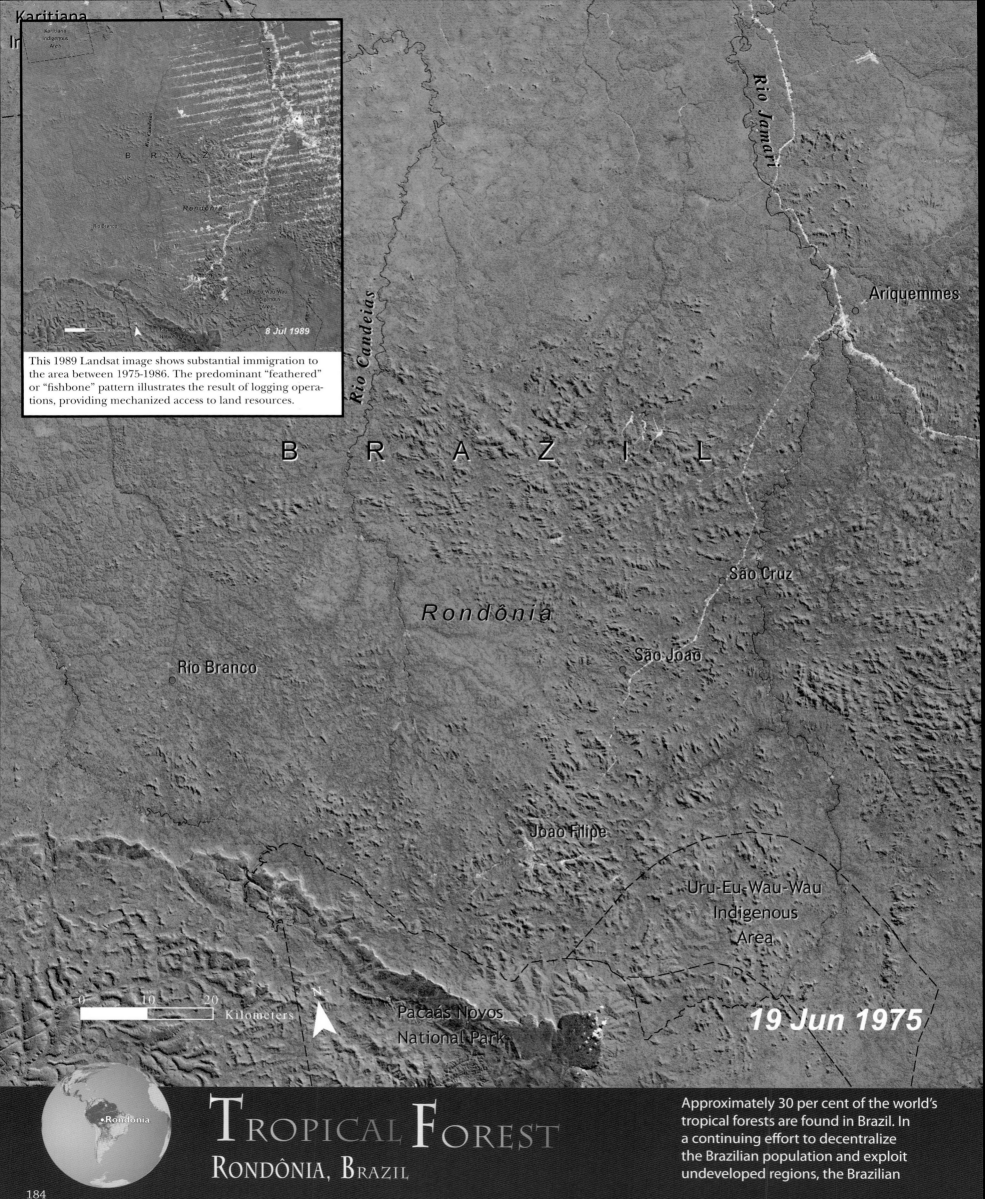

Karitiana
In

This 1989 Landsat image shows substantial immigration to the area between 1975-1986. The predominant "feathered" or "fishbone" pattern illustrates the result of logging operations, providing mechanized access to land resources.

8 Jul 1989

Rio Jamari

Aríquemmes

Rio Candeias

B R A Z I L

São Cruz

Rondônia

São João

Rio Branco

Joao Filipe

Uru-Eu-Wau-Wau
Indigenous
Area

0   10   20
Kilometers

N

Pacaás Novos
National Park

19 Jun 1975

Rondônia

TROPICAL FOREST
RONDÔNIA, BRAZIL

Approximately 30 per cent of the world's tropical forests are found in Brazil. In a continuing effort to decentralize the Brazilian population and exploit undeveloped regions, the Brazilian

Karitiana
Indigenous
Area

Credit: Ron Levy/UNEP/Topham

Primary land uses are cattle ranching and annual crop farming. More sustainable perennial crops like coffee and cacao occupy less than 10 per cent of the agricultural land areas.

Rio Jamari

Ariquemmes

B R A Z I L

Rondônia

São Cruz

São Joao

Rio Branco

Joao Filipe

Uru-Eu-Wau-Wau
Indigenous
Area

0   10   20   Kilometers

N

Pacaás Novos
National Park

19 Sep 2001

government constructed the Cuiaba-Port Velho highway through the province of Rondônia. Completed in 1960, the road serves as the access route for infrastructural development in the region, previously occupied solely by indigenous people. In 1975, the region was

distinctive fishbone pattern of forest exploitation had appeared and by 2001 had expanded dramatically. The highway has become a major transportation route for immigrant farmers seeking income-producing opportunities. Migration into the area continues unabated.

Credit: David P. Shorthouse/UNEP/Forestry Images
Boreal forests cover large parts of Alaska, Canada, Scandinavia, and western Russia.

Oleniy Nat.
Preserve
Val

Sea of Okhotsk

Nogliki

Katangli

Aleksandrovskiy
Nat. Preserve

Nysh

Gorki

Tatarskiy Proliv

Gorki

0   10   20
Kilometres

N

17 Jun 1989

Sakhalin

BOREAL FOREST
SAKHALIN, RUSSIA

186

Mixed deciduous and evergreen needle-leaf trees dominate the boreal forests of Sakhalin Island, just off the eastern coast of Russia. The tremendous natural reserves of the boreal forests serve as "carbon sinks" that help to regulate global climate. Boreal forests are also home to a unique collection of plants and animals, including rare and endangered species such

Credit: David J. Moorhead/UNEP/Forestry Images
Excessive, out-of-control fire can destroy tree roots, even though trees remain standing.

Oleniy Nat. Preserve

Val

Sea of Okhotsk

Nogliki

Katangli

Aleksandrovskiy Nat. Preserve

Nysh

Gorki

Tatarskiy Proliv

Gorki

0   10   20   Kilometres

N

3 Oct 1999

as the Amur Tiger. Fire is a natural and often vital component in maintaining the health of boreal forests. But since the 1950s, the frequency of fires has increased on Sakhalin Island as its forests have been subjected to rapid exploitation and disturbance in the acquisition of lumber, oil, coal, and peat. As people have moved into the region in greater numbers, the risk of fires started by trains, cars, trash fires, and wood stoves has increased greatly.

These satellite images show the impact of forest fires on Sakhalin Island. In 1998, roughly 300 intensely hot fires burned an area nearly the size of Luxembourg. Three people died and nearly 600 were made homeless by a very rapidly moving crown fire that consumed the town of Gorki within a few hours. The 1999 image very clearly shows the extent of the fire damage to the island's forests near the end of that year.

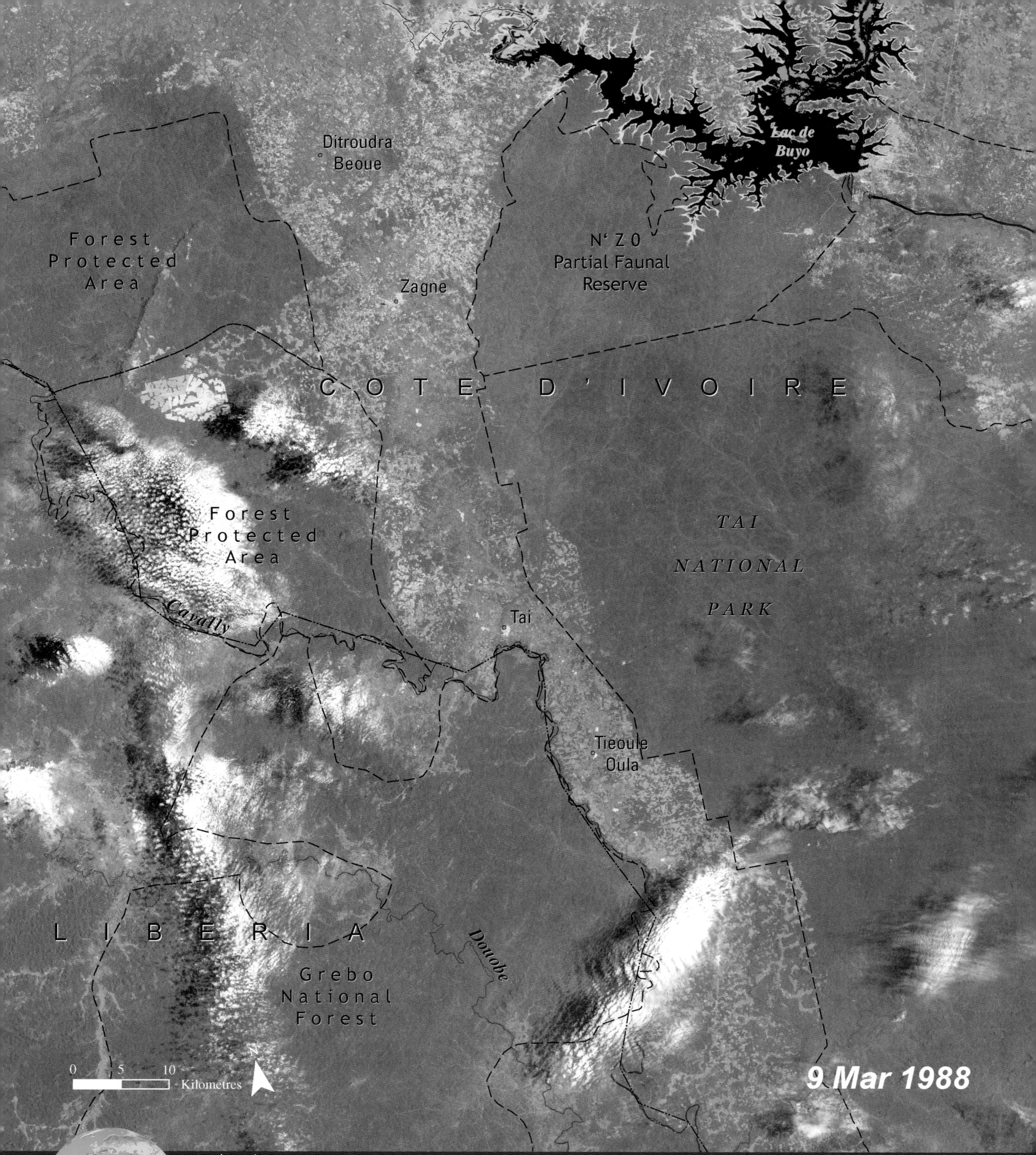

Ditroudra
Beoue

Lac de
Buyo

Forest
Protected
Area

N'ZO
Partial Faunal
Reserve

Zagne

C Ô T E   D ' I V O I R E

Forest
Protected
Area

TAÏ

NATIONAL

Cavally

Taï

PARK

Tieoule
Oula

L I B E R I A

Douobe

Grebo
National
Forest

0   5   10
Kilometres

9 Mar 1988

Taï National Park

TROPICAL FOREST
TAÏ NATIONAL PARK, CÔTE D'IVOIRE

The deforestation rate in Côte d'Ivoire is
thought to be one of the highest in tropical
regions worldwide. Conservation of large
forested areas, such as those within the
boundaries of the Taï National Park, is

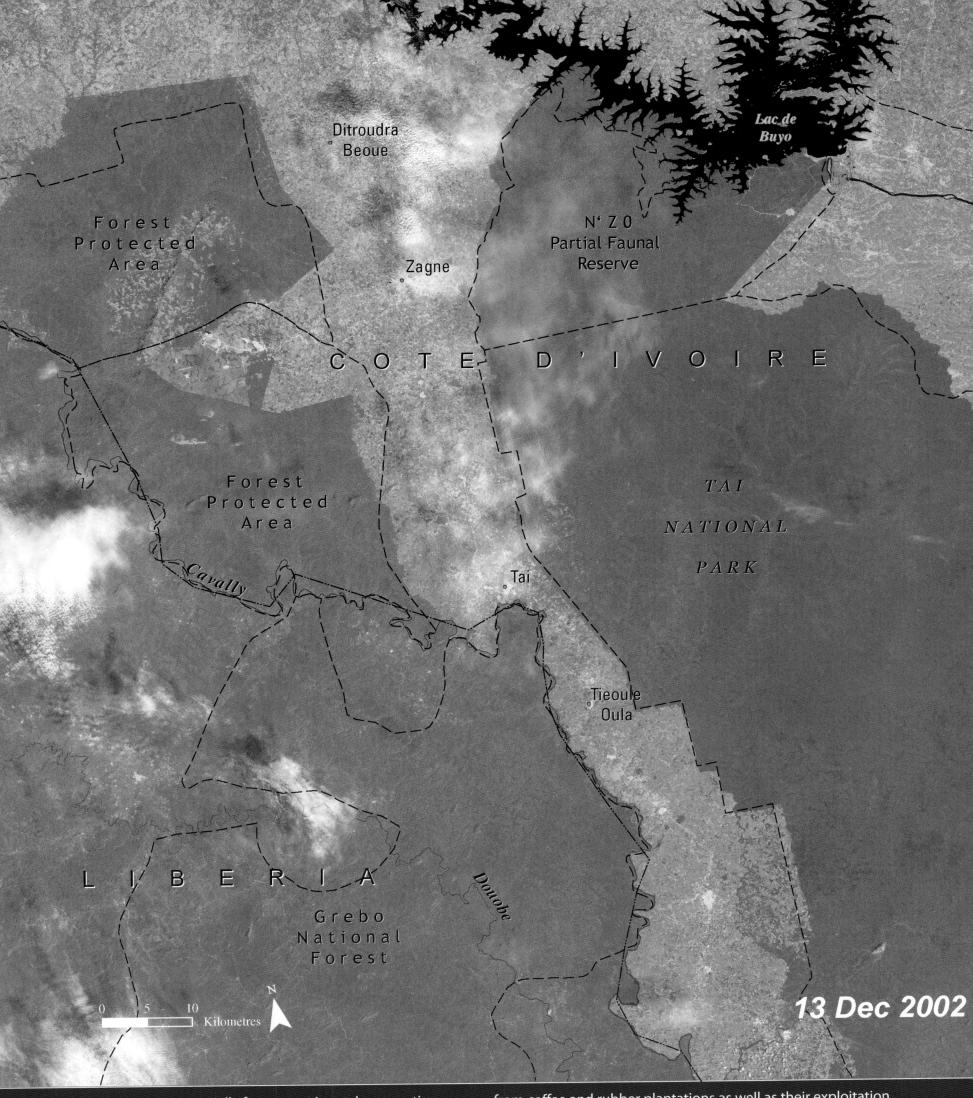

Ditroudra
Beoue

Forest
Protected
Area

Zagne

Lac de
Buyo

N'ZO
Partial Faunal
Reserve

C O T E   D ' I V O I R E

Forest
Protected
Area

Cavally

TAI

NATIONAL

PARK

Tai

Tieoule
Oula

L I B E R I A

Douobe

Grebo
National
Forest

0   5   10
Kilometres

N

13 Dec 2002

189

of primary importance, especially from a continental perspective. Conservation of smaller forested areas is also essential, both for biological conservation purposes and to meet the needs of rural communities.

A comparison of these satellite images from 1988 and 2002 shows the destruction of small forest fragments due to increased pressure from coffee and rubber plantations as well as their exploitation for fuel wood. The lighter green strip bisecting the images is the result of extensive deforestation and intensive cultivation between the protected Grebo National Forest and the Tai National Park. Encouragingly, the boundaries of the protected areas have remained relatively intact.

Chanquiuque

C H I L E

Tanguao

Credit: Juan Schlatter/UNEP/Instituto de Silvicultura, UACH

Temporal sequence of land-use changes from agricultural land (1975) to *Pinus radiata* plantation (1981).

0   5   10
Kilometres

N

22 Mar 1975

BOREAL FOREST
VALDIVIAN, CHILE

Valdivian

Chile has been recently considered as one of the most economically competitive countries in Latin America. Rapid growth in Chile's production and export of forest products is based on the expansion and management of exotic species forest plantations in the last 30 years. However, some studies have demonstrated that such

Chanquiuque

C H I L E

Tanguao

Credit: Claudio Donoso/UNEP/Instituto de Silvicultura, UACH

Mixed forest of native (*Nothofagus glauca*) and exotic (*Pinus radiata*) tree species.

0   5   10
Kilometres

N

29 Jan 2001

expansion of forest plantation has produced a decrease in native forests in the south-central region of the country. These two Landsat, MSS and ETM scenes taken in 1975 and 2001, respectively, show changes in land use during the last 30 years. Many endangered tree and shrub species have been affected by this change, which has also led to a dramatic reduction of landscape diversity as well as goods and services from forestlands. The traditional land-use practices of small-scale logging of native forests, livestock and agriculture have been replaced by large-scale timber production that puts endemic endangered tree and shrub species at risk.

3.5 Cropland

The success of the human race can, in many respects, be attributed to the development of agriculture. The ability to raise crops and therefore control a large portion of our food supply has enabled humankind to expand and flourish as a species, and to grow in numbers far beyond the natural carrying capacity of the environment. It is also through agriculture that people have brought about some of the greatest changes to the global environment.

The Food and Agriculture Organization of the United Nations defines cropland as "land used for cultivation of crops" (FAO 2002). The foods and fibers we grow on croplands around the world are many and diverse. They include: annual crops such as maize, rice, cotton, wheat, and vegetables; crops harvested after more than a year such as sugar cane, bananas, sisal, and pineapple; and perennial crops such as coffee, tea, grapes, olives, palm oil, cacao, coconuts, apples, and pears.

The total area devoted to crops worldwide increased from 1 350 million hectares (3 336 million acres) in 1961 to 1 510 million hectares (3 731 million acres) in 1998, an annual increase of about 0.3 per cent. Most of this expansion took place in developing countries, where cropland expanded 1.0 per cent annually (Wiebe 2003). According to FAO estimates, the 1 500 million hectares (3 706 million acres) of land currently used for growing crops represents just 35 per cent of the 4 200 million hectares (10 378 million acres) of the world's land judged to be suitable for crop production. Nevertheless, much of the undeveloped arable land has marginal productivity due to costs for sustainable development and use.

Food production has more than kept pace with global population growth (WRI 2000). On average, food supplies are now 24 per cent higher per person than in 1961, and food prices are 40 per cent lower. It is estimated that world population will be in-

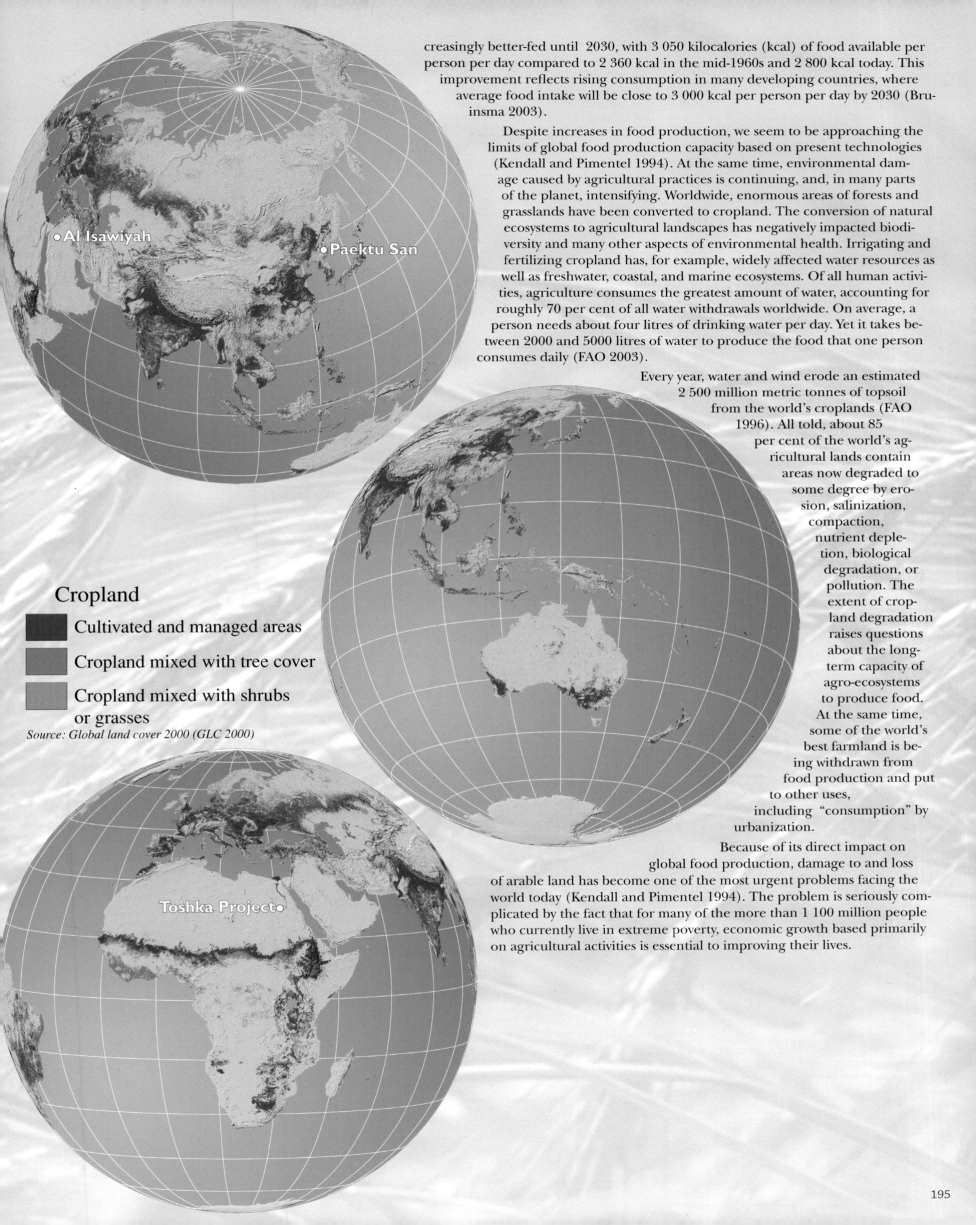

creasingly better-fed until 2030, with 3 050 kilocalories (kcal) of food available per person per day compared to 2 360 kcal in the mid-1960s and 2 800 kcal today. This improvement reflects rising consumption in many developing countries, where average food intake will be close to 3 000 kcal per person per day by 2030 (Bruinsma 2003).

Despite increases in food production, we seem to be approaching the limits of global food production capacity based on present technologies (Kendall and Pimentel 1994). At the same time, environmental damage caused by agricultural practices is continuing, and, in many parts of the planet, intensifying. Worldwide, enormous areas of forests and grasslands have been converted to cropland. The conversion of natural ecosystems to agricultural landscapes has negatively impacted biodiversity and many other aspects of environmental health. Irrigating and fertilizing cropland has, for example, widely affected water resources as well as freshwater, coastal, and marine ecosystems. Of all human activities, agriculture consumes the greatest amount of water, accounting for roughly 70 per cent of all water withdrawals worldwide. On average, a person needs about four litres of drinking water per day. Yet it takes between 2000 and 5000 litres of water to produce the food that one person consumes daily (FAO 2003).

Every year, water and wind erode an estimated 2 500 million metric tonnes of topsoil from the world's croplands (FAO 1996). All told, about 85 per cent of the world's agricultural lands contain areas now degraded to some degree by erosion, salinization, compaction, nutrient depletion, biological degradation, or pollution. The extent of cropland degradation raises questions about the long-term capacity of agro-ecosystems to produce food. At the same time, some of the world's best farmland is being withdrawn from food production and put to other uses, including "consumption" by urbanization.

Because of its direct impact on global food production, damage to and loss of arable land has become one of the most urgent problems facing the world today (Kendall and Pimentel 1994). The problem is seriously complicated by the fact that for many of the more than 1 100 million people who currently live in extreme poverty, economic growth based primarily on agricultural activities is essential to improving their lives.

Cropland

- Cultivated and managed areas
- Cropland mixed with tree cover
- Cropland mixed with shrubs or grasses

Source: Global land cover 2000 (GLC 2000)

Al Isawiyah

Paektu San

Toshka Project

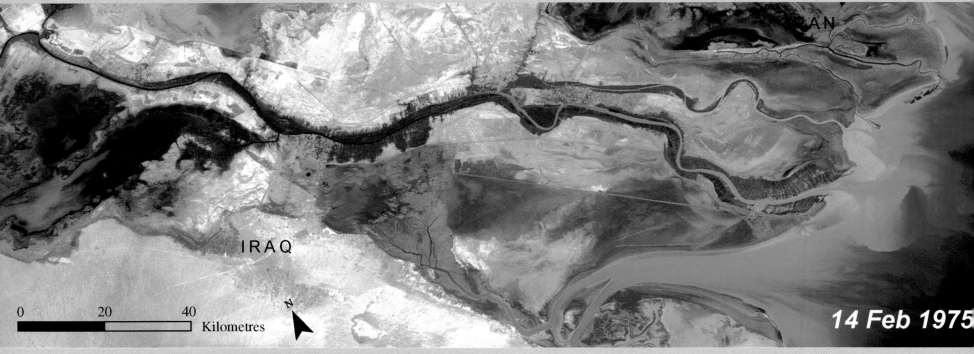

0   20   40
                Kilometres

N

14 Feb 1975

Healthy vegetation is characterized by a distinctively strong reflectance in the near infrared and appears red. In the infrared Landsat images above, the date palm belt skirting the Shatt al-Arab appears as a dark red hue in 1975. In 2002, the intensity of infrared emittance in the date belt is considerably diminished; the pallid red brown indicates stressed and dead vegetation, and the replacement of palms by reeds and desert scrub.

Case Study: Shatt al-Arab Palm Forest Destruction
1975–2002

(By Hassan Partow, UNEP/DEWA/GRID-Geneva & GRID-Sioux Falls)

Lining the 193-km-long (120-mile-long) Shatt al-Arab estuary, formed by the confluence of the Tigris and Euphrates Rivers, is the largest date palm forest in the world. Stretching back from the riverbanks towards the desert, date plantations extend for distances varying from a few hundred metres to almost six kilometres (4 miles). In the mid-1970s, the region counted some 17-18 million date palms or a fifth of the world's 90 million palm trees. By 2002, more than 14 million, or 80 per cent, of the palms were wiped out.

Destruction of the palm forest is due to a variety of factors. War has had the most direct impact, but salinisation and pest infestation have also caused long-term damage. The livelihoods of millions of people dependent on dates for food and income are in ruins, including a regional trade with export earnings ranked second only to oil.

Impact of War

Most of the Shatt al-Arab is in Iraq. But roughly about the last half of its course, near its juncture with the Karun River, forms the border between Iraq and Iran. Demarcation of the borderline has been disputed by the two countries and was invoked as a cause in the outbreak of hostilities in 1980. The conflict, which lasted for eight years, was the longest conventional war of the twentieth century, claiming an estimated one million human lives and causing extensive environmental damage. With the Shatt al-Arab waterway recast into a major theatre of war, the palm forest was unavoidably caught in the prolonged and intense crossfire. The destructive power unleashed by modern weapons in ground battles and aerial bombardments as well as deliberate felling reduced the palm forest to an emaciated shadow of what it was in its lustrous past.

Salt and Pests

Date cultivation along the banks of the Shatt al-Arab is a rare example of extensive tidal irrigation. Under the influence of the strong twice-daily tidal action of the Gulf, upper layers of fresh estuary water are swept into the creeks, irrigating date palm groves on the flood and draining them on the ebb.

Credit: Nik Wheeler/UNEP/UNEP-GRID Geneva

Al Qurnah

Shatt al-Arab

Karun

Khorramshahr

14 Feb 1975

IRAN

Abadan

Bahmanshir

Al Basrah

Abadan Island

IRAQ

Al Faw

Persian Gulf

0   20   40 Kilometres

N

Analysis of Landsat satellite imagery shows that of the 52 000 hectares (128 494 acres) of date farms fringing the Shatt al-Arab in 1975 only 11 000 hectares (27 181 acres), or 21 per cent, remained in 2002. In total, war, salt and pests have destroyed approximately 14 million palms—around 9 million in Iraq and 5 million in Iran. Moreover, many of the 3-4 million remaining palms are in poor condition.

Alarming signs of salinisation in the Shatt al-Arab region began emerging in the late 1960s. The situation rapidly deteriorated as dam construction intensified throughout the

Credit: Hassan Partow/UNEP/UNEP-GRID Geneva

Tigris-Euphrates basin, considerably reducing freshwater flows and eliminating periodic flooding of the Shatt al-Arab that formerly washed out accumulated salts. The supply and quality of water reaching the estuary dipped further with the desiccation of the vast Mesopotamian wetlands immediately above it and the diversion of marsh waters. Moreover, decreased inland discharge has stimulated deeper seawater penetration into the Shatt al-Arab, and water quality is steadily worsening due to polluted backflow from expanding irrigation projects in the watershed. Despite the date palm's high salt tolerance, excessive salinity has triggered large-scale palm dieback, with those nearest to the sea most affected but with the process continuing unabated inland. Finally, abandonment of date farms during the war and overall deterioration in palm vigour has rendered the trees susceptible to ravaging pest infestations, which have been particularly severe in the 1990s.

The Phoenix Factor

The date palms, whose botanical name is *Phoenix dactylifera L.*, resemble the mythical Phoenix bird that sprang from the ashes in that date palms are also able to regenerate from fire damage. Biotechnology may be the modern phoenix that will help replace the millions of palms that have been destroyed along the Shatt al-Arab. Iran is using a new cloning technique to accelerate mass date production, as dates are naturally slow to propagate. Already, thousands of palm plantlets have been introduced. Biosafety regulations, however, will need to be observed to ensure that the Iran-Iraq treasure grove of 800 plus date varieties, representing more than a quarter of world date diversity, is not jeopardised by a broad dissemination of cloned palms.

Al 'Isawiyah

Wadi as Sirhan

JORDAN

Tubarjal

S A U D I   A R A B I A

Al 'Isawiyah

Wadi as Sirhan

Tubarjal

JORDAN

S A U D I   A R A B I A

24 Feb 1991

This image shows the region shortly after the introduction of center pivot irrigation.

0   10   20

Kilometres

N

2 Feb 1986

Al Isawiyah

CROPLAND
AL' ISAWIYAH, SAUDI ARABIA

198

Rich in oil but lacking abundant water resources, Saudi Arabia has used oil revenues to adopt some of the best technologies available for farming in arid and semi-arid environments. One such technology is the center-pivot irrigation system (CPI). In satellite images, CPI-irrigated fields appear as green dots.

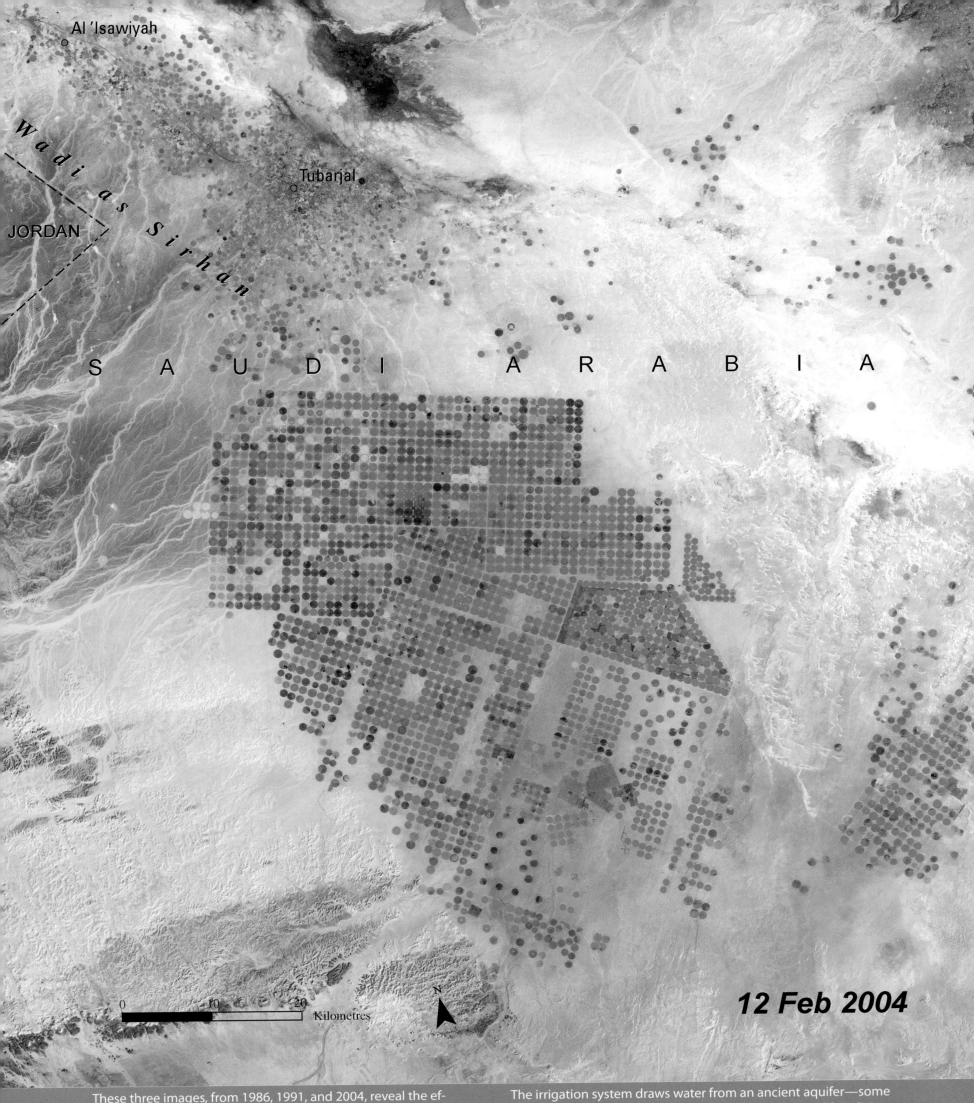

Al 'Isawiyah

W a d i   a s   S i r h a n

Tubarjal

JORDAN

S A U D I   A R A B I A

0   10   20
Kilometres

N

12 Feb 2004

These three images, from 1986, 1991, and 2004, reveal the effects of this irrigation strategy in a vast desert region in Saudi Arabia known as Wadi As-Sirhan. This region was once so barren that it could barely support the towns Al'Isawiyah and Tubarjal that can be seen in the upper left of each image. Following the introduction of center-pivot irrigation, however, barren desert was gradually transformed into a greener, food-producing landscape.

The irrigation system draws water from an ancient aquifer—some of the water it contains may be as much as 20 000 years old. Judicious use of water resources, and climate-appropriate technology, has in this situation helped improve food production without being detrimental to the environment.

S P A I N

Dalias

A l m e r i a

El Ejido

Roquetas
de Mar

La Mojonera

*C a m p o   d e
D a l i a s*

Balerma

Almerimar

Punta Entinas-Sabinar

A l b o r a n   S e a

N

0   3   6
                        Kilometres

24 Jan 1974

Almeria

CROPLAND
ALMERIA, SPAIN

This pair of satellite images shows the impact of massive and
rapid agricultural development in Almeria Province along
Spain's southern coast. In the earlier image, the landscape
reflects rather typical rural agricultural land use. In the 2000
image, much of the same region—an area covering roughly
20 000 hectares (49 421 acres)—has been converted to inten-

S P A I N

Dalias

A l m e r i a

El Ejido

Roquetas
de Mar

La Mojonera

*C a m p o   d e
D a l i a s*

Balerma

Almerimar

Punta Entinas-Sabinar

N

A l b o r a n   S e a

0   3   6
Kilometres

30 Apr 2000

sive greenhouse agriculture for the mass production of market pro-
duce. (Greenhouse-dominated land appears as whitish gray patches.)
In order to address increasingly complex water needs throughout
Spain, the government adopted the Spanish National Hydrological
Plan (SNHP) in 2001. Initially, this water redistribution plan involved
the construction of 118 dams and 22 water transfer projects that

would move water from parts of the country where it was relatively
abundant to more arid regions. In 2004, the Spanish government
announced it would begin exploring more environmentally friendly
water-saving technologies, such as wastewater recycling and seawater desalinization.

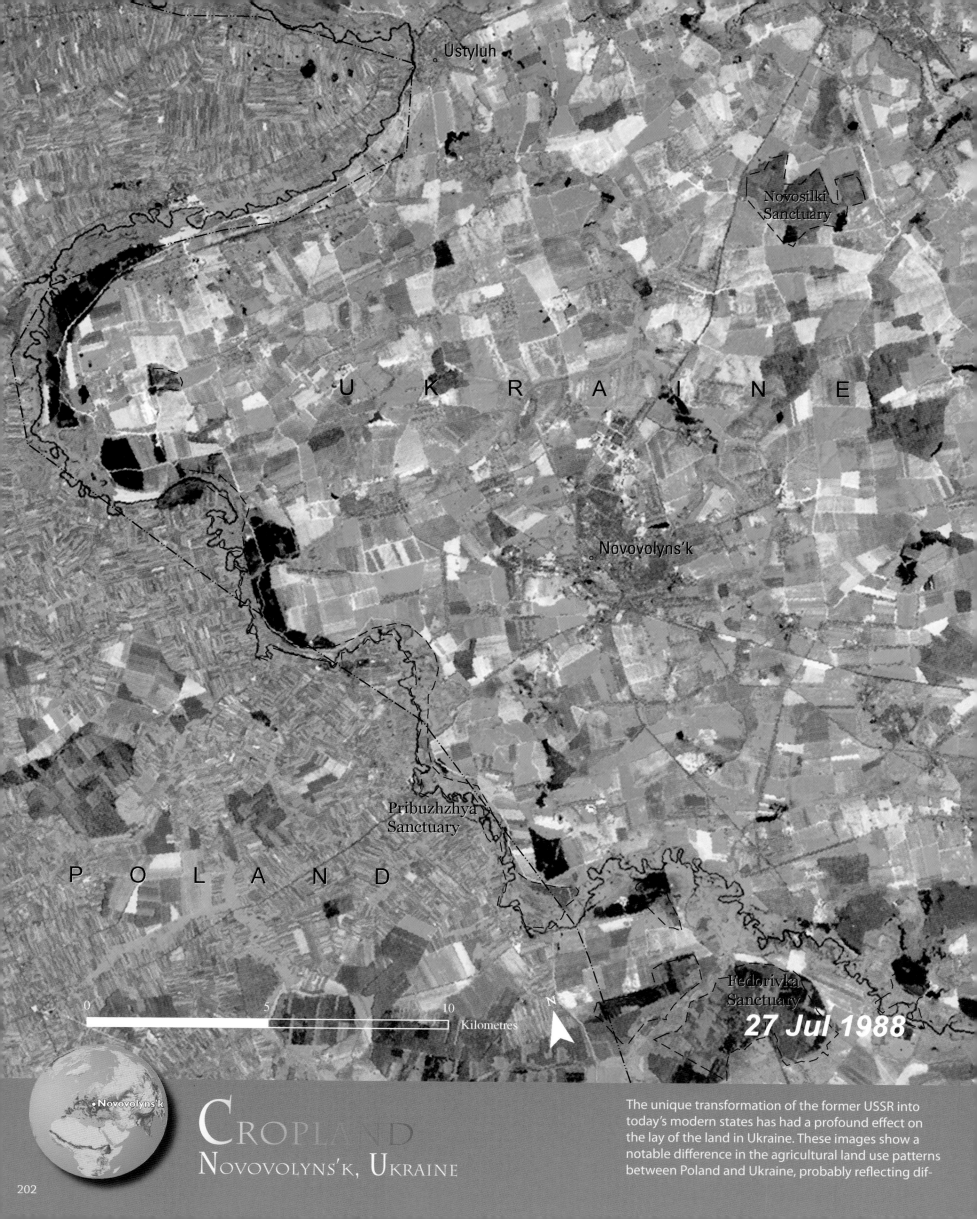

Ustyluh

Novosilki
Sanctuary

U K R A I N E

Novovolyns'k

Pribuzhzhya
Sanctuary

P O L A N D

Fedorivka
Sanctuary

27 Jul 1988

0   5   10
|▬▬▬▬▬▬▬▬▬▬▬▬▬▬▬▬▬▬▬▬| Kilometres

N

• Novovolyns'k

CROPLAND
NOVOVOLYNS'K, UKRAINE

The unique transformation of the former USSR into
today's modern states has had a profound effect on
the lay of the land in Ukraine. These images show a
notable difference in the agricultural land use patterns
between Poland and Ukraine, probably reflecting dif-

Ustyluh

Novosilki
Sanctuary

U K R A I N E

Novovolyns'k

Pribuzhzhya
Sanctuary

P O L A N D

Fedorivka
Sanctuary

10 Jun 2000

0   5   10
Kilometres   N

ferent policies and approaches to land use. Of particular interest are the sizes and patterns of the fields in the two countries; while Poland the farms are comparably much smaller, those in Ukraine are larger.

Though the town of Novovolyns'k has not changed appreciably in size, an apparent change in the approach to land use in Ukraine has taken place;

in the 2000 image, larger fields have been divided, following the pattern in Poland. The satellite images reveal quite vividly the contrast in land-use practices between the individual farms of Poland and Ukraine's former state farm plan—and how the latter has changed over time.

CHINA

Paektu San

△

Samjiyŏn

Changbai Shan
Natural Reserve

DEMOCRATIC
PEOPLE'S
REPUBLIC OF
KOREA

0   10   20
Kilometres

N

26 Sep 1977

Paektu San

CROPLAND
PAEKTU SAN, NORTH KOREA

Situated on the border between China and
North Korea, the mountain Paektu San is a sym-
bol of patriotism for the Korean people and an
embodiment of their national spirit. The mountain's rich volcanic soils and its dry, relatively
cool climate make it suitable for agriculture.

CHINA

Paektu San △

Samjiyŏn

DEMOCRATIC PEOPLE'S REPUBLIC OF KOREA

Changbai Shan
Natural Reserve

0   10   20
Kilometres

N

2 Sep 1999

These two satellite images reveal the degree to which agricultural activities have expanded on and around Paektu San, particularly on the North Korean side of the border, where intensive land development has served to both increase food production and underscore North Korea's territorial claims. In these images, green represents natural vegetation while grayish-brown areas are bare agricultural lands in which crops have not yet emerged from the soil. Areas of deforestation and other types of land clearing appear pink and are dissected by the fine lines of mountain streams. Near the center of the more recent image there is further evidence of land-cover change along the border between the two countries where a dam has been constructed.

San Ramon

Río San Pedro

Los Cafes

B O L I V I A

Grande (Guapay)

La Esperanza

Santa Cruz

Cotoca

Santa Cruz

0   10   20
Kilometres

N

2 Jun 1986

Lack of bridges and roads offered only limited access. By 1986, roads were established and clearing for agriculture had began in earnest.

17 Jun 1975

Santa Cruz

Santa Cruz, Bolivia

CROPLAND

Santa Cruz is situated in Bolivia's rich, fertile lowlands, a region highly suitable for agriculture. In the 1975 satellite image, the region's forested landscape appears as a dense, essentially un-broken expanse of deep green that extends

San Ramon

Río San Pedro

Los Cafes

B O L I V I A

Grande (Guapay)

La Esperanza

Santa Cruz

Pozo del
Tigre

Cotoca

anta Cruz

0   10   20
|————————————————| Kilometres

N

6 May 2003

to the Rio Grande (Guapay) River. By 1986 (inset image), roads had been built that linked the region to other population centers. As a result, large numbers of people migrated to the area. A large agricultural development effort (the Tierras Baja project) led to widespread deforestation as forests were clear-cut and converted to pastures and cropland. By 2003, almost the entire region had been converted to

agricultural lands, including the area east of La Esperanza across the river. In the area north and west of Los Cafes (upper left), notice the grid of squares on the landscape, each with an internal star-shaped pattern. At the center of each square is a small community.

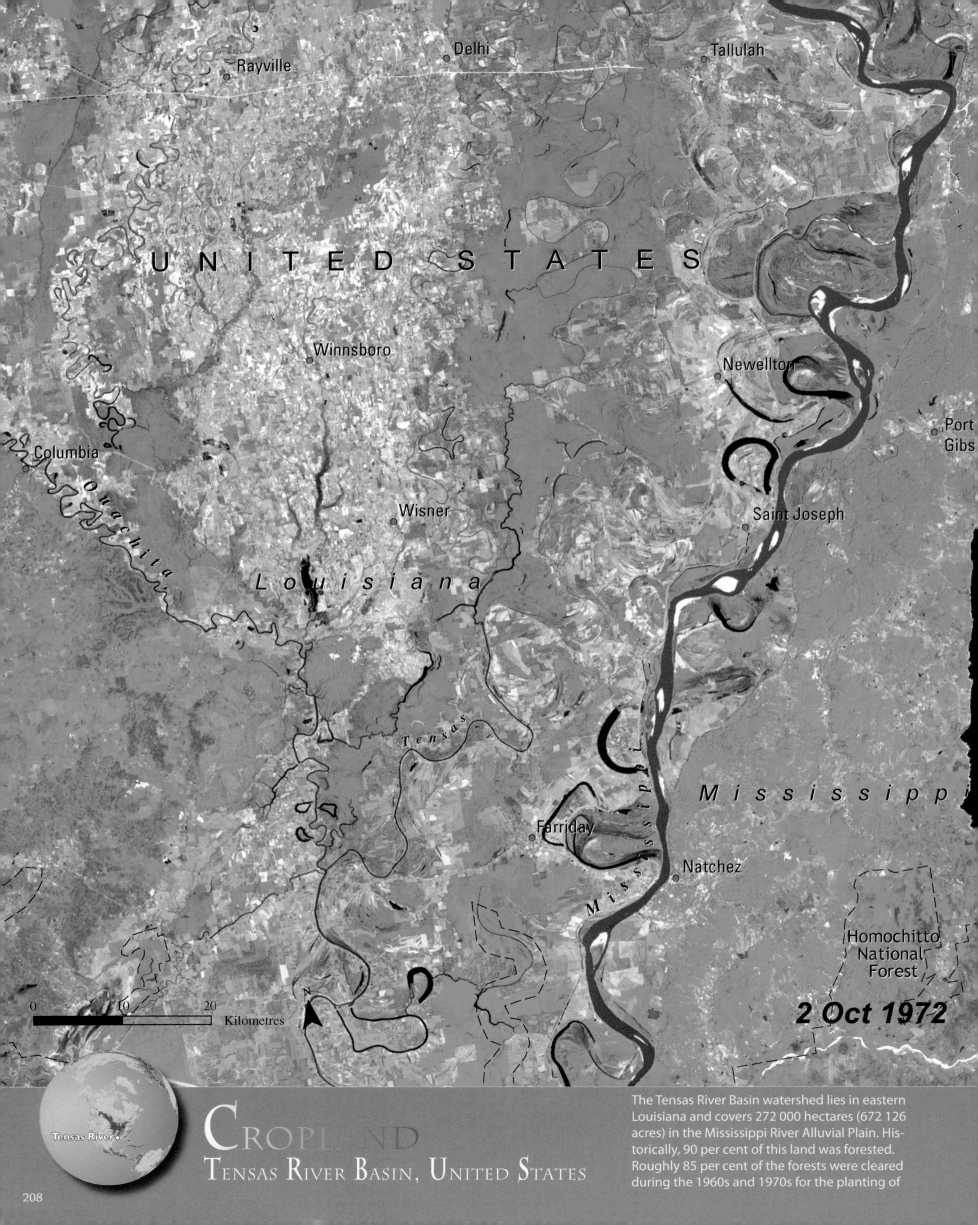

Rayville

Delhi

Tallulah

U N I T E D   S T A T E S

Winnsboro

Newellton

Ouachita

Columbia

Port
Gibs

Wisner

Saint Joseph

L o u i s i a n a

Tensas

M i s s i s s i p p i

Farriday

Mississippi

Natchez

N

Homochitto
National
Forest

0   10   20

Kilometres

2 Oct 1972

Tensas River

CROPLAND

TENSAS **R**IVER **B**ASIN, **U**NITED **S**TATES

The Tensas River Basin watershed lies in eastern
Louisiana and covers 272 000 hectares (672 126
acres) in the Mississippi River Alluvial Plain. Historically, 90 per cent of this land was forested.
Roughly 85 per cent of the forests were cleared
during the 1960s and 1970s for the planting of

soybeans. Clearing of the forest has exacerbated flooding problems and increased erosion.

As this pair of images reveals, intensive agricultural development has continued in the Tensas River Basin over time. Croplands appear in shades of tan; forests are green. The only remaining large tracts of hardwood forests in the watershed are in isolated wildlife refuges and management areas. Small forest remnants also occur on some private lands. The contrast between the amount of land cover change that has occurred on opposite sides of the Mississippi River in these images is striking. In the state of Mississippi, the forests remain largely intact, possibly due to the absense of lands suitable for cultivation.

La Campana

Tlahualilo de Zaragoza

M E X I C O

Bermejillo

Coahuila

Durango

Dinamita

Gómez Palacio

Ciudad Lerdo

Torreón

Matamoros

N

9 Oct 1972

0   10   20   Kilometres

CROPLAND
TORREÓN, MEXICO

Torreon

The city of Torreón is located in the State of Coahuila in central Mexico. Founded in 1893, Torreón is a modern industrial city that is home to flour mills, textile plants, iron foundries, a rubber factory, and various other industries.

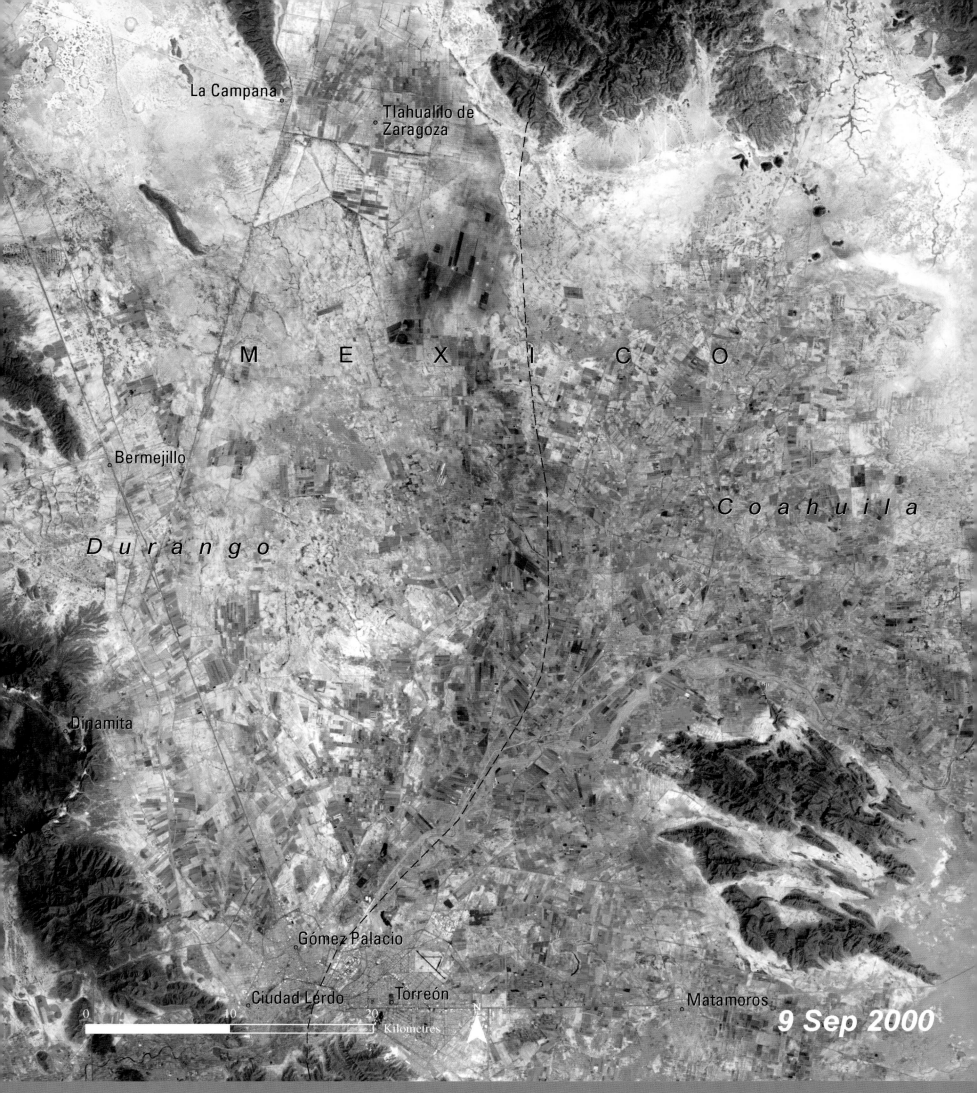

La Campana

Tlahualilo de
Zaragoza

M E X I C O

Bermejillo

C o a h u i l a

D u r a n g o

Dinamita

Gómez Palacio

Ciudad Lerdo   Torreón   Matamoros

9 Sep 2000

0   10   20

Kilometres   N

Torreón is also situated in a rich agrarian region noted for its cotton and wheat farms and cattle ranches.

Since the 1970s, however, there has been a significant decrease in cropland in the Torreón region due to drought and subsequent extraction of ground water from aquifers. In 1992, the Mexican government passed the Federal Water Law, in which the government sought to shift responsibility for some water management rights issues from federal to local governments, or even individuals. This left farmers in a position to negotiate their own water rights. At the same time, however, prices of water for irrigation were also raised. The amount of land around Torreón on which crops are raised continues to decrease.

EGYPT

29 Sep 1987

13 Sep 1984

Lake
Nasser

0   10   20 Kilometres

N

29 Sep 1987-13 Sep 1984

Toshka Project

Cropland
Toshka Project, Eygpt

Egypt's Toshka Project has transformed part of
the country's scorching hot southern desert
into a region dotted by lush, neatly tended
vegetable plots that are supplied with water
and fertilizer by drip irrigation systems. These
images, from 1984 and 2000, document the

E G Y P T

Toshka

Lake Nasser

0 — 10 — 20 Kilometres

N

23 Aug–1 Sep 2000

changes and success Egypt has had in this desert reclamation project, which was begun in the mid-1990s and aimed to double the size of Egypt's arable land in fifteen years' time.

The project created four new lakes in the desert by drawing water through a concrete-lined canal from Lake Nasser, which was formed by damming the Nile River at Aswan. The water flows through the

canal into the Toshka Depression, where it forms the lakes visible in the 2000 image. The faint blue-green areas visible around some of the lakes are agricultural lands, newly created by irrigation. While providing people with new arable land on which crops can be grown, the Toshka Project's environmental impacts are still under study.

3.6 Grasslands

Grasslands cover roughly 40 per cent of the Earth's land surface. They are, as their name implies, natural landscapes where the dominant vegetation is grass. For purposes of this report shrublands are also considered grasslands. Grasslands typically receive more water than deserts, but less than forested regions. Worldwide, these ecosystems provide livelihoods for nearly 800 million people. They are also a source of forage for livestock, wildlife habitat, and a host of other resources (White et al. 2000).

Most of the world's meat comes from animals that forage on grasslands. World meat production has nearly doubled since 1975, from 116 million metric tonnes to 233 million metric tonnes in 2000 (UNEP 2002b). Grasslands and their soils store about one-third of the global stock of carbon in terrestrial ecosystems. These lands also are habitat for diverse and biologically important plants and animals.

Most of the world's original grasslands that receive enough rainfall to support the growing of crops have been converted to agricultural lands. In other areas, irrigation using imported water or groundwater has been implemented on traditional rangeland areas (SRM n.d.). Precise measurements of area changes are difficult to come by as there is no international organization tracking grasslands and because of the difficulty in identifying what is grassland and what is not. However, it has been estimated that there were over seven million km² (three million square miles) of grassland and scrubland lost between the development of agriculture and 1982 (Mathews 1983). In addition, it is known that all croplands were developed either from forests or grasslands. In that respect, since cropland areas are expanding, it can be assumed that on the whole, grassland areas are continuing to decline. On the other hand, large areas of tropical rainforests are being cleared to provide pasture for livestock. Therefore, grasslands—at least in the form of pastures—may be expanding in some localized areas.

Worldwide, the quality of surviving grasslands is declining. This is due primarily to human-induced modifications such as agriculture, excessive or insufficient fire, livestock grazing, fragmentation, and invasive plants and animals (White et al. 2000). Invasion of the world's grasslands by woody plants is

Green River

among the dominant changes in the Earth's vegetation during the last two centuries (Polley et al. 2003). Woody plants tend to displace grasses on grasslands. Invasion by woody plants traditionally has been attributed to the introduction of exotic species, overgrazing, fire suppression, and elimination of small mammals that kill woody seedlings. Fire suppression in some parts of the world is resulting in the reintroduction of woody plants into grasslands that have been controlled by fire by thousands of years. At the same time, an increase in human-made fires in other parts of the world are making it possible for grasslands to supplant forests.

The displacement of grasses by woody plants may also be related to the 30 per cent increase in the concentration of CO_2 in the Earth's atmosphere that has occurred over the last 200 years. Grasses use water more slowly as carbon dioxide levels increase. Consequently, grassland soil may retain water better during droughts when atmospheric carbon dioxide concentrations are high. Such an increase in soil water may indirectly promote the invasion of woody plants into grassland by enhancing the survival of shrub seedlings during droughts.

It is estimated that 73 per cent of the world's grazing land has so deteriorated that it has lost at least 25 per cent of its animal carrying capacity (UNEP 1999b). Even though the damage from overgrazing is spreading, the world's livestock population continues to grow in step with increases in the human population, and a growing demand for meat that accompanies increased wealth. As world population increased from 2 500 million in 1950 to 6 100 million in 2001, the world's cattle population grew from 720 million to 1 530 million. The number of sheep and goats increased from 1 040 million to 1 750 million.

With 180 million people worldwide now trying to make a living tending 3 300 million cattle, sheep, and goats, grasslands are under heavy pressure. As a result of overstocking and over-grazing, grasslands in much of Africa, the Middle East, Central Asia, the northern part of the Indian subcontinent, Mongolia, and much of northern China are deteriorating. While grazing was once a pastoral activity that involved people moving with their herds from place to place, it has become a far more sedentary undertaking. The result is an increase in grassland degradation near settlements and the creation of grassland landscapes perforated by bore holes.

Initially, overgrazing of grasslands reduces their productivity and ultimately destroys them. Worldwide, there are now 680 million hectares (1 680 million acres) of degraded grasslands (Brown 2002). Desertification is estimated to involve 3 600 million hectares (8 896 million acres) of land—roughly 25 per cent of the world's total surface area.

Grasslands

Dense

Sparse

Source: Global land cover 2000 (GLC 2000)

Wyperfeld National Park

eanut Basin • Revane

Narok

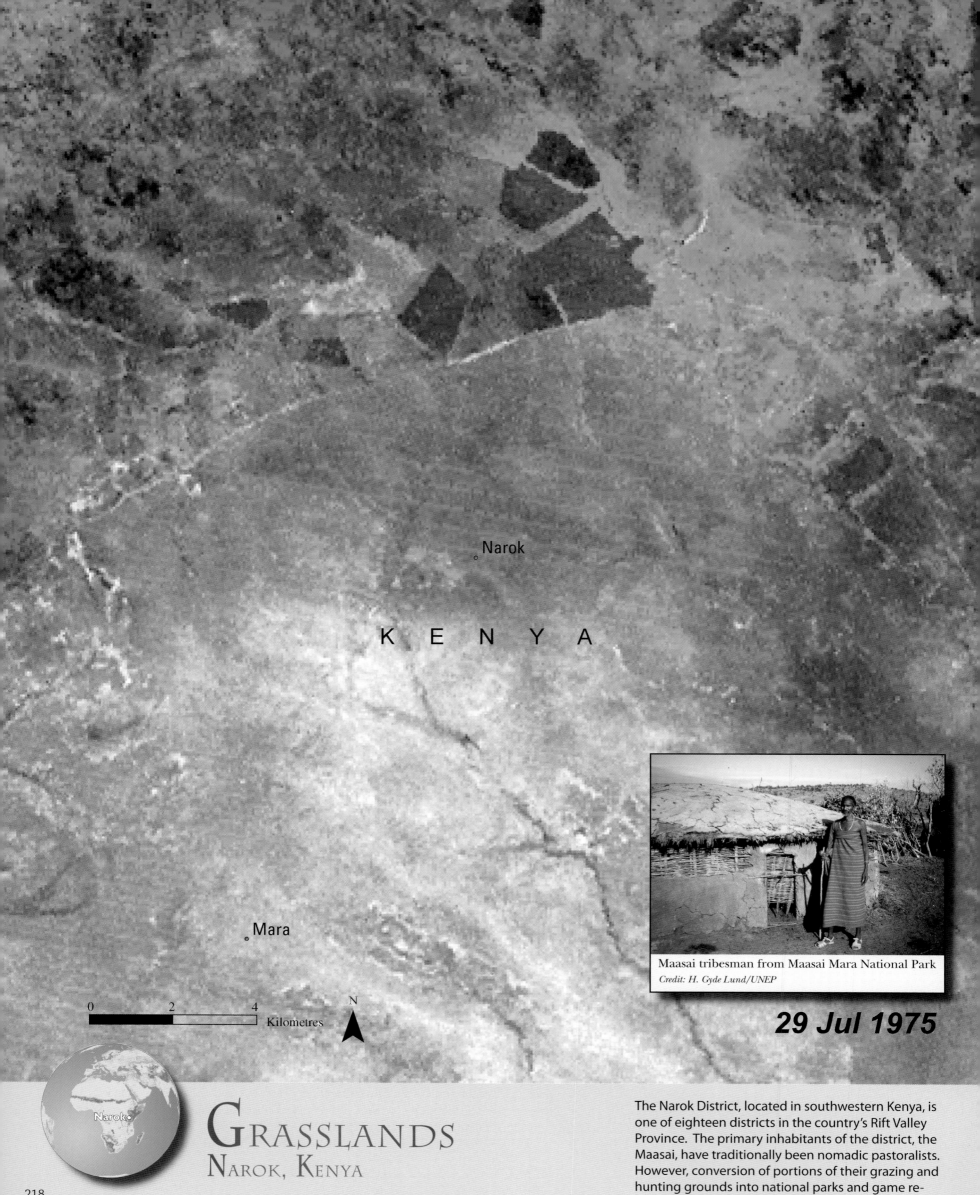

Narok

K E N Y A

Mara

Maasai tribesman from Maasai Mara National Park
Credit: H. Gyde Lund/UNEP

29 Jul 1975

0   2   4
Kilometres

N

Grasslands
Narok, Kenya

The Narok District, located in southwestern Kenya, is one of eighteen districts in the country's Rift Valley Province. The primary inhabitants of the district, the Maasai, have traditionally been nomadic pastoralists. However, conversion of portions of their grazing and hunting grounds into national parks and game re-

JAPAN

Isahaya

Isahaya
Bay

0   1   2   Kilometres

N

23 Oct 2003

of light- and dark-colored water. Behind the sea wall, tidal flats can be seen drying as water is slowly drained away. In the 2003 image, that area has been fully reclaimed from the sea.

The Isahaya Bay Reclamation project has been fraught with controversy. Environmental groups have criticized the project for its destruction of wetland habitat. The Isahaya Bay area is known for its production of nori

(seaweed), and local farmers have complained that the reclamation project has negatively impacted the quality and abundance of the nori growing in the bay. The Isahaya project prompted the formation of the Japan Wetlands Action Network, a group of grassroots and national conservation organizations who are protesting the project and recommending that the sea wall gates be opened to restore ecological balance.

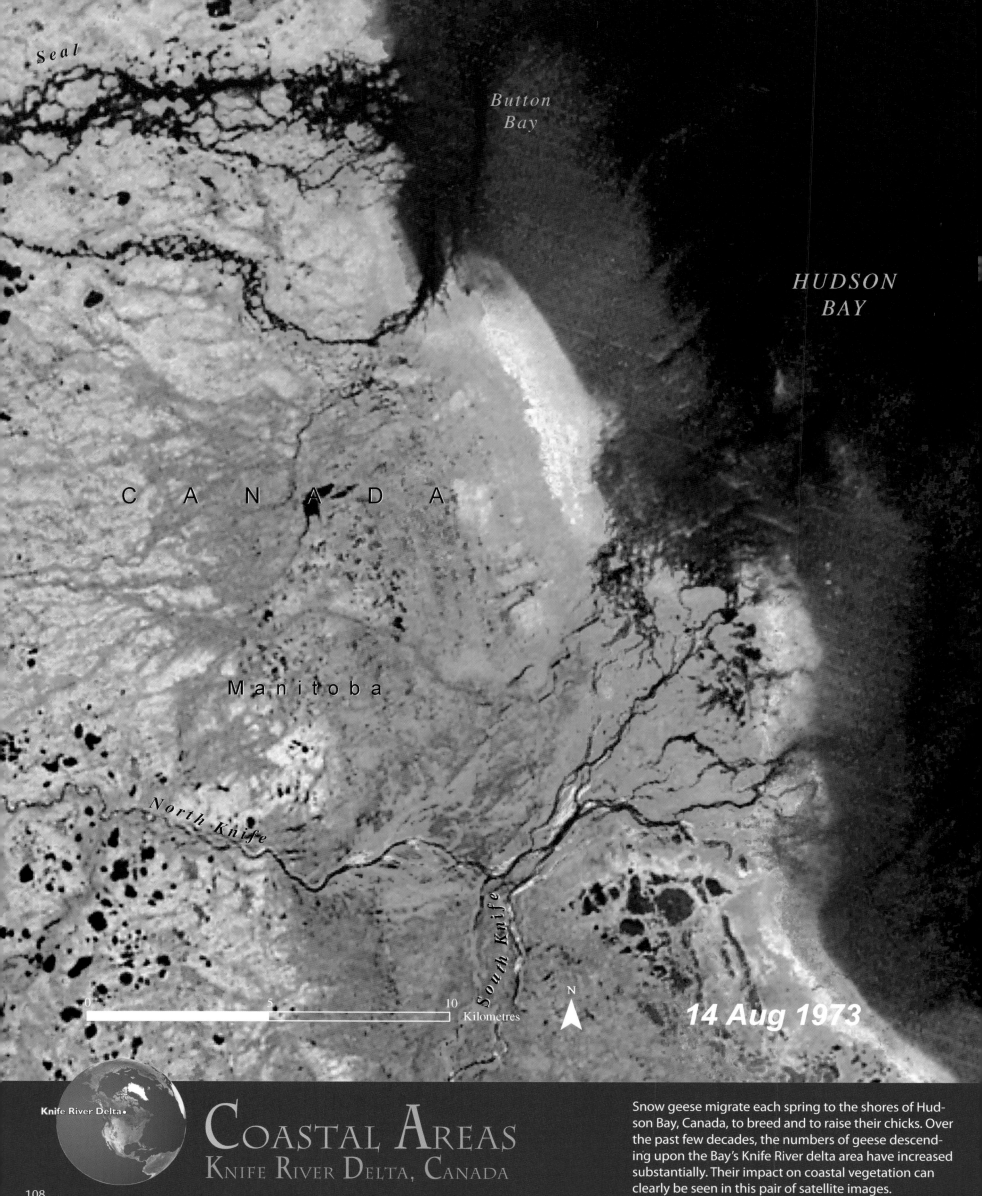

Seal

Button Bay

HUDSON BAY

C A N A D A

M a n i t o b a

North Knife

South Knife

N

0   5   10
Kilometres

14 Aug 1973

Knife River Delta

COASTAL AREAS
KNIFE RIVER DELTA, CANADA

Snow geese migrate each spring to the shores of Hudson Bay, Canada, to breed and to raise their chicks. Over the past few decades, the numbers of geese descending upon the Bay's Knife River delta area have increased substantially. Their impact on coastal vegetation can clearly be seen in this pair of satellite images.

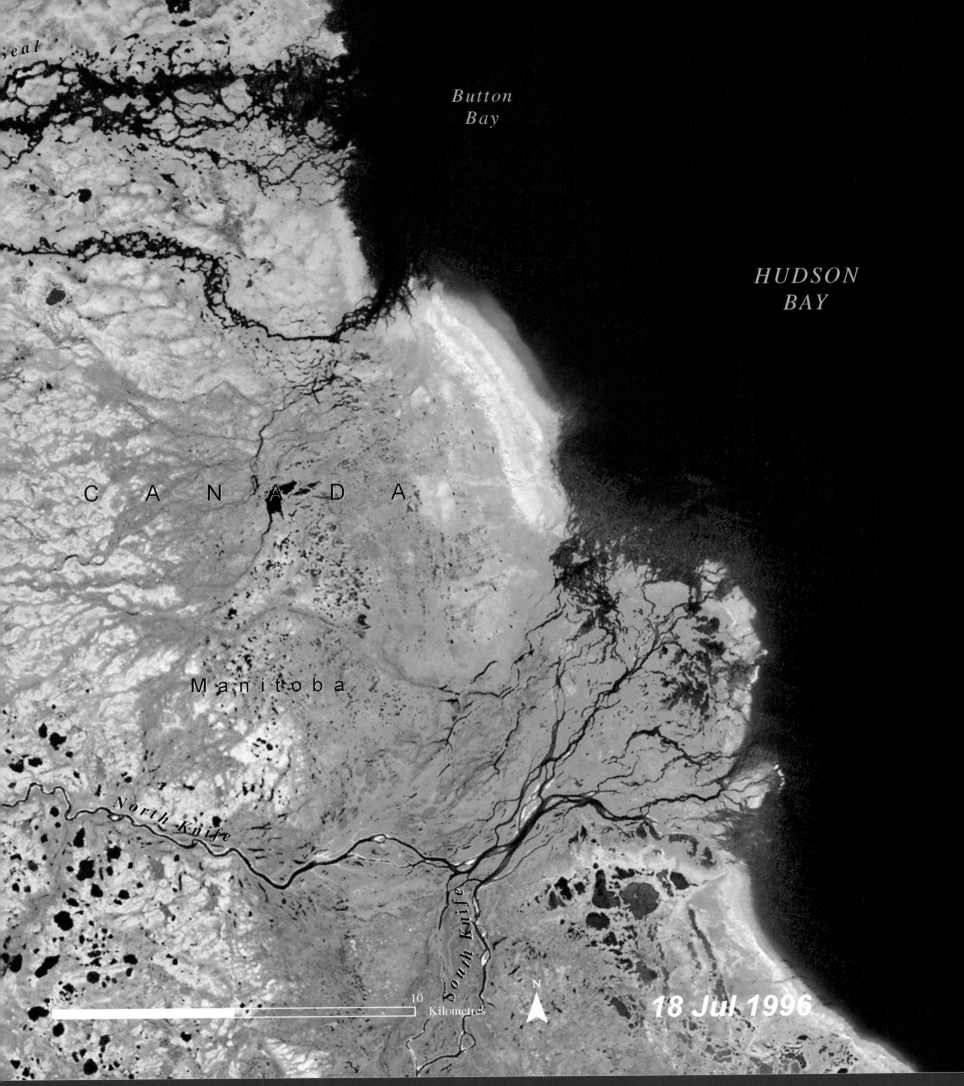

Seal

Button
Bay

HUDSON
BAY

C A N A D A

M a n i t o b a

North Knife

South Knife

0   5   10
Kilometres

N

18 Jul 1996

In the image on the right, notice how the vegetation (green) has receded from the shoreline north of the delta. Snow geese have overgrazed this area and turned the shoreline into an enormous mudflat. Having denuded the shoreline of vegetation, the geese have also moved inland in search of food on the tundra, where overgrazed soil quickly becomes barren and develops a crust of salt due to evaporation. The salty layer prevents the regrowth of plants, and ultimately leads to erosion. Some researchers have suggested lifting restrictions on the hunting of snow geese in an attempt to reduce their numbers and control the overgrazing problem. Others believe such measures are "too little, too late."

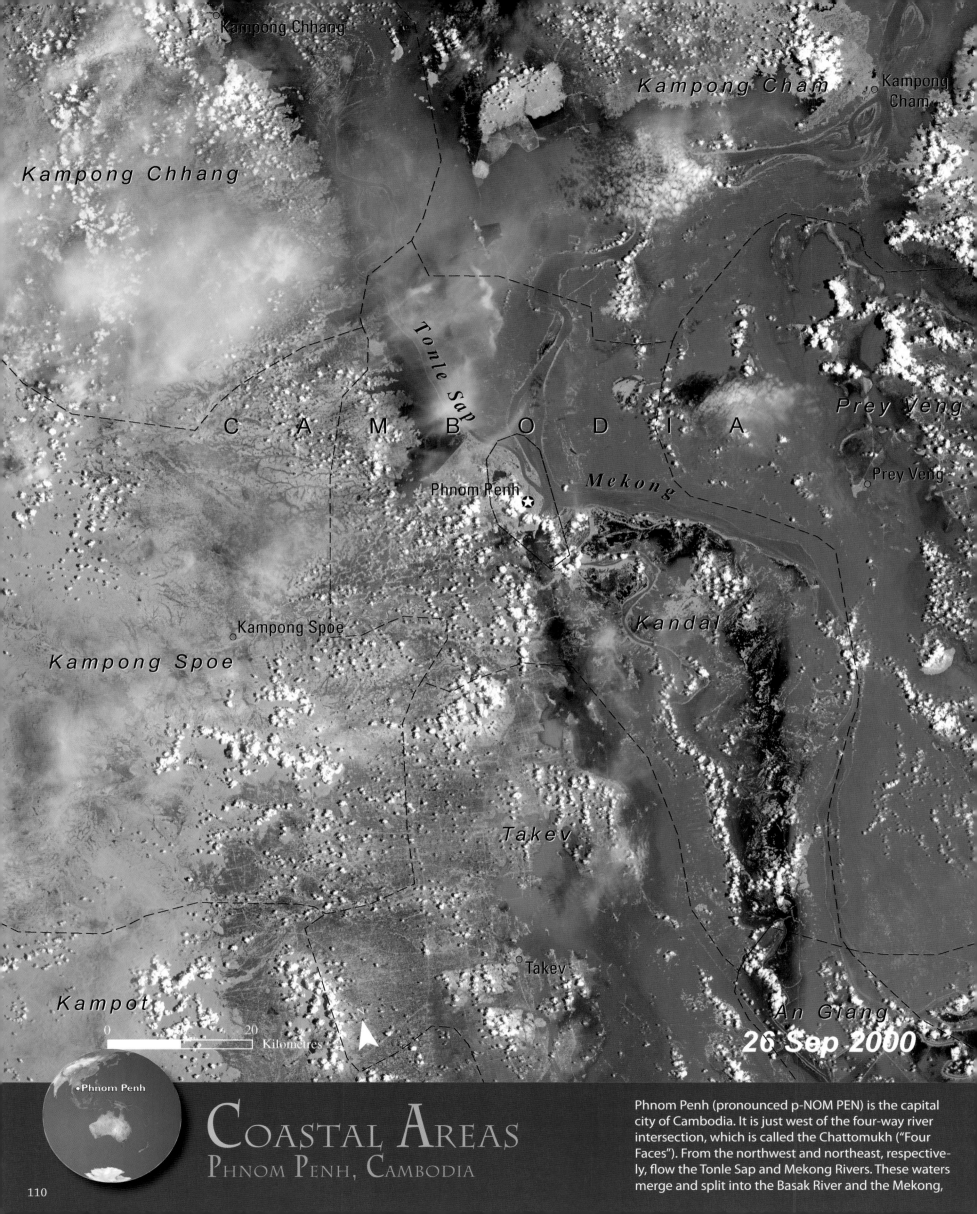

Kampong Chhang

Kampong Cham

Kampong
Cham

Kampong Chhang

Tonle Sap

Prey Veng

C A M B O D I A

Phnom Penh

Mekong

Prey Veng

Kampong Spoe

Kampong Spoe

Kandal

Takev

Kampot

Takev

An Giang

26 Sep 2000

N

0   20
                    Kilometres

• Phnom Penh

COASTAL AREAS
PHNOM PENH, CAMBODIA

Phnom Penh (pronounced p-NOM PEN) is the capital
city of Cambodia. It is just west of the four-way river
intersection, which is called the Chattomukh ("Four
Faces"). From the northwest and northeast, respectively, flow the Tonle Sap and Mekong Rivers. These waters
merge and split into the Basak River and the Mekong,

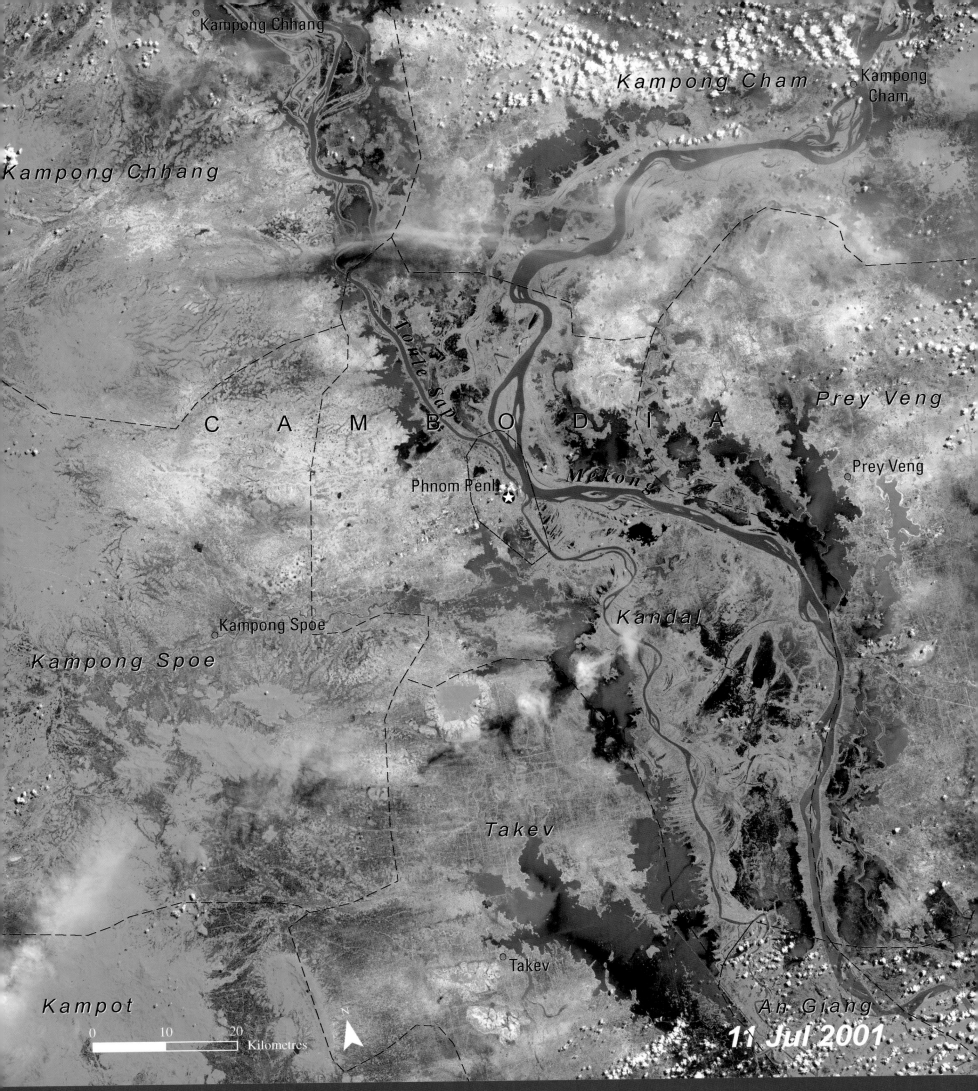

Kampong Chhang

Kampong Cham

Kampong Chhang

Kampong
Cham

Tonle Sap

C A M B O D I A

Prey Veng

Mekong

Prey Veng

Phnom Penh

Kandal

Kampong Spoe

Kampong Spoe

Takev

Kampot

Takev

An Giang

0   10   20
Kilometres

N

11 Jul 2001

which flow southeast to the South China Sea. The Mekong River is the 12th longest in the world, flowing from western China to the Mekong Delta in southern Vietnam. Every autumn, monsoon rains are too great for the Mekong to carry, and it floods a large area of Cambodia. This flood even reverses the flow of the Tonle Sap River, northward to the Tonle Sap ("Great Lake") which can expand to ten times its normal size.

This pair of images show the extent of flooding associated with the two rivers. The 2000 image was taken during a period of flooding while the 2001 image was taken after the flood waters had receded. Visible also in the images, especially in the south-central area of the 2001 image, are extensive ditches and canals that are used in irrigation.

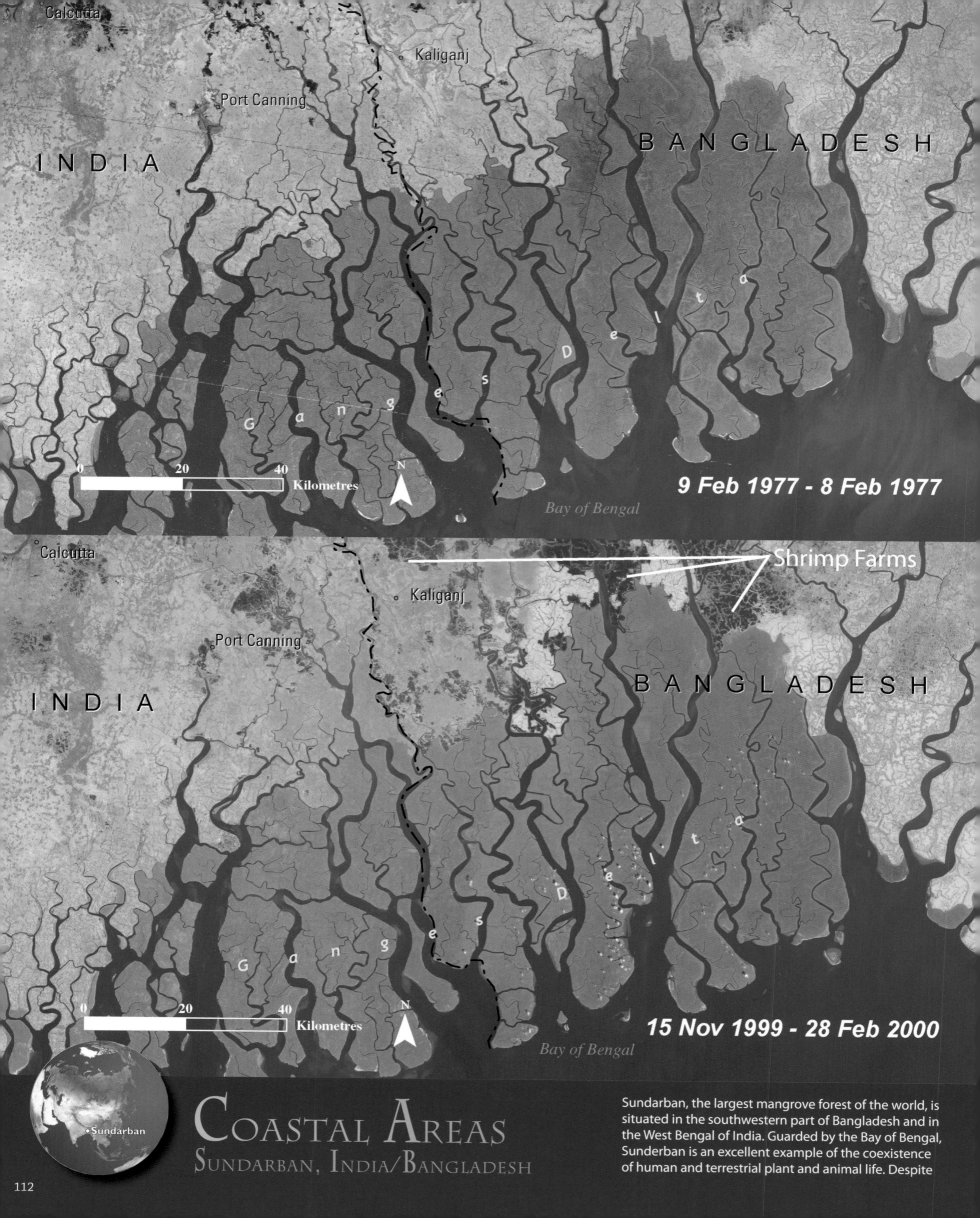

Calcutta

Kaliganj

Port Canning

I N D I A

B A N G L A D E S H

G
a
n
g
e
s
D
e
l
t
a

0   20   40
Kilometres

N

9 Feb 1977 - 8 Feb 1977

Bay of Bengal

Calcutta

Shrimp Farms

Kaliganj

Port Canning

I N D I A

B A N G L A D E S H

G
a
n
g
e
s
D
e
l
t
a

0   20   40
Kilometres

N

15 Nov 1999 - 28 Feb 2000

Bay of Bengal

•Sundarban

COASTAL AREAS
SUNDARBAN, INDIA/BANGLADESH

Sundarban, the largest mangrove forest of the world, is situated in the southwestern part of Bangladesh and in the West Bengal of India. Guarded by the Bay of Bengal, Sunderban is an excellent example of the coexistence of human and terrestrial plant and animal life. Despite

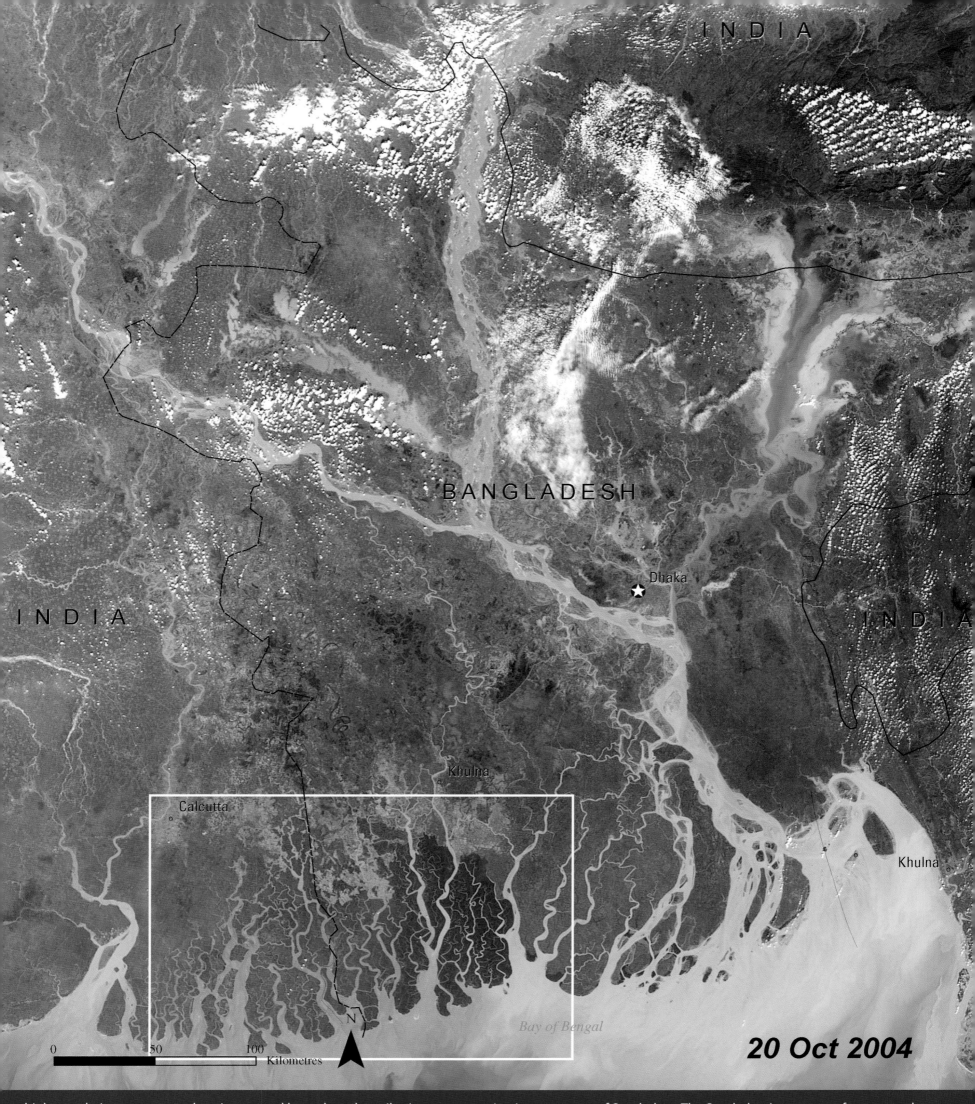

20 Oct 2004

high population pressure and environmental hazards such as siltation, cyclone flooding and sea level rise, the aerial extent of the mangrove forest has not changed significantly in the last 25 years. In fact, with improved management, the tiger population has increased from a mere 350 in 1993 to 500-700 in 2000 and ecotourism is progressing well. However, while sufficient data is not available, several reports suggest that forest degradation has been occurring in many parts of Sundarban. The Sundarban's mangrove forests are also becoming more vulnerable due to the significant rise of shrimp farming in the region. The increase of shrimp farming has negatively affected agriculture and also contributed to the loss of mangrove forests during the past two decades.

Krung Thep

★ Bangkok
(Krung Thep)

Chachoengsao

Thon Buri

T H A I L A N D

Phra
Pradaeng

Chao Phraya

Samut Prakan

Bang Pakong

Samut Prakan

Chon
Buri

Shrimp farm
Credit: H. Gyde Lund/UNEP

Chon
Buri

*Gulf of
Thailand*

0   5   10
Kilometres

N

6 Jan 1973-16Dec 1978

Si Racha

Thon Buri

COASTAL AREAS
THON BURI, THAILAND

As the city of Bangkok, Thailand, has grown, the need to provide food and an additional economic base for its burgeoning population has been a primary concern. Parts of the Thai coastline, including those near Bangkok, offer conditions favorable to aquaculture,

Murrayville

Wyperfeld
National park

Lake
Albacutya

Rainbow

A U S T R A L I A

Lake
Hindmarsh

Bunns Springs

Jeparit

Victoria

Yanac

Nhill

South
Australia

Kaniva

Bordertown

0   10   20
Kilometres

N

31 May 2004

when the river overflows its banks. Much of the park's vegetation is mallee, a type of shrubland dominated by several sparse, tall varieties of eucalyptus. Over 450 species of plants, 200 species of birds, and a variety of mammals and reptiles live within the park.

Fires set by people have been used to maintain the Australian bush for thousands of years. Fires also occur naturally and occur in the park and sur-rounding area nearly every year, leaving huge fire scars on the landscape that are easily seen in satellite images (light green areas). Remote sensing is used to document the extent of burn areas, and to help land managers plan controlled burns that help maintain the native vegetation and habitat for na-tive wildlife. Wyperfeld staff currently set fuel-reduction fires along the park's edges but fight all accidental fires.

3.7 Urban Areas

An urban area is a geographical unit of land constituting a town or city. Urbanization is the process by which large numbers of people become permanently concentrated in relatively small areas to form towns or cities.

During the course of human history, urbanization has accelerated worldwide. Between 1975 and 2000, urban population increased from 1 500 million people to over 2 800 million, or about 45 per cent of the world's population (UNEP 2002b). By 2020, it is estimated that 60 per cent of the world's population will be urban (Anon 2003).

For many people, urban living represents a better lifestyle. On average, individuals living in urban areas have higher incomes and live healthier, easier lives than their rural counterparts. They have greater access to clean water and sanitation than those in rural areas. Concentrations of people also tend to strengthen infrastructures by consolidating transportation services, utilities, and roads.

It is also true that not all urban dwellers benefit from urban living. In 2001, 924 million people, or roughly 31.6 per cent of the global urban population, lived in slums (UN Habitat n.d.). A slum household is one in which a group of individuals living under the same roof lack one or more fundamental necessities, including access to clean water, access to sanitation, secure tenure, durability of housing, and sufficient living area (Warah 2003). In the next thirty years, as many as 2 000 million people will be living in urban slums unless substantial policy changes are put into place.

Wherever people are concentrated in large numbers, as they are in urban areas, the risk of disease and other health concerns have the potential to become extremely urgent issues. Overcrowding fosters epidemics of tuberculosis, influenza, and many other communicable diseases (Myers and Kent 1995). Urban areas also tend to be polluted. According to some estimates, industrialized countries exhaust 3 146 kg (6 936 lbs) of fossil fuels and produce 200 kg (440 lbs) of air pollutants every year. Fossil fuel use adds both pollutants and greenhouse gases to the atmosphere, the latter of which contribute to global warming. Temperatures in heavily urbanized areas may be 0.6-1.3°C (1.1-2.3°F) warmer than

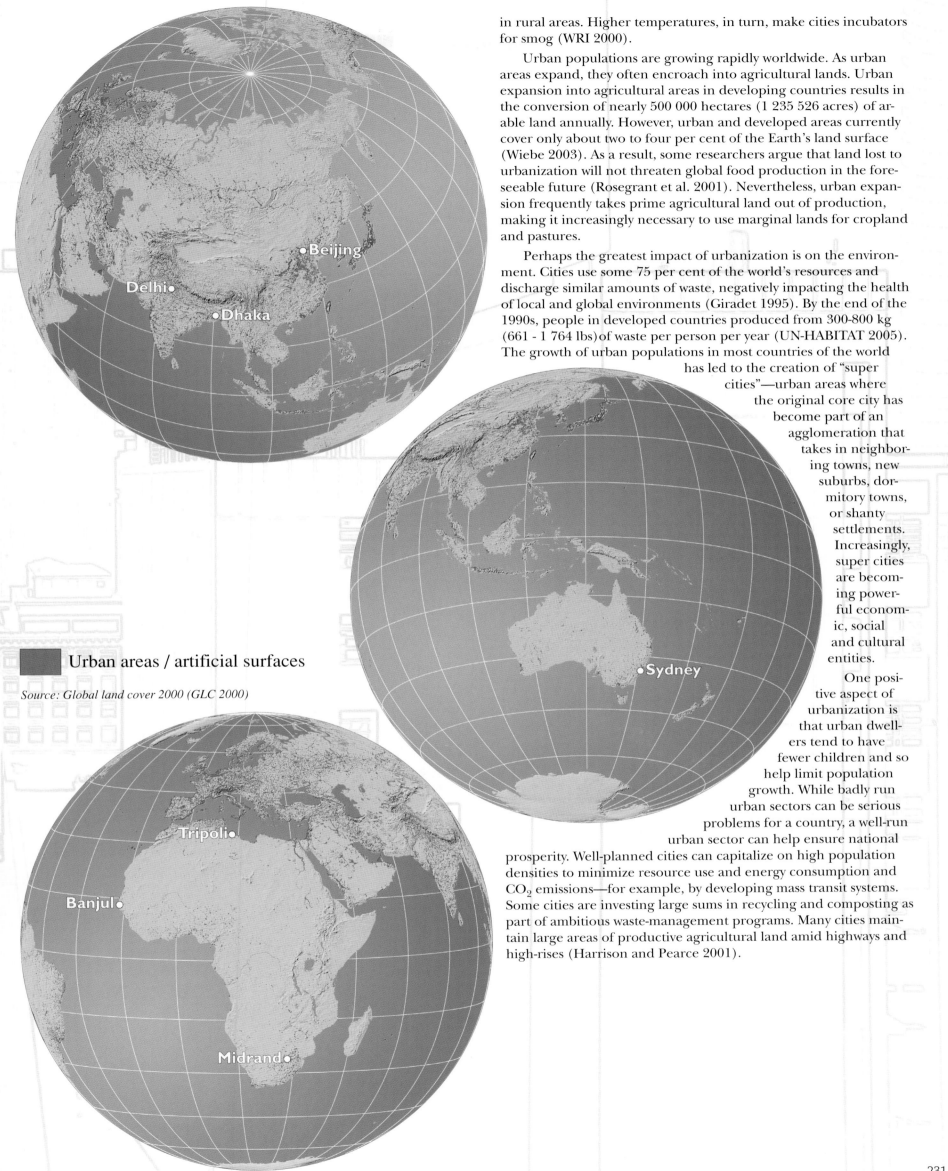

Urban areas / artificial surfaces

Source: *Global land cover 2000 (GLC 2000)*

in rural areas. Higher temperatures, in turn, make cities incubators for smog (WRI 2000).

Urban populations are growing rapidly worldwide. As urban areas expand, they often encroach into agricultural lands. Urban expansion into agricultural areas in developing countries results in the conversion of nearly 500 000 hectares (1 235 526 acres) of arable land annually. However, urban and developed areas currently cover only about two to four per cent of the Earth's land surface (Wiebe 2003). As a result, some researchers argue that land lost to urbanization will not threaten global food production in the foreseeable future (Rosegrant et al. 2001). Nevertheless, urban expansion frequently takes prime agricultural land out of production, making it increasingly necessary to use marginal lands for cropland and pastures.

Perhaps the greatest impact of urbanization is on the environment. Cities use some 75 per cent of the world's resources and discharge similar amounts of waste, negatively impacting the health of local and global environments (Giradet 1995). By the end of the 1990s, people in developed countries produced from 300-800 kg (661 - 1 764 lbs) of waste per person per year (UN-HABITAT 2005). The growth of urban populations in most countries of the world has led to the creation of "super cities"—urban areas where the original core city has become part of an agglomeration that takes in neighboring towns, new suburbs, dormitory towns, or shanty settlements. Increasingly, super cities are becoming powerful economic, social and cultural entities.

One positive aspect of urbanization is that urban dwellers tend to have fewer children and so help limit population growth. While badly run urban sectors can be serious problems for a country, a well-run urban sector can help ensure national prosperity. Well-planned cities can capitalize on high population densities to minimize resource use and energy consumption and CO_2 emissions—for example, by developing mass transit systems. Some cities are investing large sums in recycling and composting as part of ambitious waste-management programs. Many cities maintain large areas of productive agricultural land amid highways and high-rises (Harrison and Pearce 2001).

Atlantic
Ocean

Barra

Essau

Bakau

Banjul

Serekunda

Sukuta

—Abuko
Nature
Reserve

T H E   G A M B I A

Gambia River

Brikama

0   2.5   5
Kilometres

N

21 Feb 1973

Urban Areas
Banjul, The Gambia

The Gambia is a small—11 295 km² (4 361 square
miles)—country in West Africa. It is surrounded by
Senegal on all sides except on its coast. The capital
city of Banjul lies at the end of a small peninsula that
protrudes into the Atlantic Ocean.

Atlantic
Ocean

Barra

Essau

Bakau

Banjul

Serekunda

Sukuta

—Abuko
Nature
Reserve

THE GAMBIA

Gambia River

Brikama

0   2.5   5
Kilometres

N

6 Dec 1999

Credit: David McKee/UNEP/Gambia Tourist Support Website
The mangroves that lie on the northeastern edge of Banjul have largely escaped destruction, as urban development has mostly progressed westward. Increasing population and human encroachment remain a threat to the mangroves.

The country's population is increasing at a rate of about 4.2 per cent annually. For the past three decades, western Gambia has undergone considerable urban growth, particularly in Banjul and some of its neighboring cities, including Serekunda, Bakau, Sukuta, and Brikama. The population of the greater Banjul area, for example, more than tripled during this time. These two satellite images, taken in 1973 and 1999 respectively, show this urban sprawl and its impact. Urban growth and the accompanying expansion of cropland around urban areas have led to a significant decline in woodland areas (dark green). The Abuko Nature Reserve, located in the center of the images, was once surrounded by woodlands. It now stands out as an isolated patch of green in an otherwise developed landscape.

Yanggezhuang

Shunyi

Beijing Shi

Wenyu

Ginghe

Credit: Simon Tsuo/UNEP/NREF
Market vendors selling fruits on a
Beijing street.

Hebei

TongXian

C H I N A

Beijing
(Peking)

Shijingshan

Fengtai

Nanyuan

Feng

Liangxiangzhen

Daxing
(Huangcun)

Liangxiang

Hebei

Tia

0   5   10
Kilometres   N

12 Jun 1978

Anci
(Langfang)

URBAN AREAS
BEIJING, CHINA

•Beijing

Beijing, the capital city of the People's Republic of
China, is located in the country's northeastern corner,
in the transition zone between the Inner Mongolia
Plateau and the North China Plain. It is a city that has
undergone tremendous change and explosive urban
growth, since the start of economic reforms in 1979.

Yanggezhuang

Langfang

Shunyi

Beijing Shi

Shahezhen

Wenyu

Hebei

Ginghe

C   H   I   N   A

TongXian

Beijing
(Peking)

Shijingshan

Credit: *Law Chun Wah/UNEP/Topfoto*

Home to 13 million people, Beijing has experienced very
rapid urban growth in the past several decades.

Fengtai

Nanyuan

Feng

Liangxiangzhen

Daxing
(Huangcun)

Liangxiang

Tianj

0   5   10 Kilometres

N

Hebei

30 Apr 2000

Anci
(Langfang)

The left-hand satellite image shows Beijing in 1978, just prior to the reforms. The light blue-gray area in the center of the image is the urban landscape of the city. The hills to the west are covered with deciduous forest, which appears green. The agricultural lands that lie around the city appear as muted red, orange, and golden yellow, depending on the crop (rice, winter wheat, or vegetables) and its stage of development. Beijing's explosive growth is very obvious in the 2000 image. The city has expanded from its original center in all directions. Prime agricultural lands that once lay outside the city are now suburbs dominated by institutional, industrial, and residential buildings. In 2000, Beijing's population was 13 million.

Goiás

Districto Federal

B R A Z I L

Planaltina

Braslândia

Sobradinho

Parque
Nacional
de Brasília

Paranoá

*Lago do
Rio Descoberto*

★ Brasília

Ceilândia

Taguatinga

*Lago
do
Paranoá*

Gama

Goiás

0   10   20
Kilometres

N

1 Aug 1973

Brasilia

Urban Areas
Brasilia, Brazil

Inaugurated on 21 April 1960, Brazil's new capital of Brasilia began with a population of 140 000 and a master plan for carefully controlled growth and development that would limit the city to 500 000. Urban planner Lucio Costa and architect Oscar Niemeyer intended that every element—from the layout of the residential

Goiás

Distrícto Federal

B R A Z I L

Planaltina

Braslândia

Sobradinho

Parque
Nacional
de Brasília

Paranoá

*Lago do
Rio Descoberto*

Brasília

Ceilândia

Taguatinga

Credit: Lee Chui Yee/UNEP/Topfoto

New housing replaces natural forest.

Gama

Goiás

0   10   20
Kilometres

N

6 Aug 2001

and administrative districts to the symmetry of the buildings themselves—should act in harmony with the city's overall design. This consisted of a bird-shaped core with residential areas situated between the encircling "arms" of Lake Paranoá. The city was a landmark in town planning and was recognized as a World Heritage site in 1987.

As these images reveal, unplanned urban developments arose at Brasilia's fringes resulting in a collection of urban "satellites" around the city. Several new reservoirs have been constructed since Brasilia's birth, but the National Park of Brasilia stands out as a densely vegetated expanse of dark green that has remained relatively unchanged. In 1970, the population of Brasilia and its satellites was roughly 500 000. The population now exceeds 2 000 000.

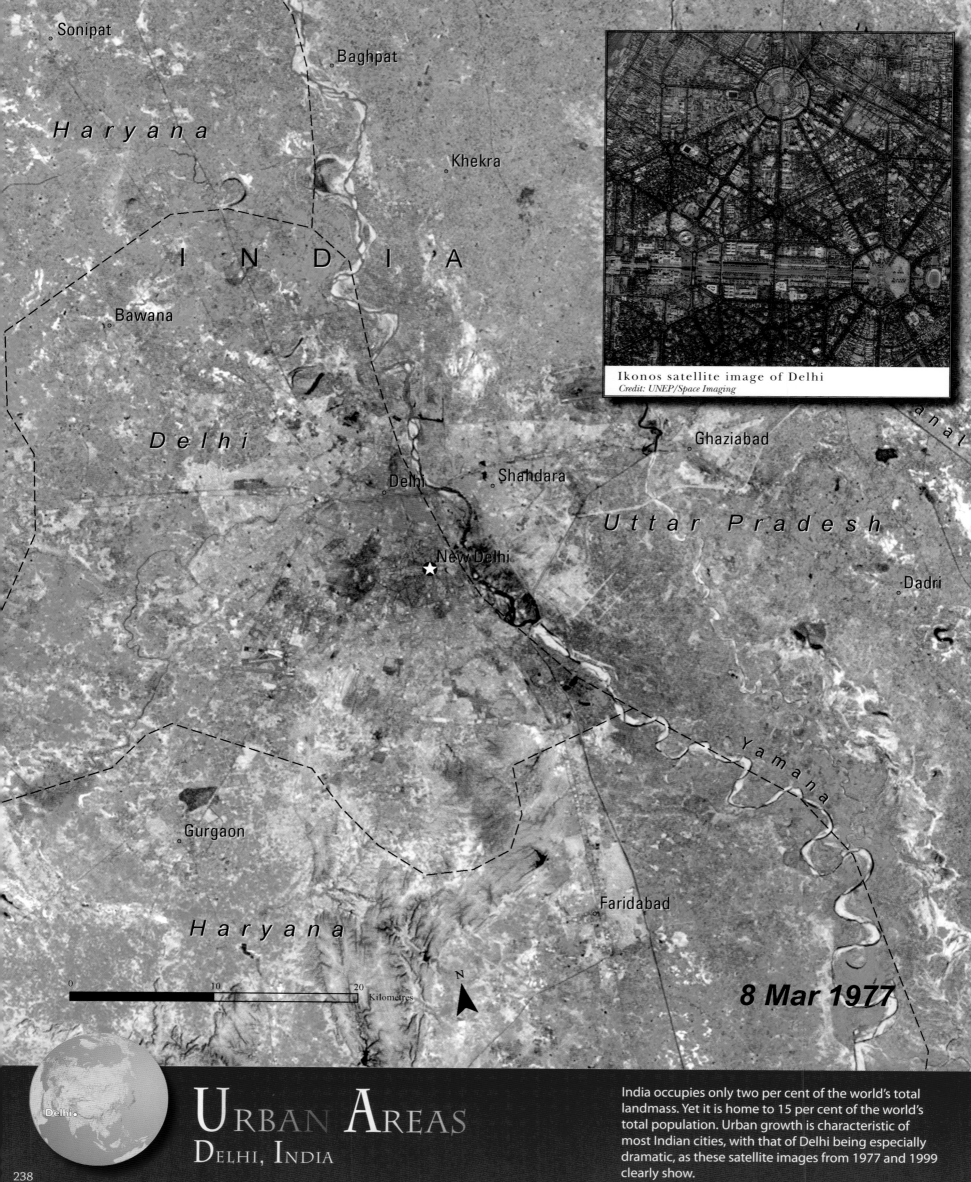

Haryana

Sonipat

Baghpat

Khekra

I N D I A

Bawana

Ikonos satellite image of Delhi
Credit: UNEP/Space Imaging

Delhi

Ghaziabad

Delhi

Shahdara

New Delhi

Uttar Pradesh

Dadri

Yamana

Gurgaon

Faridabad

Haryana

0   10   20
Kilometres

N

8 Mar 1977

Delhi

Urban Areas
Delhi, India

India occupies only two per cent of the world's total landmass. Yet it is home to 15 per cent of the world's total population. Urban growth is characteristic of most Indian cities, with that of Delhi being especially dramatic, as these satellite images from 1977 and 1999 clearly show.

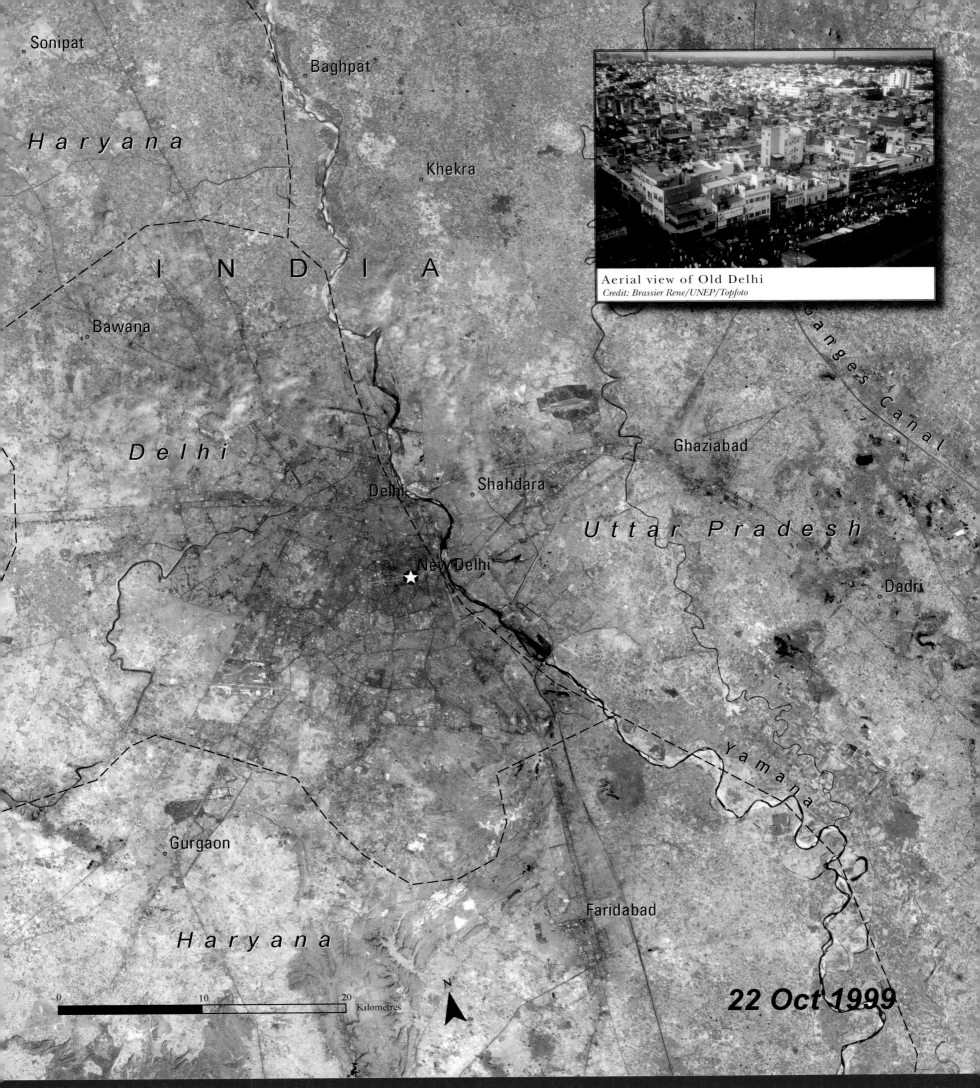

Sonipat

Baghpat

H a r y a n a

Khekra

I N D I A

Bawana

D e l h i

Delhi

Shahdara

Ghaziabad

U t t a r   P r a d e s h

New Delhi

Dadri

G a n g e s   C a n a l

Gurgaon

Y a m u n a

Faridabad

H a r y a n a

0   10   20
Kilometres

N

22 Oct 1999

Aerial view of Old Delhi
Credit: Brassier Rene/UNEP/Topfoto

In 1975, Delhi had a population of 4.4 million people or 3.3 per cent of India's entire urban population. In 2000, the city had 12.4 million inhabitants, or more than 4.5 per cent of the country's urban population. Of the world's 30 largest urban agglomerations, Delhi ranked 24th in 1975 and tenth in 2000. By 2015, Delhi's population is expected to be 20.9 million.

In these images, urban areas appear in shades of gray and purple. Growth is especially noticeable in the suburbs and areas surrounding Delhi such as Ghaziabad, Faridabad, and Gurgaon. Rapid urbanization has placed tremendous pressure on land and water resources in and around Delhi.

Turag

Tongi

B A N G L A D E S H

Meghna

Vegetable vendor in Dhaka
Credit: Jim Welch/UNEP/NREL

★ Dhaka
(Dacca)

Turag

Dhaleswari

Narayangan

0   5   10
Kilometres

N

8 Feb 1977

•Dhaka

Urban Areas
Dhaka, Bangladesh

Dhaka, the capital of Bangladesh, has undergone phe-
nomenal growth since the country gained independence in 1971. It has grown from a city of 2.5 million
inhabitants to one with a population of more than ten
million. This increase represents an average population

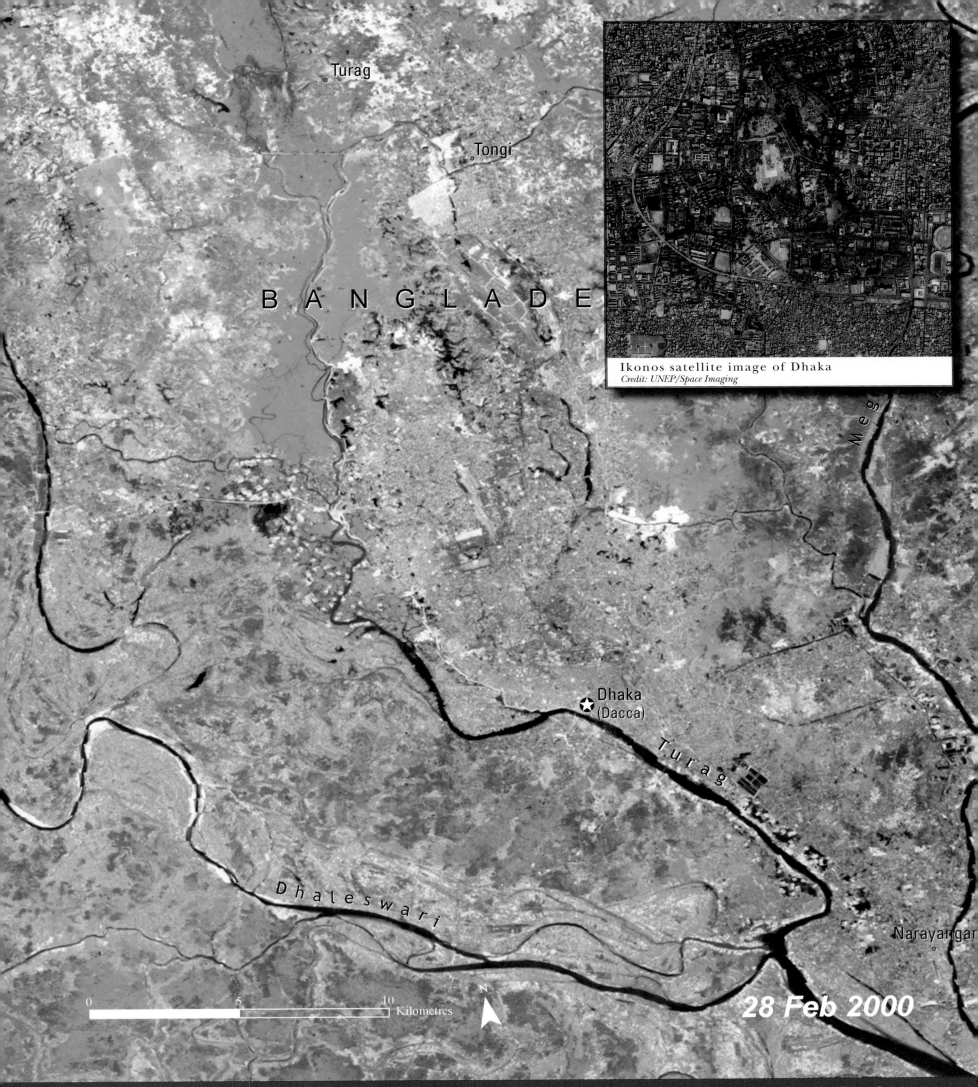

Turag

Tongi

B A N G L A D E S

Meg

Dhaka
(Dacca)

T u r a g

D h a l e s w a r i

Narayangan

Ikonos satellite image of Dhaka
Credit: UNEP/Space Imaging

0   5   10 Kilometres

N

28 Feb 2000

growth rate of about eight per cent annually. Dhaka is one of the poorest and most densely populated cities on the planet, with 6 545 people per square kilometre.

Following independence, urban areas expanded rapidly as they sought to become hubs of production and modernization. In the process, land use changed dramatically, as these images from 1977 and 2000 reveal. Dhaka is visible in the central portion of each image along the Turag River. Green areas represent forests and agricultural lands. White spots are planned areas of infrastructure. Urban areas are light purple. The 2000 image shows how, over time, lowlands and agricultural lands have been converted to urban areas where Dhaka has expanded to the north.

Quail Spring
Wilderness Study Area

Desert National Wildlife Range

Nellis A   Nellis B

Nellis Air Force
Range

Las Vegas Paiute
Indian Reservation

Nellis C

Floyd R. Lamb
State Park

Nellis Air Force
Range

U N I T E D   S T A T E S

Nevada

Las Vegas

Lake Mead
Nat. Rec.
Area

Credit: Lynn Betts/UNEP/USDA-NRCS
A girl waters the yard in Clark County, Nevada.

0   5   10
                              Kilometres

N

13 May 1973

Urban Areas
Las Vegas, United States

Las Vegas is the fastest growing metropolitan area in
the United States. Its growth was fairly slow during the
first half of the 20th century, but as the gaming and
tourism industry blossomed the population increased
more rapidly. In 1950, Las Vegas was home to 24 624
people. Today, the population of the Las Vegas Valley

Las Vegas

Quail Spring
Wilderness Study Area

Desert National Wildlife Range

Nellis A   Nellis B

Nellis C

Nellis Air Force
Range

Las Vegas Paiute
Indian Reservation

Floyd R. Lamb
State Park

Nellis Air Force
Range

U N I T E D   S T A T E S

Nevada

Las Vegas

Lake Mead
Nat. Rec.
Area

Credit: Lynn Betts/UNEP/ USDA-NRCS

New housing and a golf
course in Nevada replace
natural desert.

0   5   10
Kilometres

N

3 May 2000

tops one million, not including the tourists. According to one estimate, it may double by 2015. This population growth has put a strain on water supplies.

Satellite imagery of Las Vegas provides a dramatic illustration of the spatial patterns and rates of change resulting from the city's urban sprawl. Las Vegas is shown in the central portion of these images from 1973 and 2000. Note the profound modifications to the landscape—specifically the prolifera-

tion of asphalt and concrete roads and other infrastructure, along with the displacement of the few vegetated lands. By 2000, Las Vegas' growth had sprawled in every direction, with the greatest expansion to the northwest and southeast. As the city expanded, several new transportation networks emerged to serve the city's inhabitants.

Cuautitlán

Presa de
Guadalupe

Coacalco

Nicolás Romero

Ecapetec

Ciudad Lopez Mateos

El Caracol

M E X I C O

Texcoco

Naucalpan de
Juarez

★ Mexico City

Ixtapaluca

1910 **1929** **1941** **1959** **1970**

The red fill shows the historical urban boundaries of Mexico City.

Chalico

0   5   10
Kilometres   N

21 Apr / 23 Nov 1973

URBAN AREAS
MEXICO CITY, MEXICO

Mexico City is one of the fastest growing megalopolis
cities in the world. These satellite images show the
transformation Mexico City underwent between 1973
and 2000. Areas of urban infrastructure appear as

Mexico City

Cuautitlán

Coacalco

Nicolás Romero

Presa de
Guadalupe

Ciudad Lopez Mateos

A street view of historical Mexico City
Credit: José de Jesús Campos Enríquez/UNEP/CentroGeo

M E X I C O

El Carneol

Texcoco

Naucalpan de
Juarez

★ Mexico City

Ixtapaluca

Desierto
De Los Leones
Nat. Park
Miguel
Hidalgo
Nat. Park

Chalico

Cumbres Del
Ajusco Nat. Park

0   5   10
                              Kilometres   N

21 Mar 2000

shades of purple while natural vegetation is shown in green. In 1973 Mexico City had a population of about 9 million. In the ensuing years, the city expanded into surrounding areas. The forests in the mountains west and south of the city suffered significant deforestation as the urban sprawl progressed.

By 1986, Mexico City's population had soared to 14 million. In 1999, Mexico City had a population of 17.9 million, making it the second largest metropolitan area in the world behind Tokyo, Japan. The Mexican megalopolis is expected to reach 20 million in the next few years.

Pretoria

Lyttelton

S O U T H   A F R I C A

Midrand

Sandton

Alexandra

G a u t e n g

Benoni

14 Dec 1978

Johannesburg

0   5   10
|_____|_____| Kilometres

N

URBAN AREAS
MIDRAND, SOUTH AFRICA

Midrand is located approximately halfway between the major urban centers of Johannesburg and Pretoria in South Africa. The major highway that connects these two large cities dissects the city of Midrand into east

Downtown Johannesburg
Credit: Stephan Volz/UNEP/Africa Focus

Pretoria

Lyttelton

SOUTH AFRICA

Midrand

Sandton

Alexandra

Gauteng

Benoni

Johannesburg

0   5   10
Kilometres

N

7 Jan 2002

and west halves. Since 1978, the city has been rapidly transformed as a result of population growth, agriculture, mining, and industry.

In the 1978 image, the area surrounding Midrand consists largely of agricultural lands and rural residential zones, with some evidence of commercial development. The 2002 image reveals high-density urban development throughout. Rapid growth of Midrand's economy is expected to continue. Current development trends and population growth rates indicate that if effective environmental management strategies are not adopted soon, significant deterioration in the quality of the environment can be expected.

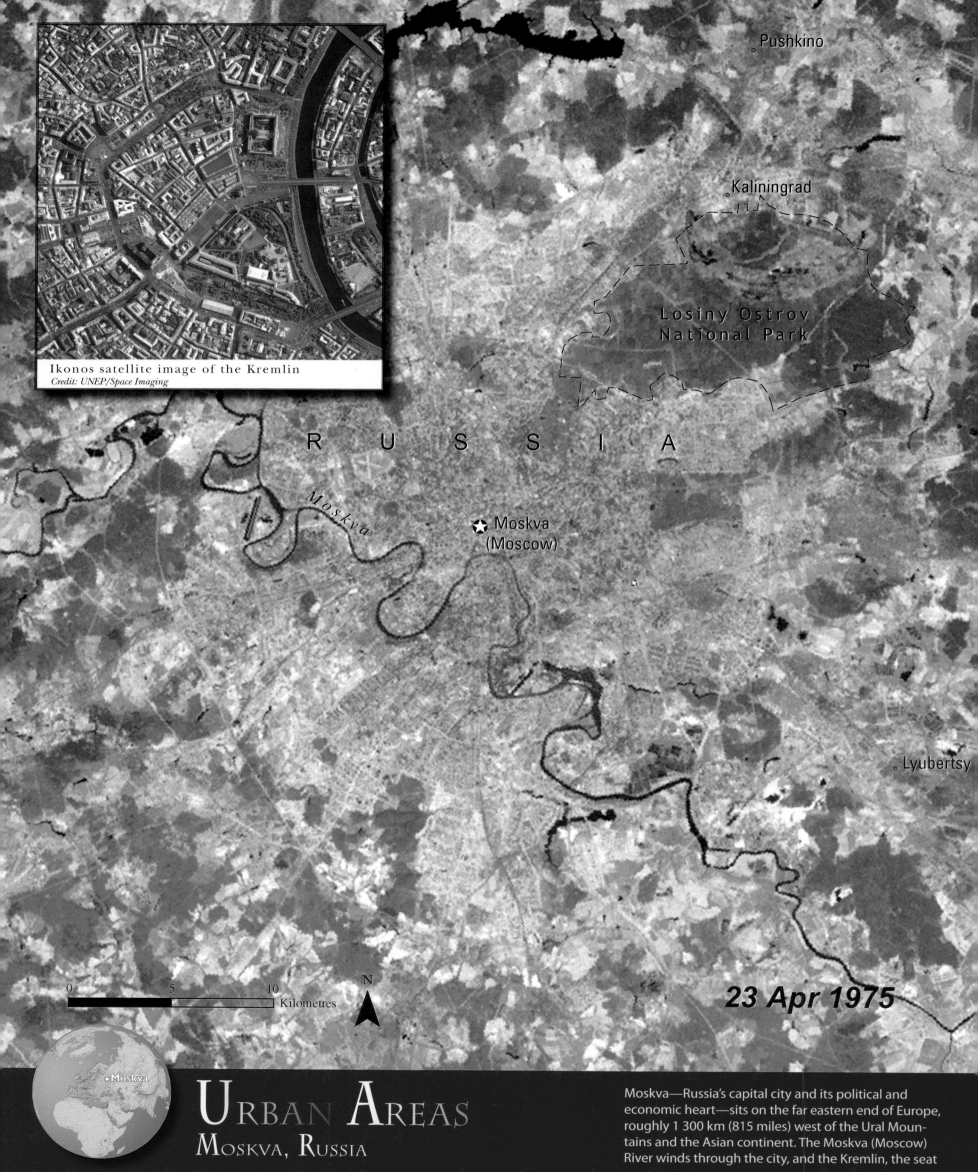

Ikonos satellite image of the Kremlin
Credit: UNEP/Space Imaging

Pushkino

Kaliningrad

Losiny Ostrov
National Park

R U S S I A

Moskva
Moskva
(Moscow)

Lyubertsy

0   5   10
Kilometres

N

23 Apr 1975

Moskva

URBAN AREAS
MOSKVA, RUSSIA

Moskva—Russia's capital city and its political and
economic heart—sits on the far eastern end of Europe,
roughly 1 300 km (815 miles) west of the Ural Mountains and the Asian continent. The Moskva (Moscow)
River winds through the city, and the Kremlin, the seat

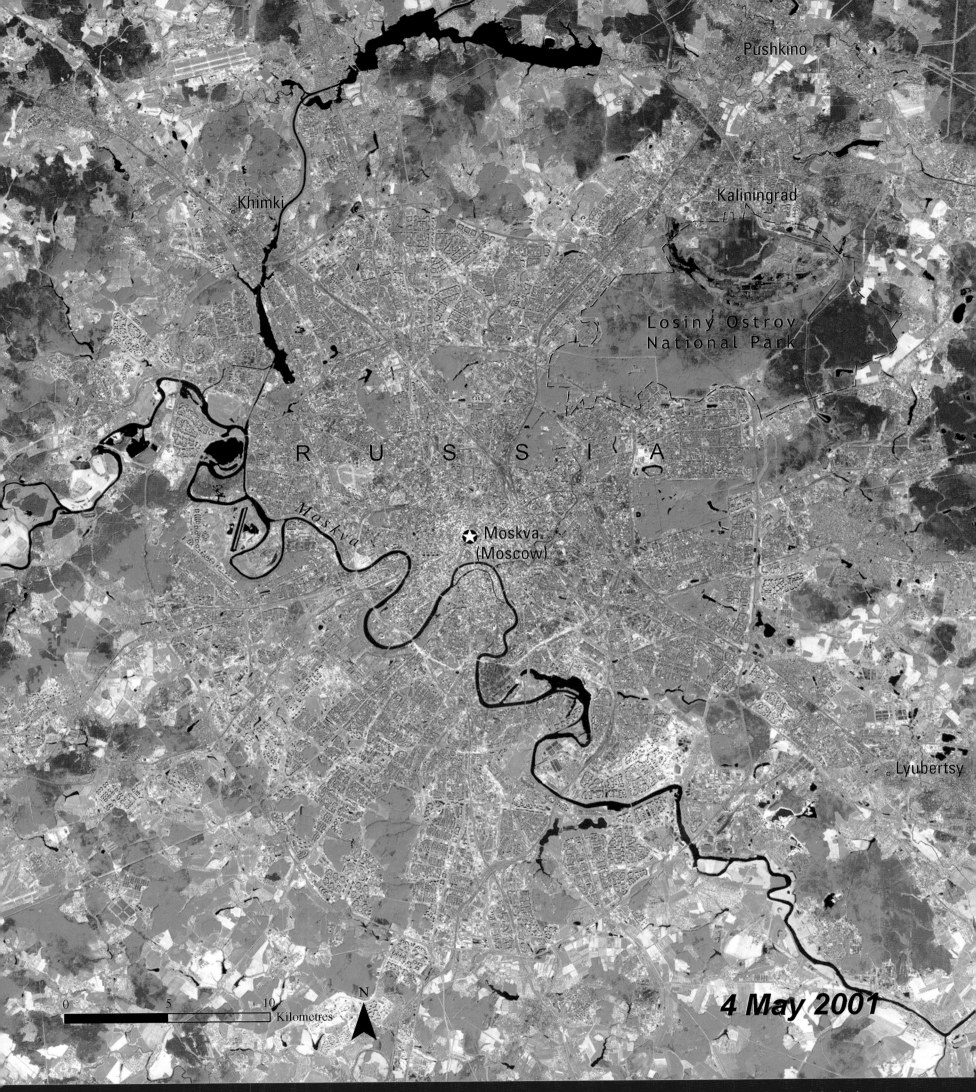

Khimki

Pushkino

Kaliningrad

Losiny Ostrov
National Park

R U S S I A

★ Moskva
(Moscow)

Moskva

Lyubertsy

0   5   10 Kilometres

N

4 May 2001

of the Russian government, lies at its center. With a population close to 9 million and an area of 1 035 km² (405 square miles), Moskva is believed to be the largest of all European cities.

These two images show the urban expansion Moskva experienced during the last 25 years of the 20th century. The blue-gray patches are urban areas.

The light green areas surrounding the city are farms while the brown areas are regions of sparse vegetation

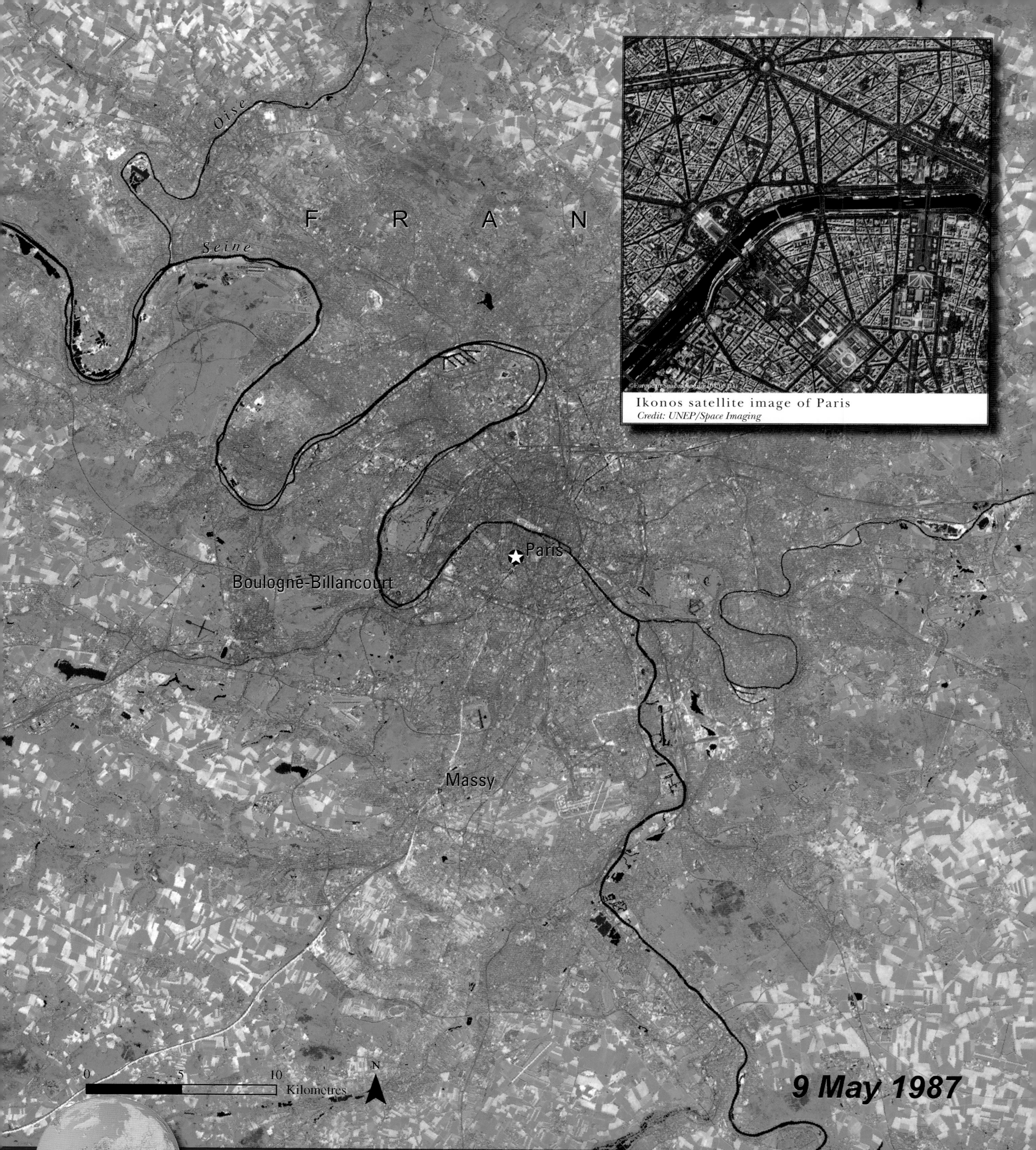

F R A N

Oise

Seine

Ikonos satellite image of Paris
Credit: UNEP/Space Imaging

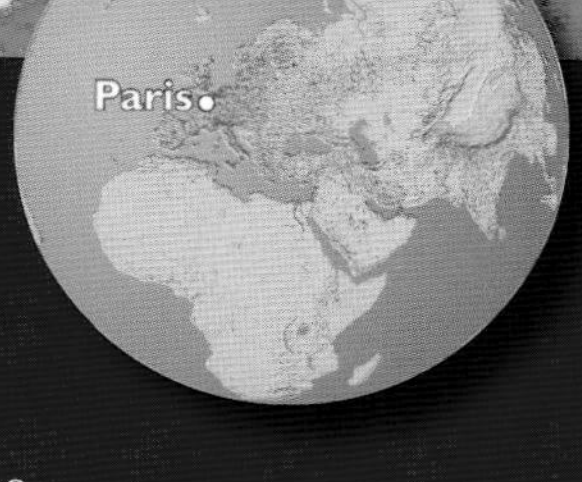

★ Paris

Boulogne-Billancourt

Massy

9 May 1987

0   5   10
Kilometres

N

Paris

URBAN AREAS
PARIS, FRANCE

France is a large country with relatively few large metropolitan areas. Only 16 French towns and cities have populations of more than 150 000 people. Paris, the capital city of France, is the largest of these and home

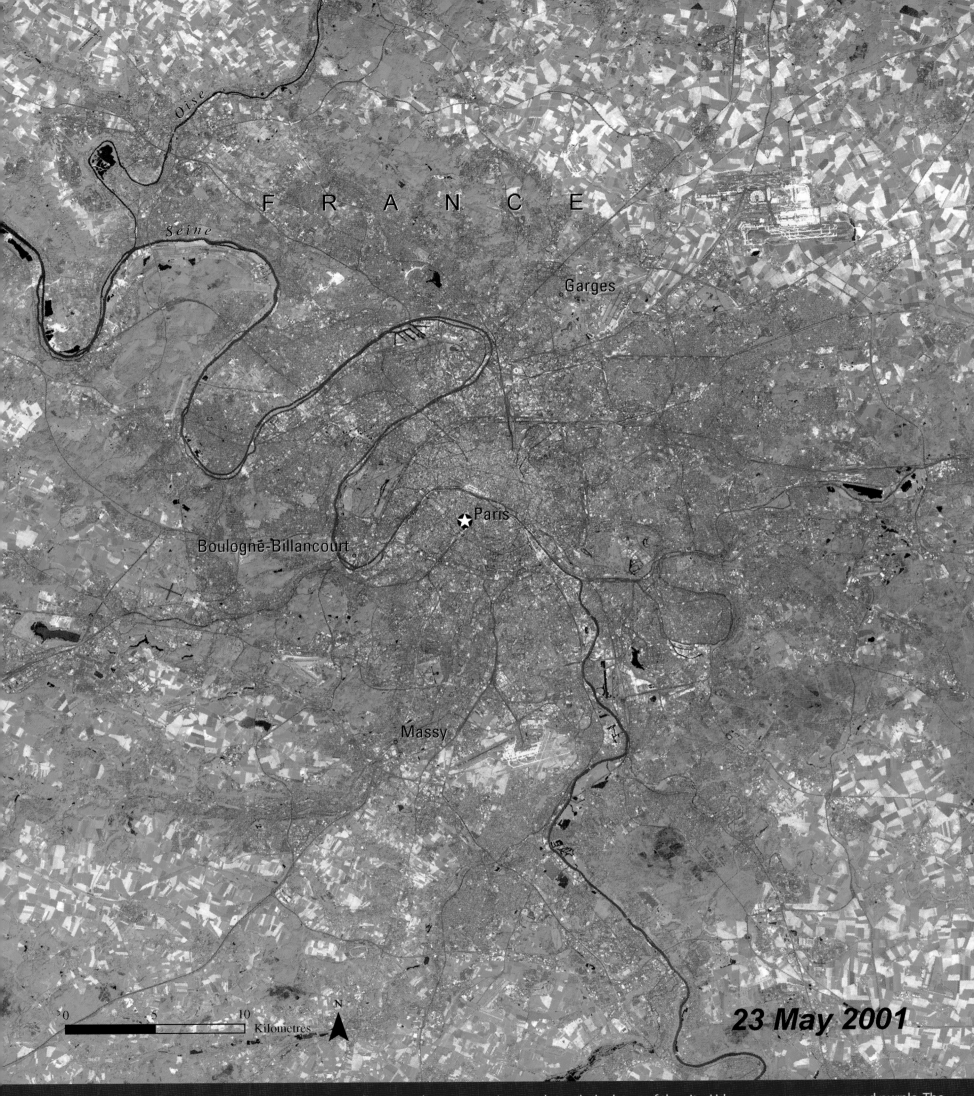

FRANCE

Oise

Seine

Garges

Paris

Boulogne-Billancourt

Massy

Seine

0   5   10
Kilometres

N

23 May 2001

to roughly 2 million inhabitants. The entire Paris metropolitan area, however, includes more than 11 million people.

Lying roughly 160 km (100 miles) southeast of the English Channel in northern France, Paris is considered by many to be one of the most beautiful cities in the world. In the images above, the Seine River can be seen winding its way through the heart of the city. Urban areas appear gray and purple. The patchwork of green, brown, tan and yellow around the city is primarily farmland. Note how the city has expanded in the years between 1987 and 2001, reaching ever-further into the surrounding rural areas.

C H I L E

☆ Santiago

Padre Hurtado

San Bernardo

Puente Alto

San Jose de Maipo

Talagante

Río Maipo

N

0   5   10
Kilometres

22 Mar 1975

URBAN AREAS
SANTIAGO, CHILE

Santiago, the capital of Chile, is home to more than one-third of the country's total population of 15 million. Santiago's rapid growth is part of a national trend, but it is also a reflection of the large numbers of immigrants who are moving into the city.

Santiago

Downtown Santiago, Chile
Credit: Unknown/UNEP/Earthshots

C H I L E

★ Santiago

Padre Hurtado

San Bernardo

Puente Alto

San José de Maipo

Río Maipo

Talagante

0   5   10
Kilometres

N

31 Mar 2000

Santiago's population growth has led to a horizontal expansion of the city, principally towards the south and southeast. Chilean urban scholars speak of this expansion as the "urban stain" that continually exceeds and expands the limits of the Metropolitan Region of Santiago (MRS) while incorporating previously rural areas into it. Characteristics of Santiago's urban sprawl are haphazard growth, low-density housing, poor transportation, and air pollution. In the time frame illustrated by these images, Santiago's population has nearly doubled.

Richmond

Napean

A U S T R A

New South Wales

Parramatta

Prospect Reservoir

Sydney

Blue
Mts.
N.P.

Mulgoa

Botany Bay

Narellan

Tasm Sea

Camden

Sydney Opera House is one of the architectural
wonders of the world, with its design and construction involving countless innovative design ideas and
construction techniques.

Credit: DTCreations/UNEP/Morguefile.com

0   10   20
Kilometres

N

12 Oct 197

URBAN AREAS
SYDNEY, AUSTRALIA

Sydney

Australia is the sixth largest country in the world.
is roughly the same size as the conterminous Uni
States and 50 per cent larger than Europe. Yet Au
has the lowest population density of any country
the world. With 4 million inhabitants, Sydney is

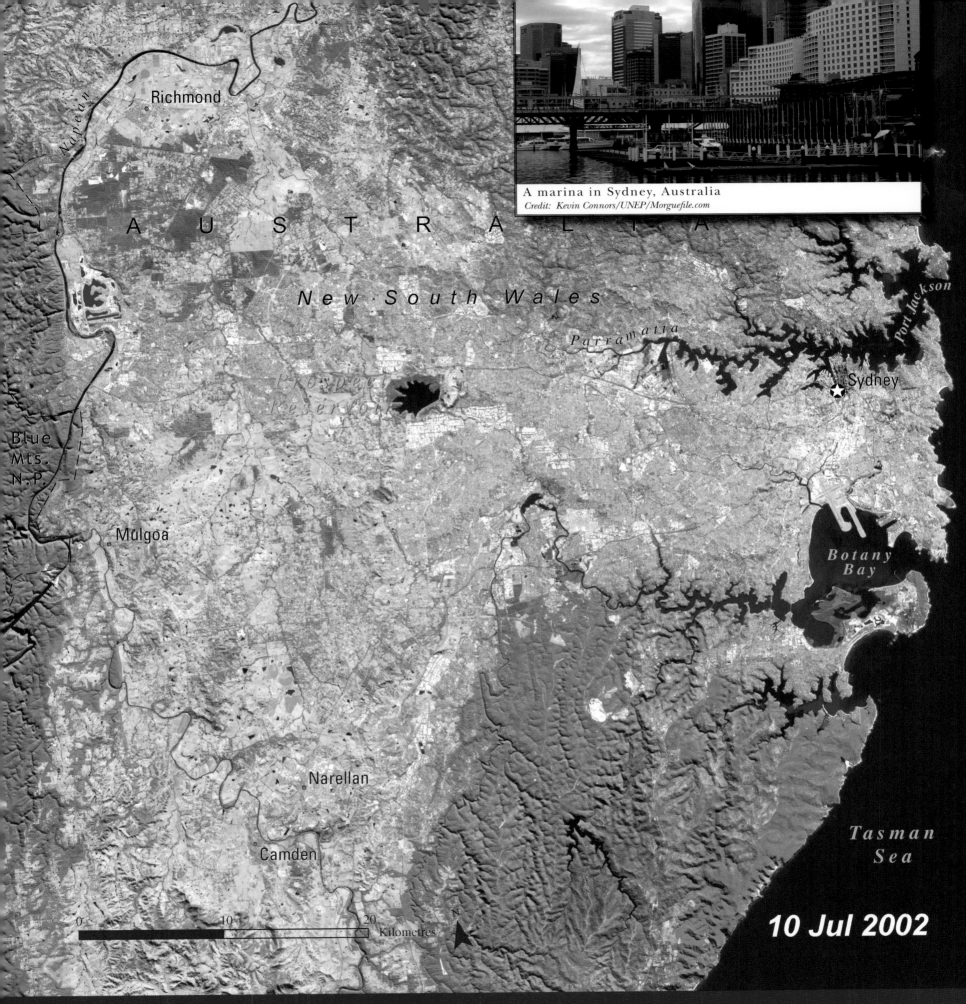

Richmond

Napean

A U S T R A L I A

New South Wales

Parramatta

Prospect
Reservoir

Blue
Mts.
N.P.

Mulgoa

Port Jackson

Sydney

Botany
Bay

Narellan

Tasman
Sea

Camden

0   10   20
                    Kilometres

N

10 Jul 2002

A marina in Sydney, Australia
Credit: Kevin Connors/UNEP/Morguefile.com

Australia's largest city. It is also the capital of New South Wales, the country's most densely populated state. Sydney is bounded by the Pacific Ocean to the east, national parks and deep-water inlets to the north and south, and the spectacular Blue Mountains far to the west. These natural boundaries have influenced Sydney's urban growth patterns. Over the past several decades, the city's expansion has been largely westward toward the Blue Mountains, as can be seen in these two satellite images. As suburbs sprawl into bushland, they become vulnerable to summer bush fires.

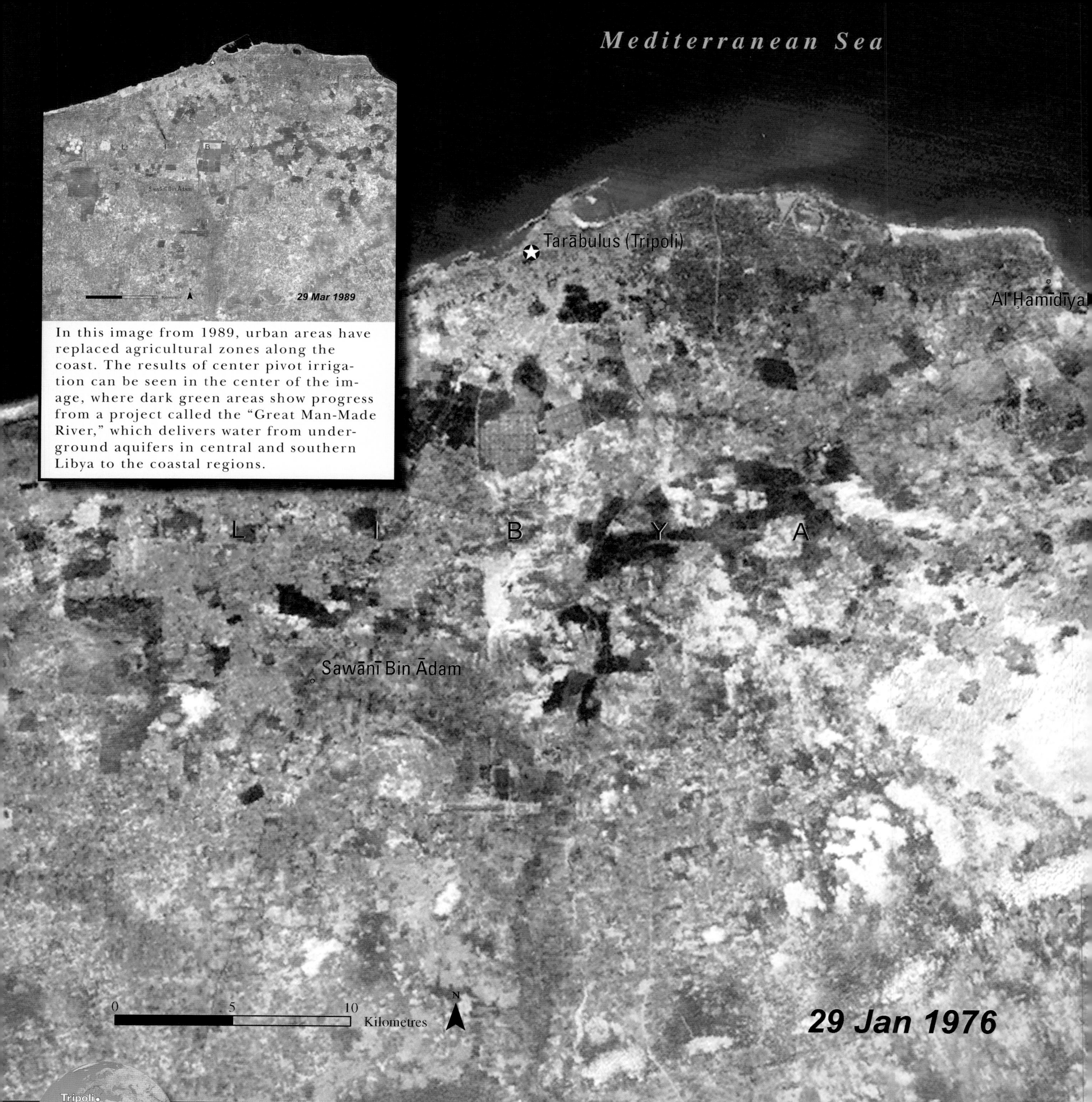

Mediterranean Sea

29 Mar 1989

In this image from 1989, urban areas have replaced agricultural zones along the coast. The results of center pivot irrigation can be seen in the center of the image, where dark green areas show progress from a project called the "Great Man-Made River," which delivers water from underground aquifers in central and southern Libya to the coastal regions.

★ Tarābulus (Tripoli)

Al Ḩamīdīya

L I B Y A

Sawānī Bin Ādam

0   5   10 Kilometres

N

29 Jan 1976

Tripoli

URBAN AREAS
TRIPOLI, LIBYA

Tripoli, the capital city of Libya, is located on the country's Mediterranean coast along a narrow band of fertile lowlands that quickly give way to a vast interior of arid, rocky plains and seas of sand. Tripoli has undergone steady urban growth over the past thirty years.

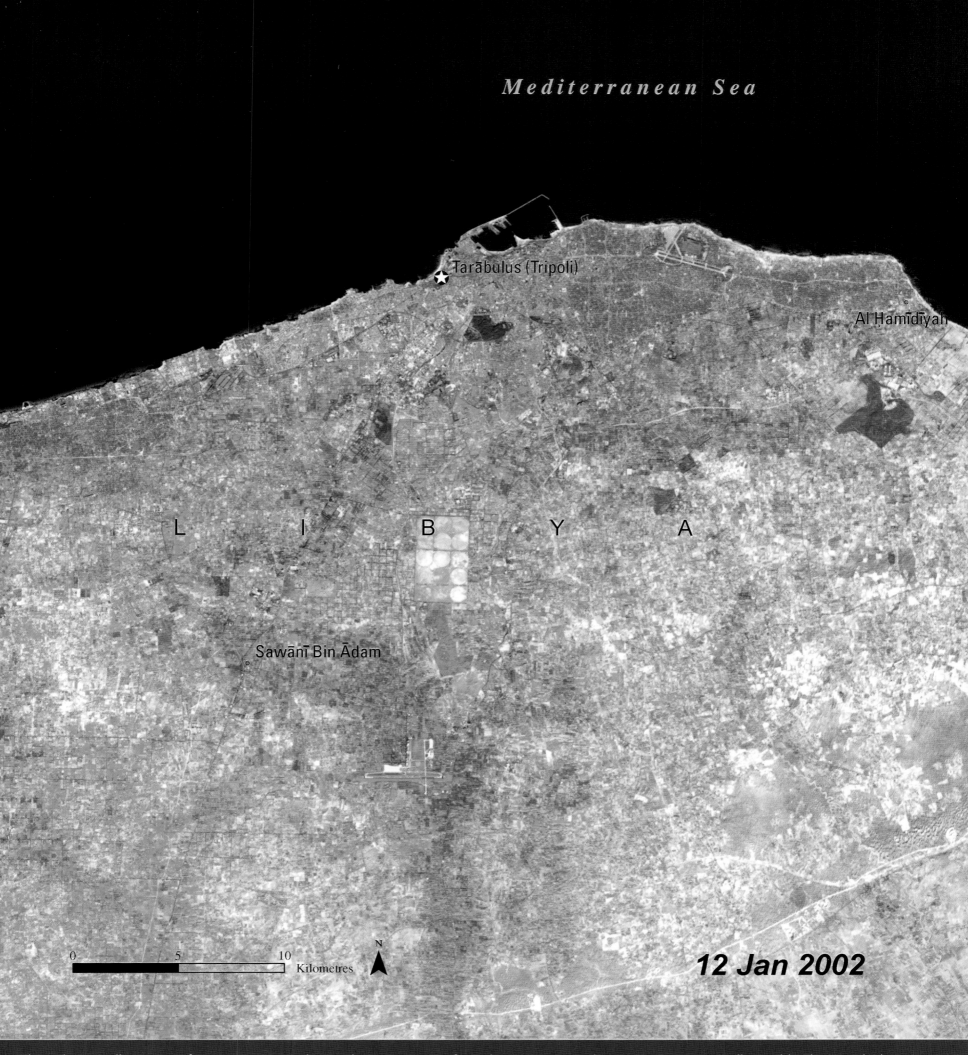

Mediterranean Sea

Tarābulus (Tripoli)

Al Ḩamīdīyah

L I B Y A

Sawānī Bin Ādam

0   5   10   Kilometres

N

12 Jan 2002

These three satellite images, from 1976, 1989, and 2002, document some of the major changes.

Urban areas appear as shades of grey. Darker patches south of the city, visible in both the 1976 and 1989 images, represent grasslands that have been converted to agricultural fields. Bright green areas are planted croplands. In the 2002 image, urban expansion is especially notable. The irregular brown patch in the upper far right of this image, south of Al Hamidiyan, is perhaps the last remaining vestige of natural vegetation in the Tripoli region.

Credit: Andrew Magor/UNEP/Topfoto

3.8 Tundra and Polar Regions

Of all the terrestrial biomes, tundra is the coldest. Tundra comes from the Finnish word *tunturia*, which means treeless plain (Pullen 1996). There are two distinct types of tundra: the vast Arctic tundra and high-altitude alpine tundra on mountains.

Arctic tundra is located in the Northern Hemisphere on lands encircling the North Pole and extending south to the coniferous boreal forests of the taiga and covering approximately 5.6 million km^2 (2 million square miles) Wookey 2002). Arctic tundra is characterized by cold, desert-like conditions. Although somewhat variable from place to place, precipitation on the Arctic tundra, including melted snow, is roughly 15 to 25 cm (6 to 10 inches) annually. The average winter temperature is -34° C (-30° F); the average summer temperature is 3 to 12° C (37 to 54° F). Winters are long and summers brief, with the growing season only 50 to 60 days long. During summer, only the top few centimeters of the soil thaw. Beneath the surface is a layer of permanently frozen subsoil called permafrost. Because the topsoil is so shallow and underlaid by permafrost, it becomes quickly saturated with water. Lakes, ponds, and bogs dot the surface of the Arctic tundra throughout the brief summer months, providing moisture for plants and nesting and feeding habitats for huge numbers of waterfowl and other animals (Pullen 1996).

Alpine tundra is found on mountains throughout the world, at high altitudes—above the tree line—where conditions are too cold and too dry for trees to grow. The growing season in alpine tundra is approximately 180 days. Nighttime temperatures are usually below freezing. Unlike soils in the Arctic tundra, soils in alpine tundra are usually well-drained (Pullen 1996). Alpine tundra is also characterized by relatively high biodiversity.

The Earth's polar regions are high-latitude zones above the Arctic Circle in the Northern Hemisphere and the Antarctic Circle in the Southern Hemisphere (EEA n.d.). Although similar in many ways, the two polar regions differ in that the Arctic is a frozen ocean surrounded by land, whereas the Antarctic is a frozen continent surrounded by ocean.

Most of the world's fresh water is locked up in polar ice caps. Large glaciers and ice sheets cover Arctic islands and Greenland in the north and the continent of Antarctica in the south. Where ice sheets and glaciers meet the ocean, huge chunks of ice continually break off, in a process known as calving, to give birth to icebergs. Icebergs are found in both Arctic and Antarctic polar oceans. In the north, most icebergs are calved from ice sheets along the western coast of Greenland. In the south, the vast ice sheets and glaciers that cover Antarctica give rise to icebergs in polar seas.

The Earth's tundra and polar regions are unique and vital parts of the global environment. They are the world's least

The Arctic region is often defined as that area where the average temperature for the warmest month is below 10ºC (50ºF)

Map of the Arctic *Source: Modified from http://www.lib.utexas.edu/maps/islands_oceans_poles/arctic_region_pol02.jpg*

Credit: Budd Christman/UNEP/NOAA

populated regions. Antarctica has no permanent residents. The Arctic has approximately 3.7 million inhabitants from eight countries. Sparsely populated and relatively undisturbed, tundra and polar regions therefore contain the world's largest remaining wilderness areas. They also possess a surprising range of natural resources, from marine life to oil and gas. Yet despite their rugged appearance, tundra and polar regions are fragile ecosystems that are extremely sensitive to the effects of resource exploitation. Managing these regions and their resources effectively places huge demands on both technical and political capacities (SPRI n.d.).

Tundra and polar regions also exert a profound effect on global climate. Variations in the extent of sea ice, for example, affect the Earth's surface radia-

tion balance by changing average surface albedo (albedo is the fraction of sunlight reflected). During the peak of the last Ice Age, one-third of the Earth's land surface was covered by thick sheets of ice that extended from polar regions toward the equator. The high albedo of these ice sheets reflected a great deal of sunlight out into space, which cooled the Earth and allowed the ice sheets to grow. Large changes in sea ice extent are also thought to influence deep-ocean convection and global ocean currents (Jezek 1995).

Many climate and biogeochemical studies indicate that carbon cycling in the Arctic tundra and boreal forests strongly influences global climate as well. Cold tundra soils contain huge amounts of stored organic carbon. They are known sinks for atmospheric CO_2 through the accumula-

tion of peat, and are significant sources of CH_4 as a result of anaerobic decomposition (Christensen n.d.).

While tundra and polar regions play a major role in shaping the Earth's climate, they also are highly sensitive ecosystems that have the potential to be profoundly affected by changes in the Earth's climate (NRDC 2004). Nearly all climate models indicate that environmental changes brought about by global warming are expected to be greater in tundra and polar regions than for most other places on Earth. In that respect, tundra and polar regions form a sort of early warning system for climate change and its effects on the planet and its inhabitants. The monitoring of high-latitude and high-altitude ecosystems, then, represents a way to detect early signs of regional and global climate change. The advance or retreat of glaciers, ice sheets, and sea ice has been given particular attention by climate change researchers.

A rapid warming trend in the Arctic polar region over the last 25 years has dramatically reduced the region's sea ice. Scientists have been monitoring ongoing changes in Arctic sea ice for decades, both firsthand through fieldwork and remotely through the use of satellite imagery. In 2002, the extent of multi-year Arctic sea ice was the lowest on record since satellite observations began in 1973. There

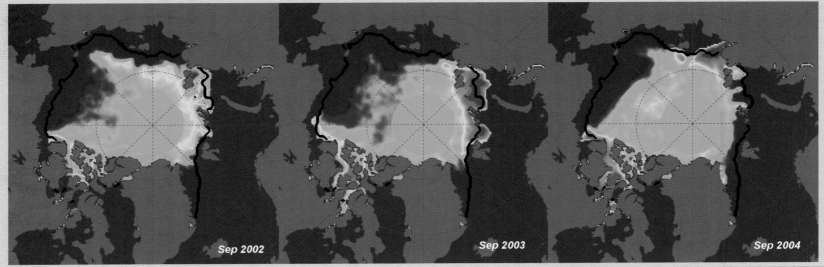

Sep 2002 Sep 2003 Sep 2004

Case Study: Arctic Seas

The extent of Arctic sea ice in September– the end of the summer melt period–is the most valuable indicator of the state of the ice cover. On average, sea ice in September covers an area of about seven million km², an area roughly equal in size to the continent of Australia.

In the images above, the Sea Ice Concentration Anomaly scale indicates the per cent by which the local sea ice extent

differs above or below the average for the period 1979-2000. The median ice edge for 1979-2000 is indicated by the black outer line. In 2002, total September ice extent was 15 per cent below this average. This represents a reduction equivalent to an area roughly twice the size of Texas or Iraq. From caparisons with records prior to the satellite era, this was probably the least amount of sea ice that had covered the Arctic over the past 50 years.

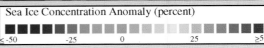

Sea Ice Concentration Anomaly (percent)

< -50   -25   0   25   ≥50

Quite often, a "low" ice year is followed by recovery the next year. However, September of 2003 was also extreme, with 12 per cent less ice extent than average. Caculations performed for 30 September 2004 show a sea ice extent loss of 13.4 per cent, especially pronounced north of Alaska and eastern Siberia. *Source: NSIDC*

was only slightly more sea ice present in 2003. According to one study, perennial sea ice—sea ice that survives the summer and remains year round—is melting at the alarming rate of 9 per cent per decade (NASA 2003d). If this trend continues, Arctic sea ice may be gone by the year 2100.

Researchers also documented temperature increases in different regions within and near the Arctic Circle, north of 66°. Average temperatures increased by 0.3°C (0.5°F) per decade over sea ice and by 0.5°C (0.9°F) per decade over the northernmost land areas of Europe and Asia. Temperatures over northern North

America experienced the highest regional warming, increasing by 1.06°C (1.9°F) per decade. Greenland cooled by less than one-tenth of a degree C per decade. The cooling found over Greenland was mainly at high elevations, while warming trends were observed around its periphery. These results are consistent with a National Snow

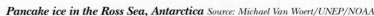

Pancake ice in the Ross Sea, Antarctica Source: Michael Van Woert/UNEP/NOAA

and Ice Data Center study that found record loss of sea ice around Greenland's periphery in 2002 (NSIDC n.d.).

As sea ice melts, Arctic waters warm. Less ice means more heat gain by polar waters, which creates a positive feedback leading to further ice melting and increased warming. The loss of Arctic sea ice, and the warming of Arctic polar waters, have enormous implications for both regional and global climate patterns. One major concern is that the disappearance of Arctic sea ice may cause changes in ocean circulation leading to unexpected and rapid shifts in climate worldwide (SPRI n.d.).

Over the past 30 years, Antarctic ice has also undergone changes. Ice sheets and glacier tongues are among the most dynamic and changeable features along the coastal regions of Antarctica. Seaward of a line where these masses of ice are grounded, the floating ice margins are subject to frequent and large calving events. These events lead to annual and decadal changes in the position of ice edge varying from several to many kilometres.

Yet ice events are also occurring in Antarctica that appear to be out of the or-dinary. Along the Antarctic Peninsula, for instance, the Wordie Ice Shelf has practically disappeared. In 2002, a section of the Larsen B Ice Shelf collapsed—the largest such event in the last 30 years.

In other parts of Antarctica, however, ice cover has actually increased (UPI 2003). What is happening with the vast West Antarctic ice sheet is not yet clear. Some studies seem to indicate that it is getting thicker (NCPPR n.d.). Other studies indicate that this mammoth ice sheet is shrinking in size. If the West Antarctic ice sheet melts, global sea levels would rise by many metres. Such a change would severely impact densely populated coastal regions around the world, forcing people to move to higher elevations.

Although the details may be still unclear, there is no doubt that the Earth's tundra and polar regions are undergoing many changes. Some are related to climate change and long-distance pollution. Some are the result of on-site human activities. On a positive note, many of the human-induced environmental threats in the Arctic have not yet occurred in the largely unpopulated Antarctic (Harrison and Pearce

Between 2000 and 2002, scientists observed the formation of a crack in the Ward Hunt Ice Shelf on the northern shore of Canada's Ellesmere Island. The crack allowed the waters of a rare freshwater Arctic lake to empty into the Arctic Ocean. Rising temperatures also brought about the thinning of this 3,000-year-old shelf, which is the Arctic's largest. *Credit: V. Sahanatien/UNEP/Parks Canada*

2001). Activities in Antarctica are carried out under the Antarctic Treaty, a model of international cooperation. In the Arctic, the common needs of indigenous peoples living in remote areas are addressed through the Arctic Council and other circumpolar institutions. Thus, the polar regions offer hope that nations can cooperate in addressing the changes taking places in these and other parts of the world (SPRI n.d).

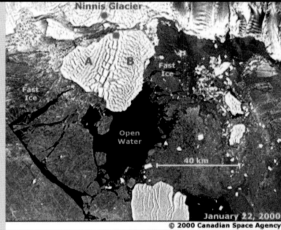

Images courtesy of Dr. Rob Massom, Antarctic CRC © 2000 Canadian Space Agency

Case Study: Ninnis Glacier, Antarctica 2000

To better understand the Antarctic Ice Sheet's potential response to global climate change and its effect on global sea level, it is important to detect and monitor the calving of large icebergs. The series of images shown here depict the 2000 disintegration of the Ninnis Glacier tongue into two sections. Each image is a sub-section of a SCANSAR scene of the Ninnis Glacier Tongue region.

22 January 2000. This image captures the Ninnis Glacier Tongue region soon after the initial calving. The resultant iceberg (sections A and B) had an area of approximately 900 km^2 (347 square miles). NOTE: Purple dots indicate the area where the iceberg broke away from the glacier.

5 February 2000. Roughly two week after calving, the iceberg split into two sections (A and B). When this image was taken Berg A had drifted 20 km (about 12.5 miles) to the west, Berg B had drifted to the northeast, and a smaller section (C) remained grounded in front of the Ninnis Glacier.

20 February 2000. At this point Bergs A and B had almost totally separated, rotated counterclockwise, and drifted to the north. Note that both sections are now well away from the Ninnis Glacier.
Source: USGS 1999; Schmidt 2000

Case Study: Recession of Gangotri Glacier 1780-2001

Gangotri Glacier is situated in the Uttar Kashi District of Garhwal Himalaya, northern India. With its tributary glaciers, it is one of the largest glaciers in the Himalayas. It has been receding since 1780, although studies show its retreat quickened after 1971. It is currently 30.2 km (18.8 miles) long and between 0.5 and 2.5 km (0.3 and 1.6 miles) wide. The blue contour lines drawn in the image show the recession of the glacier's terminus over time. They are approximate, especially for the earlier years. Over the last 25 years, Gangotri Glacier has retreated more than 850 m (2 788 ft) with an accelerated recession of 76 m (249 ft) from 1996 to 1999 alone. The retreat is an alarming sign of global warming, which will impact local communities. Glaciers play an important role in storing winter rainfall, regulating water supply through the year, reducing floods, shaping landforms, and redistributing sediments.

Source: NASA 2004j

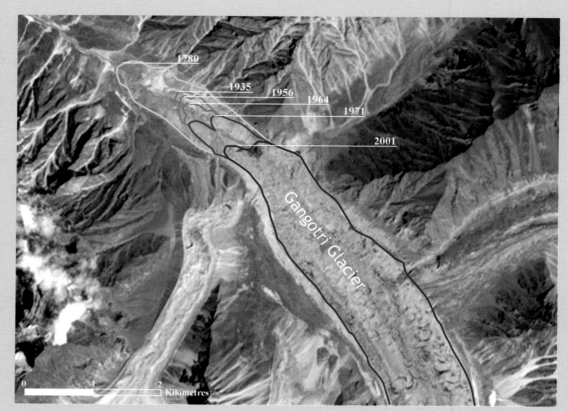

Credit: NASA 2004

Mountain peak in the Himalayas *Credit: Unknown/UNEP/Freefoto.com*

Huge icebergs are found in Antractica's regions. These icebergs influence the weather and climatic conditions. It is believed that if these icebergs melt, the sea levels will rise significantly. The study of the state of icebergs and their behaviors are very important aspects of climae change research. Today, multi-sensored remote sensing data are used in monitoring the state of icebergs. *Credit: Michael Van Woert/UNEP/NOAA*

Case Study: Drygalski Ice Tongue February 2005

The Drygalski ice tongue is located on the Scott Coast, in the northern McMurdo Sound of Antarctica's Ross Dependency, 240 km (149 miles) north of Ross Island. It stretches 70 km (43 miles) out to sea from the David Glacier, reaching the sea from a valley in the Prince Albert Mountains of Victoria Land.

The ice tongue was discovered in 1902 by Robert Falcon Scott, and is thought to be at least 4 000 years old.

This image, collected by the Advanced Synthetic Aperture Radar on the European Space Agency's ENVISAT satellite, shows the David Glacier on the 1 831-metre-high (6 007-feet-high) Mt. Joyce. As ice piles on the glacier, it slides under its own weight to the ocean. The ice doesn't break up when it reaches the ocean; rather, it floats, forming a long tongue of ice. The floating end of the David Glacier is the Drygalski Ice Tongue.

This floating spit of ice was recently menaced by the B-15A iceberg, a 120-km-long (74-mile-long) giant that had been drifting on a collision course with the ice tongue before becoming grounded. On 21 February 2005, Drygalski calved an iceberg. The five-by-ten-km (three-by-six mile) iceberg was floating off the left side of the ice tongue on 22 February when this image was acquired. The event is a normal part of the evolution of the ice tongue—pieces regularly break from the tongue as the glacier pushes more ice out over the sea. This image shows cracks, formed by time and ocean currents, which become more numerous towards the end of the tongue.

Source: NASA 2005, WIKIPEDIA

22 Feb 2005

Credit: European Space Agency—ESA

17.8.2003

Mount Dzhimarai-Khokh, elevation 4 780 metres (15 682 feet), towers above the Kolka Cirque. Rock and ice falling from the steep walls of the cirque since the end of July 2002, eventually triggered the collapse of the Kolka Glacier.

Black Sea

Mt. Kazbek

Caspian Sea

Caucasus Mountains

N 100 km

13 June 2001

Case Study: Collapse of the Kolka Glacier 20 September 2002

Rebecca Lindsey, Olga Tutubalina, Dmitry Petrakov, Sergey Chernomorets

Running east to west across the narrow isthmus of land between the Caspian Sea to the east and the Black Sea to the west, the Caucasus Mountains make a physical barricade between southern Russia to the north and the countries of Georgia and Azerbaijan to the south. In their center, a series of 5 000-metre-plus (16 000-feet-plus) summits stretch between two extinct volcanic giants: Mt. Elbrus at the western limit and Mt. Kazbek at the eastern. On the lower slopes, snow disappears in July and returns again in October. On the summit, winter is permanent. Glaciers cover peaks and steep-walled basins called cirques. The remote, sparsely populated area is popular with tourists and backpackers.

Elevations reach 5 642 metres (18 511 feet), and glaciers accumulate from heavy snowfall in the steep mountain valleys. Around Mount Kazbek, a dormant volcano, glaciers intermittently collapse, burying the landscape below under rock and ice. The latest of such collapses happened in 2002. Rebecca Lindsey, science writer with NASA's Earth Observatory, in close collaboration with Russian scientists Olga Tutubalina, Dmitry Petrakov (Moscow State University), and Sergey Chernomorets (University Centre for Engineering Geodynamics and Monitoring) compiled the details of this event.

On the evening of 20 September 2002, in a cirque just west of Mt. Kazbek, chunks of rock and hanging glacier on the north face of Mt. Dzhimarai-Khokh tumbled onto the Kolka glacier below. Kolka shattered, setting off a massive avalanche of ice, snow, and rocks that poured into the Genaldon River valley. Hurtling downriver nearly 13 km (8 miles), the avalanche exploded into the Karmadon Depression, a small bowl of land between two mountain ridges, and swallowed the village of Nizhniy Karmadon and several other settlements.

At the northern end of the depression, the churning mass of debris reached a choke point: the Gates of Karmadon, the narrow entrance to a steep-walled gorge. Gigantic blocks of ice and rock jammed into the narrow slot, and water and mud sluiced through. Trapped by the blockage, avalanche debris crashed like waves against the mountains and then finally cemented into a towering dam of dirty ice and rock, creating lakes upstream. At least 125 people were lost beneath the ice.

The Kolka Glacier collapse partially filled the Karmadon Depression with ice, mud, and rocks, destroying much of the village of Karmadon. The debris swept in through the Genaldon River Valley and backed up at the entrance to a narrow gorge. The debris acted as a dam, creating lakes upstream. Boulders, pebbles, and mud covered the surface of the debris flow, resulting in treacherous footing. The pathless maze of debris was only one of many hazards that slowed exploration of the disaster area.

Scratches on the surface of rocks of the Maili Glacier's moraine show the violence of the event. The avalanche, moving up to 180 kilometres per hour (112 mph), scoured the rocks below, leaving parallel grooves called "striations." Striations are typically observed in the bedrock underlying glaciers, created by the slow, scouring action of rocks caught beneath the ice.

Large-scale avalanches and glacial collapses are not uncommon on the slopes of Mount Kazbek and nearby peaks. The Kolka Glacier collapsed in 1902, surged in 1969, and collapsed again in 2002. Evidence, including historical accounts, indicates similar events have happened in neighboring valleys as well.

After the collapse, people speculated that something called a glacial surge had triggered the Kolka collapse. In 1902, a more significant collapse at Kolka Glacier killed 32 people. Despite a history of disasters there, routine monitoring of the Kolka Glacier cirque ended in the late 1980s.

19 October 2002 **22 May 2003** **11 July 2003** **30 August 2003**

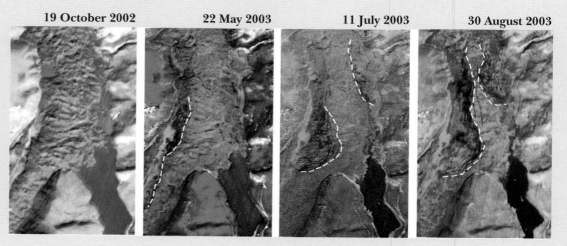

This sequence of images from the Indian Remote Sensing (IRS) satellites showed that the lakes (except Lake Saniba) were draining gradually through crevasses in the ice mass, and were not likely to cause subsequent catastrophic floods. *Credit: IRS*

Credit: Sergey Chernomorets/UNEP

The rapidly rising water was a continuing danger, threatening a sudden outburst that would cause flooding downstream.

Russian researchers evaluated the risk of future danger at the disaster site using a time-series of satellite images collected in the year following the disaster. Satellite imagery was crucial throughout the late fall and winter of 2002 and 2003, when dangerous weather prevented on-site observations of the ice-dammed lakes.

Russian scientists combined satellite data with ground observations to create maps of the Kolka Glacier Cirque. The IRS Satellite image (acquired 11 July 2003) shows details of the cirque, including scars caused by post-collapse rockfall,

a large remnant of the Kolka Glacier, ice cliffs high above the floor of the cirque, displaced porous ice, the Maili Glacier, a temporary lake, and deposits of rubble left along the path of the collapsing glacier.

There is uncertainty also about what triggered the collapse of rocks and hanging glaciers on Mount Dzhimarai-Khokh. Two small earthquakes jarred the region in the months before the collapse, probably destabilized the hanging glaciers. In the first days after the collapse, an Emercom (Russian Emergencies Ministry) crew flew to the site via helicopter, but was forced to evacuate immediately when the crew detected an overpowering smell of sulfur-containing gas. It seems there may be some

fumaroles—volcanic vents—on the face of Mount Dzhimarai-Khokh in the area where the hanging glacier collapsed.

Based on the available data and observations, the scientists say they don't expect any additional catastrophic processes within the next 10 to 20 years. The remaining lakes will likely continue to drain through crevasses and channels being cut through the ice mass, and as they drain, the risk of flooding decreases.

Published 9 September 2004
Source: http://earthobservatory.nasa.gov/Study/Kolka/ kolka.html

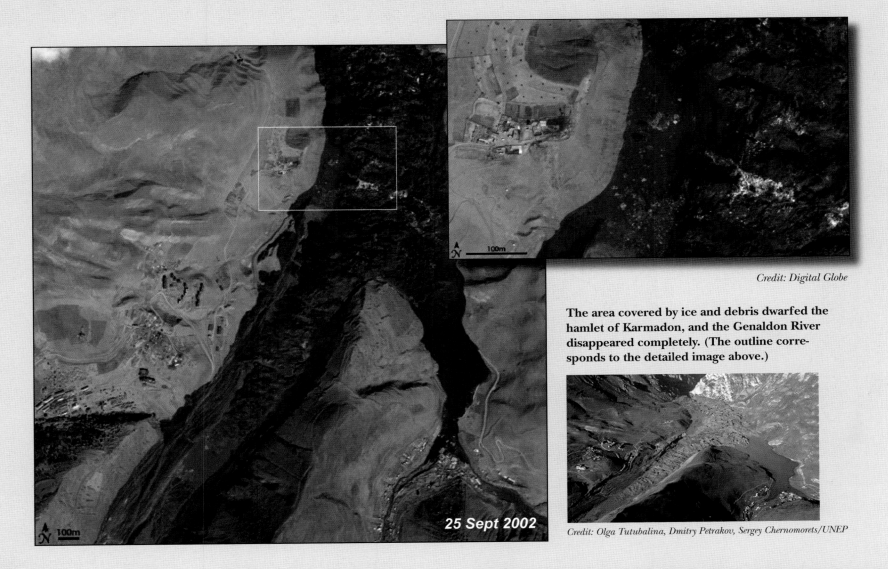

Credit: Digital Globe

The area covered by ice and debris dwarfed the hamlet of Karmadon, and the Genaldon River disappeared completely. (The outline corresponds to the detailed image above.)

Credit: Olga Tutubalina, Dmitry Petrakov, Sergey Chernomorets/UNEP

The edge of the pack ice
Source: Michael Van Woert/UNEP/NOAA

Case Study: Arctic Sea Ice

In September 2003, scientists from the United States and Canada announced that the largest ice shelf in the Arctic had broken up. The Ward Hunt ice shelf to the north of Canada's Ellesmere Island split into two main parts, with other large blocks of ice also pulling away from the main sections.

Evidence continues to emerge that average temperatures in the Arctic are rising even more rapidly than the global average. Satellite data indicate that the rate of surface temperature increase over the last 20 years was eight times the global average over the last 100 years (Comiso 2003).

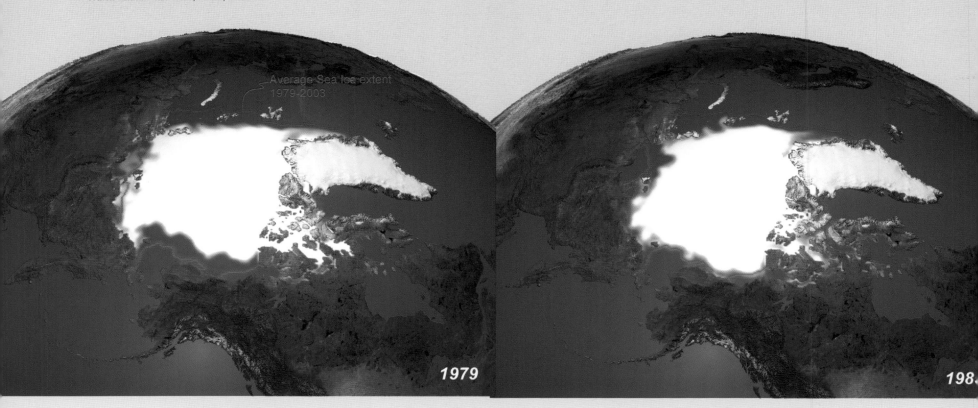

Average Sea Ice extent
1979-2003

1979

198.

Credit: Gyde Lund/UNEP

Studies report that the extent of Arctic sea ice has shrunk by 7.4 per cent over the past 25 years, with record-low coverage in September 2002 (Johannessen et al. 2003). An analysis of 30 years of satellite data suggests that the loss of Arctic sea ice is also accelerating (Cavalieri et al. 2003). There are projections that much of the sea ice, until now thought to be permanent, will melt during the summer by the end of this century if the current trend in global warming continues. This will have major direct impacts on indigenous people and Arctic wildlife such as polar bears and seals, and will also open the region to increased development pressure as access by sea to valuable natural resources becomes easier. The global impacts may also be significant as absorption of solar radiation increases, and could lead to changes in the world ocean circulation (UCL 2003; NASA 2003; Laxon et al. 2003).

Source: GEO Year Book 2003

These images reveal dramatic changes in Arctic sea ice since 1979. The loss of Arctic sea ice may be caused by rising Arctic temperatures that result from greenhouse gas build-up in the atmosphere and resulting global warning.

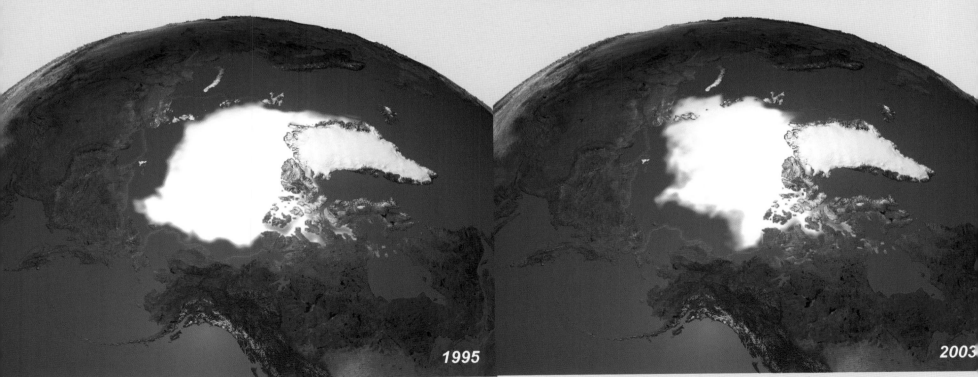

1995

2003

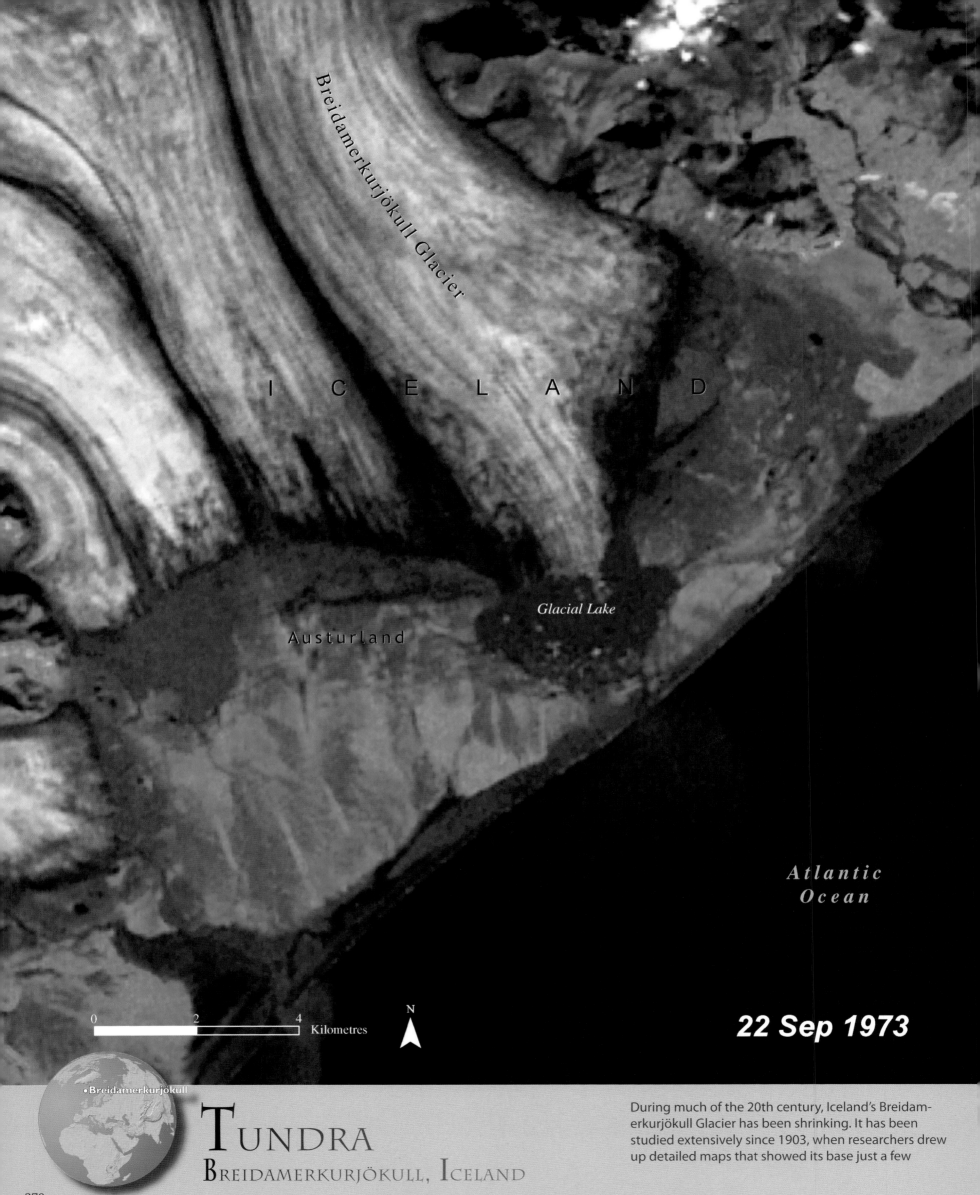

Breidamerkurjökull Glacier

I C E L A N D

Glacial Lake

Austurland

Atlantic
Ocean

0   2   4   Kilometres

N

22 Sep 1973

•Breidamerkurjökull

TUNDRA
BREIDAMERKURJÖKULL, ICELAND

During much of the 20th century, Iceland's Breidamerkurjökull Glacier has been shrinking. It has been
studied extensively since 1903, when researchers drew
up detailed maps that showed its base just a few

Breidamerkurjökull Glacier

ICELAND

Glacial Lake

Austurland

Atlantic
Ocean

23 Dec 2000

0   2   4   Kilometres

N

hundred metres from the ocean edge. Over time, the glacier has receded so that its base is now several kilometers from the coast. As the huge river of ice has pulled back across the Icelandic landscape, thousands of hectares of fertile farmland have been exposed and people are populating the area that was until relatively recently buried under tonnes of ice.

In this pair of satellite images, notice how the glacier has receded and the glacial lake at its tip has enlarged over time. Some researchers attribute the shrinking of Breidamerkurjökull to climate change and global warming. Other scientists maintain that the glacier is simply retreating from the advance it made during the Little Ice Age.

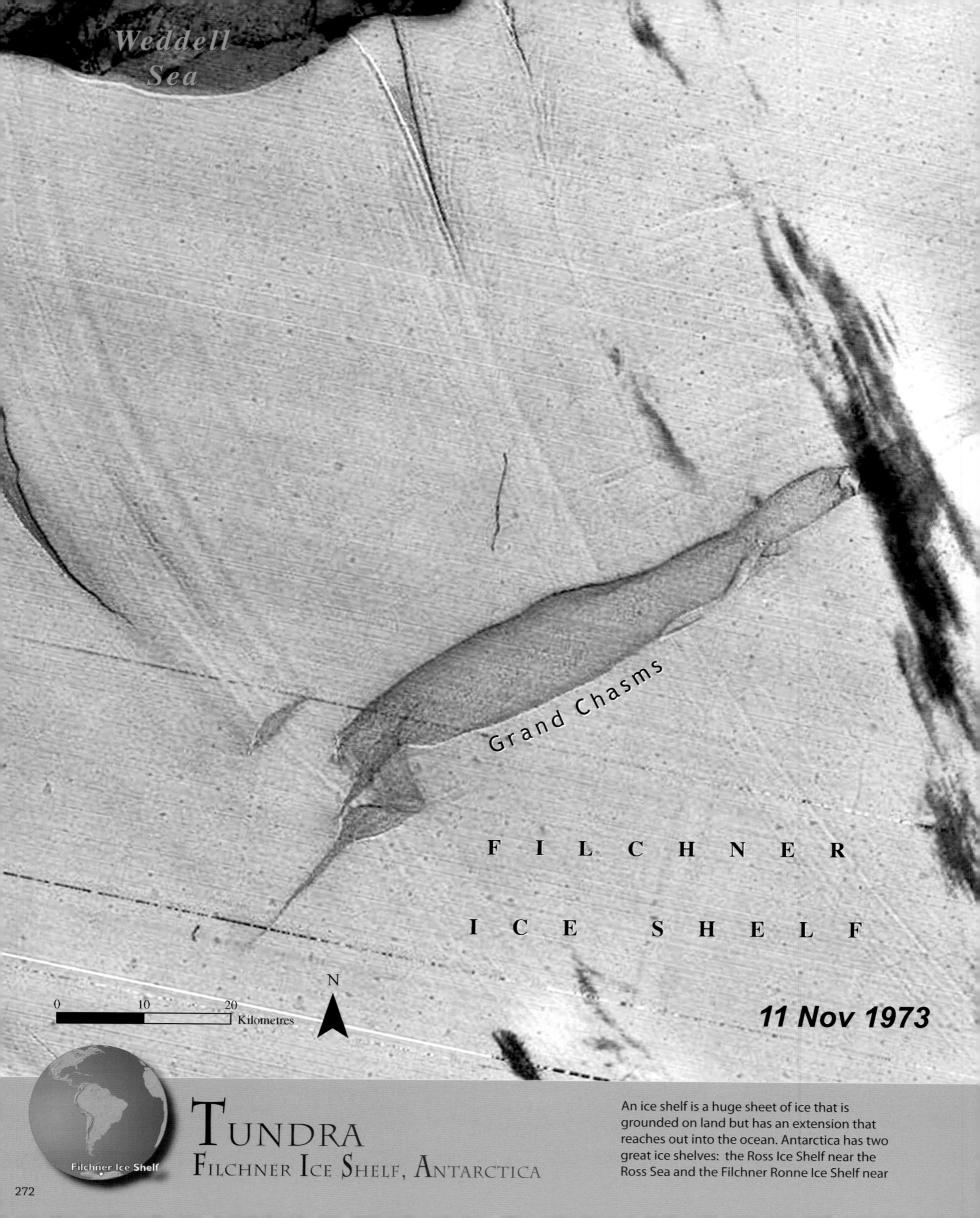

Weddell Sea

Grand Chasms

F I L C H N E R

I C E   S H E L F

11 Nov 1973

0   10   20 Kilometres

N

Tundra
Filchner Ice Shelf, Antarctica

Filchner Ice Shelf

An ice shelf is a huge sheet of ice that is grounded on land but has an extension that reaches out into the ocean. Antarctica has two great ice shelves: the Ross Ice Shelf near the Ross Sea and the Filchner Ronne Ice Shelf near

272

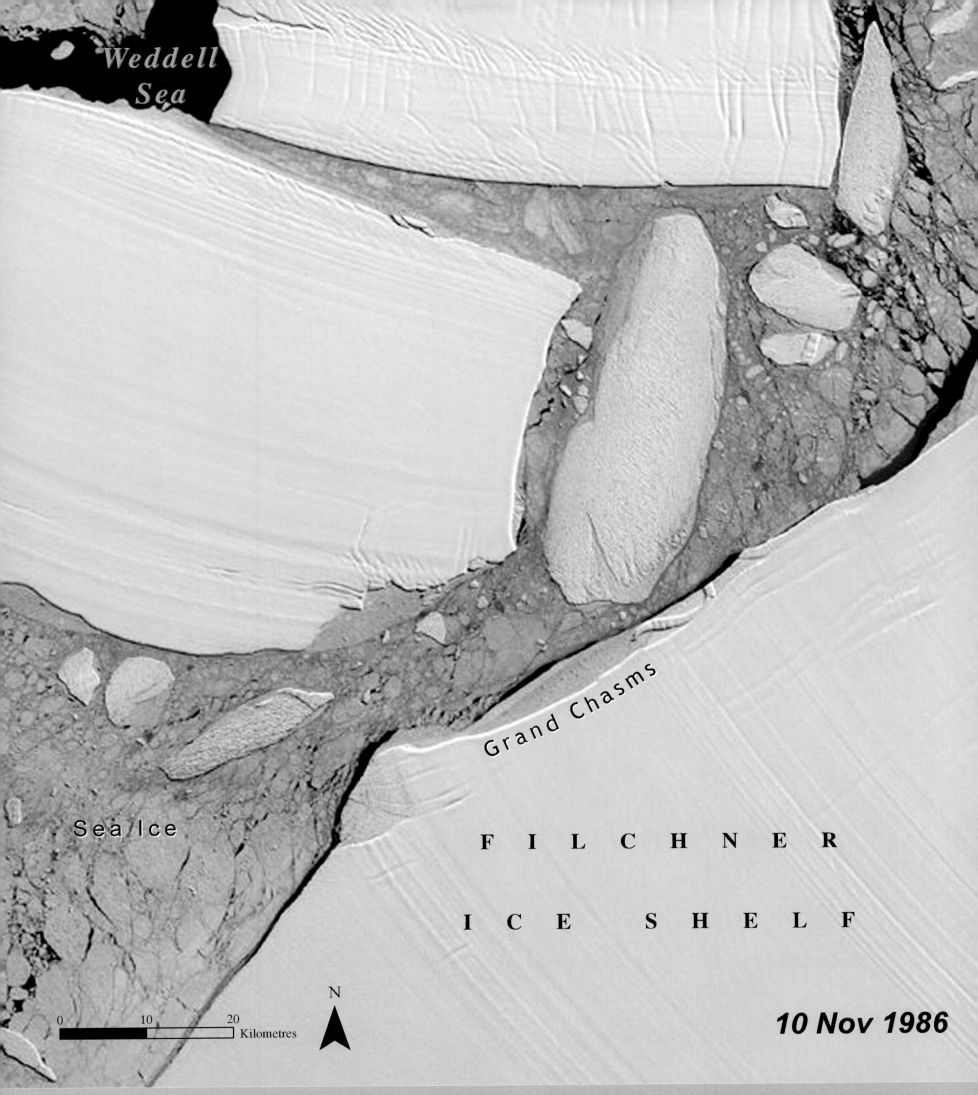

Weddell Sea

Sea Ice

Grand Chasms

F I L C H N E R

I C E   S H E L F

N

0   10   20
Kilometres

10 Nov 1986

the Antarctic Peninsula. By volume, the Filchner Ice Shelf is the largest ice shelf on the planet.

In the austral winter of 1986, the front edge of the Filchner Ice Shelf broke off into the ocean, forming three enormous icebergs named A22, A23, and A24 by glaciologists. Soon after this calving event, all three icebergs ran aground on the shallow sea floor just off

shore. In early 1990, however, A24 broke free and moved out into the open waters of the Weddell Sea, and finally, in 1991, into the southern Atlantic Ocean. In the 1986 image above, notice how the break line created when the icebergs calved has filled with sea ice. Sea ice does not contain the flow lines that are apparent in the glacial ice of the shelf that lies behind the break line.

Valerie
Glacier

Hubbard
Glacier

Disenchantment
Bay

Gilbert
Point

Russell
Fjord

7 Aug 1985

11 Sep 1986

0   2   4
Kilometres

N

14 Aug 2002

TUNDRA
HUBBARD GLACIER, UNITED STATES

•Hubbard Glacier

Hubbard Glacier, located at St. Elian National
Park near Yakutat, Alaska, is the largest calving
glacier in North America. It is currently increasing in total mass and advancing across
the entrance of 56-km-long (35-mile-long)
Russell Fjord.

Valerie Glacier

Hubbard Glacier

Photograph view ▸

Disenchantment Bay

Gilbert Point

Russell Fjord

Credit: *Unknown/UNEP/USGS, USFS, Yakutat Range District and National Park Service, Yakutat Ranger Station*

These photographs show an enlarged eastward-looking view of a small section of the Hubbard Glacier terminus and the evolution of the "squeeze-push" moraine in front of Gilbert Point that blocked the tidal exchange between Disenchantment Bay (bottom of photos) and Russell Fiord (top of photos), creating Russell Lake which rose to 18.6 metres (61 feet) above sea level over 2½ months before it finally outburst on 14 August 2002, creating the second largest glacial lake outburst worldwide in historical times.

0   2   4
Kilometres

N

20 May 2003

These images show the potential environmental disruption that a fast glacial flow is capable of producing. In 1986, the Hubbard Glacier blocked the Russell Fjord, endangering seals and porpoises by producing freshwater runoff that reduced the salinity of that body of water. Rising water levels also became a concern. By the time the ice dam eventually broke later that year, the water level of Russell Fjord had risen by 25 m (82 ft). The images show the Hubbard Glacier and surrounding area at various stages before, during, and after the formation of the ice dam.

Rift Valley

Arusha

Credit: Christian Lambrechts/UNEP/UNEP-GRID Nairobi

T A N Z A N I A

△ *Mt. Kilimanjaro*

Kilimanjaro

0 Kilometre

24 Jan 1976

Arusha

0   5   10
Kilometres

N

Mt. Kilimanjaro

T UNDRA
MT. KILIMANJARO, TANZANIA

Mt. Kilimanjaro, Africa's highest mountain, is located 300 km (186 miles) south of the equator in Tanzania. A forest belt that spans between 1 600 m (5 249 ft) and 3 100 m (10 171 ft) surrounds it. The forest has a rich diversity of eco-

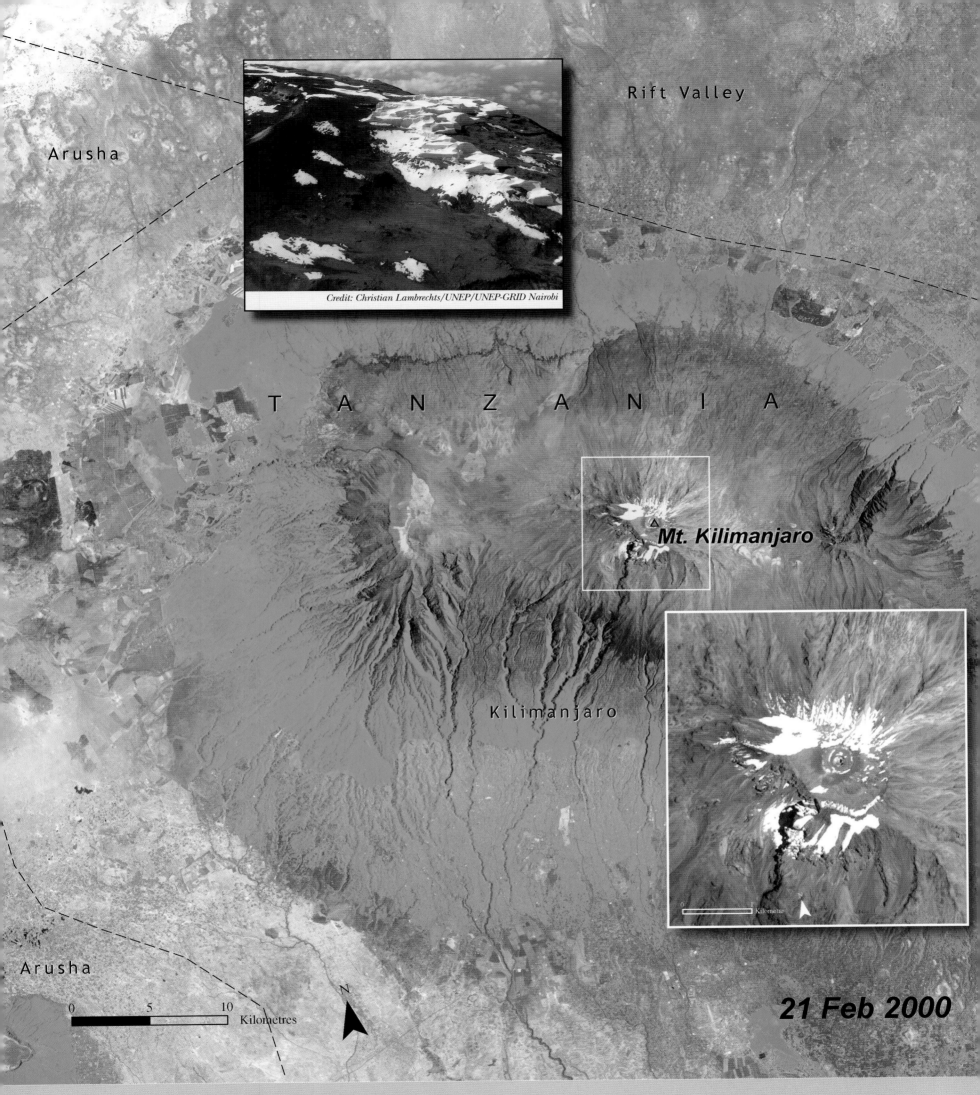

Rift Valley

Arusha

Credit: Christian Lambrechts/UNEP/UNEP-GRID Nairobi

T A N Z A N I A

Mt. Kilimanjaro

Kilimanjaro

Arusha

0   5   10
Kilometres

N

21 Feb 2000

systems, particularly of vegetation types that result mainly from the large range in altitude and rainfall of about 700 to 3 000 mm/yr (28 to 118 in/yr). It hosts a very large diversity of species, with about 140 mammal species and over 900 plant species. But of greater concern are the glaciers atop the mountain. In 1976, glaciers covered most of the summit of Mt. Kilimanjaro. By 2000, the glaciers had receded

alarmingly. An estimated 82 per cent of the icecap that crowned the mountain when it was first thoroughly surveyed in 1912 is now gone, and the remaining ice is thinning as well—by as much as a metre per year in one area. According to some projections, if recession continues at the present rate, the majority of the remaining glaciers on Kilimanjaro could vanish in the next 15 years.

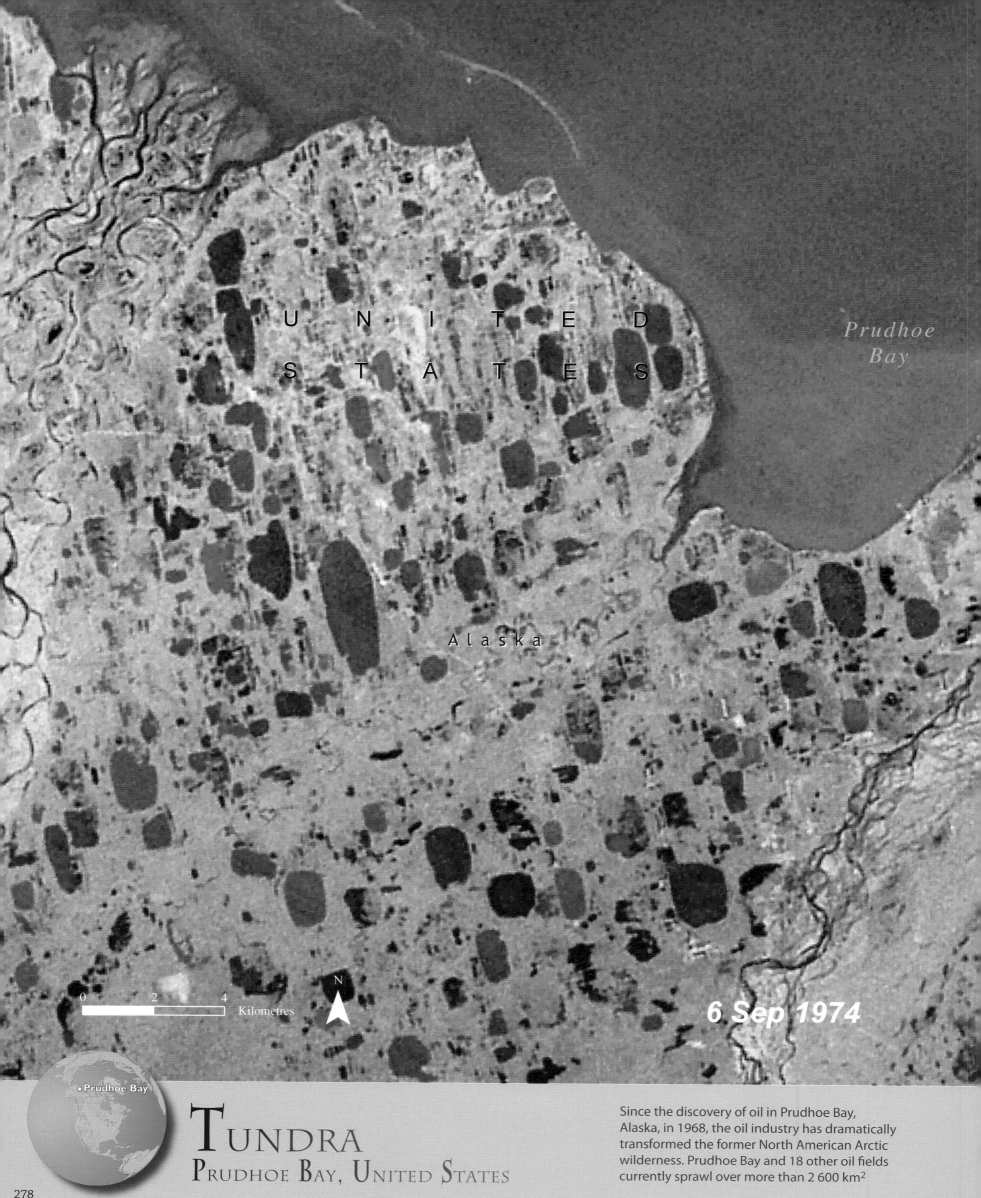

U N I T E D

S T A T E S

Prudhoe Bay

Alaska

6 Sep 1974

0   2   4   Kilometres

N

• Prudhoe Bay

TUNDRA
PRUDHOE BAY, UNITED STATES

Since the discovery of oil in Prudhoe Bay, Alaska, in 1968, the oil industry has dramatically transformed the former North American Arctic wilderness. Prudhoe Bay and 18 other oil fields currently sprawl over more than 2 600 km²

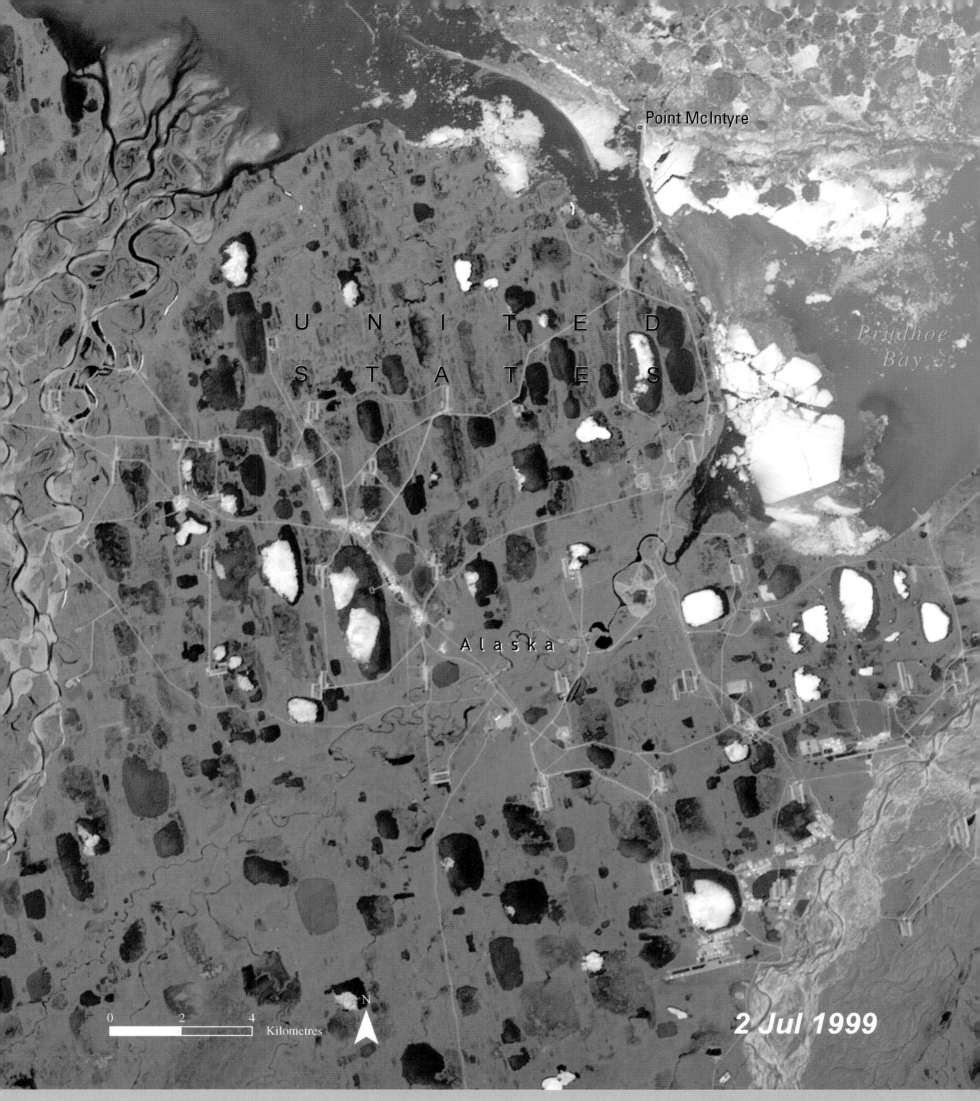

Point McIntyre

U N I T E D

S T A T E S

Prudhoe Bay

A l a s k a

0   2   4   Kilometres

N

2 Jul 1999

(1 004 square miles) in the region. North Slope oil fields now include 3 893 exploratory and producing wells, 170 production and exploratory drill pads, 800 km (497 miles) of roads, 1 769 km (1 099 miles) of trunk and feeder pipelines, two refineries, several airports, and a collection of production and gas processing facilities, seawater treat- ment plants, and power plants. These satellite images document the dramatic changes that the Prudhoe Bay region has undergone over the past 35 years.

References

Allen, J. and Simmon, R. (2003). Drought Lowers Lake Mead, NASA's Earth Observatory http://earthobservatory.nasa.gov/cgi-bin/texis/webinator/printall?/Study/LakeMead/index.html on November 13, 2003.

Anon. (2003). Water for our future: what are the trends? 2003 International Year of Freshwater Newsletter. http://www.wateryear2003.org/en/ev.php@URL_ID=3697&URL_DO=DO_TOPIC&URL_SECTION=201.html on 6 March 2004.

BBC World Service (1998). Cloned dates bear fruit for Iran. http://news.bbc.co.uk/hi/english/world/middle_east/newsid_223000/223814.stm on 28 October 2002.

Beaugrand, G., Brander, K. M., Lindley, J. A., Souissi, S., Reid, P. C. (2003). Plankton effect on cod recruitment in the North Sea, Nature, 426, 661 – 664. http://www.nature.com/cgi-taf/DynaPage.taf?file=/nature/journal/v426/n6967/abs/nature02164_fs.html&dynoptions=doi1079133113 on 6 March 2004.

Brehm, D. (2003). Nobelists sound alarm on global environmental trends. MIT News, 7 March 2003. http://web.mit.edu/newsoffice/nr/2003/nobels.html on 29 March 2004.

Brown, L.R. (2002). World's rangelands deteriorating under mounting pressure, Earth Policy Updates 5 February 2002-2. Earth Policy Institute, Washington, DC, USA. http://www.earth-policy.org/Updates/Update6.htm on 4 March 2004.

Bruinsma, J. (ed) (2003). World agriculture: towards 2015/2030: An FAO perspective. Rome, Italy, Food and Agriculture Organization of the United Nations, 444. http://www.fao.org/docrep/004/y3557e/y3557e00.htm on 31 July 2004.

Bryant, D., Nielsen, D., Tangley, L. (1997). Last Frontier Forests: Ecosystems and Economics on the Edge. World Resources Institute, Washington, DC, USA, 80. http://forests.wri.org/pubs_content.cfm?PubID=2619 on 24 February 2004.

Bryant, D., Rodenberg, E., Cox, T., Nielson, D. (1995). Coastlines at risk: An index of potential development-related threats to coastal ecosystems. WRI Indicator Brief, World Resources Institute, Washington, DC, USA.

Bryant, D., Burke, L., McManus, J. W., Spalding, M. (1998). Reefs at risk: A map-based indicator of threats to the world's coral reefs. World Resources Institute, Washington, DC, USA, 57. http://www.wri.org/wri/reefsatrisk/reefrisk.html on 12 March 2004.

Burke, L., Kura, Y., Kassem, K., Revenga, C., Spalding, M., McAllister, D. (2000). Pilot analysis of global ecosystems (PAGE): Coastal ecosystems. World Resources Institute, Washington DC, USA. http://www.wri.org/wr2000/coast_page.html on 11 March 2004.

Cavalieri, D.J., Parkinson, C.L. and Vinnikov, K.Y. (2003). 30-Year satellite record reveals contrasting Arctic and Antarctic decadal sea ice variability. Geophysical Research Letter, 30, 10.

Centre for Atmospheric Science (n.d.). The Ozone Hole Tour, Cambridge University, U.K. http://www.atm.ch.cam.ac.uk/tour/atmosphere.html on 31 March 2004.

Center for Environmental Resource Management (n.d.). What is a wetland? El Paso, TX: Center for Environmental Resource Management, University of Texas at El Paso, USA. www.cerm.utep.edu/riobosque/Education/Middle/What%20is%20a%20Wetland.htm on 28 March 2004.

Choudhury, J.K. (1997). Sustainable management of coastal mangrove forest development and social needs. World Forestry Congress. Antalya, Turkey, 13 to 22 October 1997. Volume 6, Topic 38.6. Food and Agriculture Organization of the United Nations, Rome, Italy. http://www.fao.org/montes/foda/wforcong/PUBLI/V6/T386E/1.HTM#TOP on 11 March 2004.

Christensen, T. R. (n.d.). Carbon cycling and methane emissions from wet tundra ecosystems: feedback effects on climate change, 3. http://www.ulapland.fi/home/arktinen/feed_pdf/Christensen-t.pdf on 11 July 2004.

CITEPA (1994). Air Emissions In Mainland France and Overseas Departments. Air Pollution Paris, France 2004. http://www.airparif.asso.fr/english/polluants/default.htm on 6 June 2004; Healthier Environment through the Abatement of Vehicle Emissions and Noise. http://heaven.rec.org on 6 June 2004.

Cohen, J. E., Small, C., Mellinger, A., Gallup, J., Sachs, J.D. (1997). Estimates of coastal populations, Science 278, 1209-1213. http://www.sciencemag.org/cgi/content/full/278/5341/1209c on 11 March 2004.

Comiso, J.C. (2003). Warming Trends in the Arctic from Clear Sky Satellite Observations. Journal of Climate, 16, 21, 3498-3510.

DETR (1997). Climate Change and its impacts: a global perspective. Some recent results from the UK Research Programme, United Kingdom Department of Environment, Transport and Regions: Met Office / Hadley Centre for Climate Predictions and Research. http://www.met-office.gov.uk/research/hadleycentre/pubs/brochures/B1997/index-nf.html on 31 March 2004.

Downing, R. J., Ramankutty, R., Shah, J. (1997). Rains-Asia: An assessment model for acid deposition in Asia. World Bank, Washington DC, United States.

EEA (2004). EEA multilingual environmental glossary. Copenhagen, Denmark: European Environment Agency. http://glossary.eea.eu.int/EEAGlossary on 5 April 2004.

EEA (n.d.). General Multilingual Environmental Thesaurus. European Environment Agency: European Topic Centre on Catalogue of Data Sources. http://oaspub.epa.gov/trs/trs_proc_qry.org_info?P_REG_AUTH_ID=1&P_DATA_ID=11660&p_version=1&p_list_option_cd=INFO on 11 July 2004.

FAO (1982). Date production and protection: with special reference to North Africa and the Near East. FAO, Rome, Italy.

FAO (1993). Date Palm Products. FAO, Rome, Italy.

FAO (1995). Evaluation of Food and Nutrition Situation in Iraq. FAO, Rome, Italy.

FAO (1999). Date palm cultivation. FAO, Rome, Italy.

FAO (2001). FAOSTAT (CD-ROM). FAO, Rome, Italy.FAO (1996). Food for all. United Nations Food and Agriculture Organization, Rome, Italy, 64.

FAO (2001). Forest Resources Assessment 2000. Main Report FAO Forestry Paper 140. Food and Agriculture Organization of the United Nations, Rome, Italy. http://www.fao.org/forestry/foris/webview/forestry2/index.jsp?siteId=101&langId=1 on 25 February 2004.

FAO (2002). Annex 5 - Guidelines for preparation of a Country report: State of land, water and plant nutrition resources. Land Resources Information Systems in the Near East. Regional Workshop. Cairo, 3-7 September 2001. Food and Agriculture Organization of the United Nations, Rome, Italy. http://www.fao.org/DOCREP/005/Y4357E/y4357e20.htm on 9 July 2004.

FAO (2003). Unlocking the water potential of agriculture. Food and Agriculture Organization of the United Nations, Rome, Italy. http://www.fao.org/DOCREP/006/Y4525E/Y4525E00.HTM

FAO (2005). Review of the State of World Marine Fishery Resources, FAO Fisheries Technical Paper. No. 457. Food and Agriculture Organizaton of the United Nations, Rome, Italy. ftp://ftp.fao.org/docrep/fao/007/y5852e/y5852e00.pdf on 2 April 2005.

Ghareyazie, B. (2000)."Iran: Hopes, Achievements, and Constraints in Agricultural Biotechnology" in Agricultural Biotechnology and the Poor, edited by G. Persley and M. Lantin, Consultative Group on International Agricultural Research and US National Academy of Sciences, the World Bank, Washington D.C., USA.

Girardet, H. (1995). Sustainability in the age of the city, Countdown to Habitat - 5, November 1995, 6-8. http://prog2000.casaccia.enea.it/nuovo/documenti/625.PDF on 6 March 2004.

González, E. R. and Elizabeth C. Gordon (2004). Lemna in the Lake of Maracaibo, Venezuela. Central University of Venezuela, Faculty of Science, Venezuela.

Hanley, C. J. (2004). Carbon dioxide reported at record levels. Associated Press. http://www.newsmax.com/archives/articles/2004/3/21/170709.shtml on 22 March 2004.

Harrison, P. and Pearce, F. (2001). AAAS Atlas of Population and Environment. American Association for the Advancement of Science, Washington, DC, USA, 215. http://atlas.aaas.org/ on 2 March 2004.

Health Canada (2003). Clean air champions - "getting active for cleaner air" - education resource kit. Health Canada, Ottawa, Ontario, Canada. http://www.hc-sc.gc.ca/hecs-sesc/air_quality/publications/clean_air_champions/toc.htm on 6 April 2004.

Hussain, A. (1974). Date Palms and Dates with Their Pests in Iraq. University of Baghdad, Iraq.

IPCC (2001). Houghton, J.T., Ding, Y., Griggs, D.J., Noguer, M., van der Linden, P.J., Dai, X., Maskell, K., and Johnson, C.A. (eds.): Climate Change 2001: The Scientific Basis. Contribution of Working Group I to the Third Assessment Report of the Intergovernmental Panel on Climate Change, Cambridge University Press, Cambridge, United Kingdom and New York, NY, USA, 881, 944. http://www.grida.no/climate/ipcc_tar/wg1/index.htm on 29 March 2004;
http://www.ipcc.ch/present/graphics/2001syr/ppt/00.01.ppt on 29 March 2004;

http://www.ipcc.ch/present/graphics/2001syr/ppt/02.01.ppt on 29 March 2004;

http://www.ipcc.ch/present/graphics/2001syr/ppt/06.01.ppt on 29 March 2004;

http://www.ipcc.ch/present/graphics/2001syr/ppt/04.02.ppt on 29 March 2004;

http://www.ipcc.ch/present/graphics/2001syr/ppt/05.22.ppt on 29 March 2004;

http://www.ipcc.ch/present/graphics/2001wg1/ppt/02.02.ppt on 29 March 2004;

http://www.ipcc.ch/present/graphics/2001wg3/ppt/03.02.ppt on 29 March 2004.

Jezek, K. C. (1995). Understanding the role of earth's cold regions in the global system. Organizational Profile, Byrd Polar Research Center. Geoscience and Remote Sensing Society Newsletter, 2. http://www.ewh.ieee.org/soc/grss/newsletter/Jun95prf.pdf on 8 March 2004.

Johannessen, O.M., Bengtsson, L., Miles, M., Kuzmina, S.I., Semenov, V., Alekseev, G.V., Nagumy, A.P., Zakharov, V.F., Bobylev, L.P., Pettersson, L.H., Hasselmann, K., and Cattle, H.P. (2003). Artic Climate Change – Observed and Modelled Temperature and Sea Ice, Technical Report No. 218, Bergen, Nansen Environmental and Remote Sensing Centre. http://www.nersc.no/AICSEX/rep218.pdf on 4 March 2005.

Kendall, H. W. and Pimentel, D. (1994). Constraints on the expansion of the global food supply. Ambio 23(3): 198-205. http://www.css.cornell.edu/courses/190/abstr/mahoney1.htm on 2 March 2004.

Kleypas, J.A., Buddemeier, R.W., Archer, D., Gattuso, J.P., Langdon, C., Opdyke, B.N. (1999). Geochemical consequences of increased atmospheric carbon dioxide on coral reefs. Science 284 (5411): 118-120. http://www.sciencemag.org/cgi/content/full/284/5411/118 on 13 March 2004.

Landolt, E. (1975). Morphological Differentiation and Geographical Distribution of the Lemna gibba-Lemna minor Group. Aquatic Botany 1: 345-363.

Landolt, E. (1986). The Family of Lemnaceae: A Monographic Study. Veroff. Geobot. Institute ETH, Stiftung Rubel, Volume 1, 71.

Larssen, S. (ed.) (2004). Air pollution in Europe 1990–2000. Topic report 4/2003, European Environment Agency, Copenhagen, Denmark, 77. http://reports.eea.eu.int/topic_report_2003_4/en/Topic_4_2003_web.pdf on 5 April 2004.

Laurance, W. F., Oliveira, A. A., Laurance, S. G., Condit, R., Nascimento, H. E. M., Sanchez-Thorin, A. C., Lovejoy, T. E., Andrade, A., D'Angelo, S., Ribeiro, J. E., Dick, C. W. (2004). Pervasive alteration of tree communities in undisturbed Amazonian forests, Letters to Nature, Nature 428 (6970), 171-175. http://www.heatisonline.org/contentserver/objecthandlers/index.cfm?id=4613&method=full on7 March 2005.

Laxon, S., Peacock, N., and Smith, D. (2003). High interannual variability of sea ice thickness in the Arctic region. Nature, 425, 947-950.

Lindsey, R., Simmon, R., Tutubalina,O., Petrakov, D., Chernomorets, S. (2004). Collapse of the Kolka Glacier, 9 September 2004. http://earthobservatory.nasa.gov/Study/Kolka/ on 5 December 2004.

Lund, H. G. and Iremonger, S. (1998). Omissions, commissions, and decisions - the need for integrated resource assessments. Proceedings from First International Conference on Geospatial Information in Agriculture and Forestry. Decision Support, Technology, and Applications. 1-3 June 1998, Lake Buena Vista, Florida and Ann Arbor, Michigan, USA, ERIM International, Inc. Volume I: 182-189.

Lund, H. G. (2002). When is a forest not a forest? Journal of Forestry 100(8): 21-27.

Matthews, E. (1983). Global vegetation and land-use: new high-resolution databases for climatic studies, Journal of Climate and Applied Meteorology 22, 474-487. http://www.geog.ouc.bc.ca/conted/onlinecourses/geog_210/210_6_1.html on 22 September 2004.

Myers, N. and Kent, J. (1995). Environmental exodus: An emergent crisis in the global arena. Climate Institute, Washington DC, USA.

Myers, R. A., Worm, B. (2003). Rapid worldwide depletion of predatory fish communities, Nature 423 (6937), May 2003, 280 – 283. http://www.mindfully.org/Water/2003/Predatory-Fish-Depletion15may03.htm on 7 March 2005.

NASA (1989). Upper Atmosphere Research Satellite: A program to study global ozone change. National Aeronautics and Space Administration, Washington, DC, USA. http://www.ciesin.org/docs/011-489/011-489.html on 15 March 2004.

NASA (2002). Goddard DAAC. Creeping Deadzones. http://daac.gsfc.nasa.gov/CAMPAIGN_DOCS/OCDST/dead_zones.html on 1 April 2005.

NASA (2002). Rodeo-Chediski Fire. http://earthobservatory.nasa.gov/Newsroom/NewImages/images.php3?img_id=9761 on 23 September 2004.

NASA (2003a). The Human Footprint, Scott Michon. http://earthobservatory.nasa.gov/Study/footprint/Images/fullres_footprint.gif on 4 April 2004; http://earthobservatory.nasa.gov/Study/footprint/ on 4 April 2004.

NASA (2003b). Yearly Arctic Temperature Anomaly, 16 October 2003. http://svs.gsfc.nasa.gov/vis/a000000/a002800/a002830/ on 18 June 2004.

NASA (2003c). European Heat Wave, July 2003. http://earthobservatory.nasa.gov/NaturalHazards/natural_hazards_v2.php3?img_id=11166 on 6 June 2004.

NASA (2003d). Dwindling Artic Ice. http://earthobservatory.nasa.gov/Study/ArcticIce/ on November 9, 2004

NASA (2004a). NASA Plans To Put An Aura Around The Earth. http://www.gsfc.nasa.gov/topstory/2004/0517aura.html on 6 June 2004; http://www.gsfc.nasa.gov/gsfc/earth/pictures/2003/0925ozonehole/still_hires_24Sept2003.tif on 6 June 2004; http://www.gsfc.nasa.gov/ftp/pub/ozone/ozone_still_2000_09_06.tif on 6 June 2004.

NASA (2004b). SeaWiFS: NASA Carbon Cycle Initiative, NASA/Goddard Space Flight Center Scientific Visualization Studio Scripps Institution of Oceanography (SIO). http://svs.gsfc.nasa.gov/vis/a000000/a002100/a002195/ on 15 June 2004.

NASA (2004c). Aerosols and Climate Change: Hardin, M. and Kahn, R. http://earthobservatory.nasa.gov/Library/Aerosols/ on 18 June 2004.

NASA (2004d). Terra/MOPITT CO: Full Earth NASA/Goddard Space Flight Center Scientific Visualization Studio. http://svs.gsfc.nasa.gov/vis/a000000/a002100/a002150/ on 18 June 2004.

NASA (2004e). NASA's Terra Satellite Tracks Global Pollution. http://www.nasa.gov/centers/goddard/news/topstory/2004/0517mopitt.html on 18 June 2004.

NASA (2004f). African fires during 2002. http://svs.gsfc.nasa.gov/vis/a000000/a002800/a002890/index.html on 16 December 2004.

NASA (2004g). Pollution over China and China Dust Storm Pollutes Air in the Eastern United States in April 2001 (Flatmap), NASA/Goddard Space Flight Center Scientific Visualization Studio. http://svs.gsfc.nasa.gov/vis/a000000/a002900/a002957/index.html; http://earthobservatory.nasa.gov/NaturalHazards/shownh.php3?img_id=11982 on 18 June 2004.

NASA (2004h). Mississippi Dead Zone. http://www1.nasa.gov/vision/earth/environment/dead_zone.html on 13 August 2004.

NASA (2004i). Fires in Para, Brazil. http://earthobservatory.nasa.gov/NaturalHazards/natural_hazards_v2.php3?img_id=12613 on 8 January 2005.

Deforestation and fires in Para, Brazil, http://earthobservatory.nasa.gov/Newsroom/NewImages/images.php3?img_id=5107 on 8 January 2004.

NASA (2004j). Retreat of the Gangotri Glacier. http://earthobservatory.nasa.gov/Newsroom/NewImages/images.php3?img_id=16584 on 18 February 2005.

NCAR and UCAR (2005). The National Center for Atmospheric Research and The University Corporation for Atmospheric Research Office of Programs, Pollutants in the atmosphere. http://www.ucar.edu/communications/st on 7 April 2005.

NCPPR (n.d.). Myth #7: 'Humanity is causing earth's polar regions to warm quickly resulting in unusual rates of ice melting.' Envirotruth. National Center for Public Policy Research, Washington DC, USA. http://www.envirotruth.org/myth7.cfm on 8 March 2004.

Nicholls, R.J., Hoozemans, F.M.J., Marchand, M. (1999). Increasing flood risk and wetland losses due to global sea-level rise: regional and global analyses. Global Environmental Change - Human and Policy Dimensions, 9, S69-S87. http://www.sciencedirect.com/science?_ob=ArticleURL&_udi=B6VFV-3XR2V33-6&_user=10&_handle=B-WA-A-A-D-MsSAYVA-UUA-AUYACWUVBC-AUDEAUAWBC-WEWDVUDCV-D-U&_fmt=summary&_coverDate=10%2F31%2F1999&_rdoc=5&_orig=browse&_srch=%23toc%236020%231999%23999999999.8998%23137271!&_cdi=6020&view=c&_acct=C000050221&_version=1&_urlVersion=0&_userid=10&md5=f8d6c6f54f6270a67d203f819dc25003 on 11 March 2004.

Nippon Koei Co. Ltd. (1972). Study Report on the Shatt Al-Arab Project: Iraq. Prepared by Nippon Koei Consulting Engineers for the Ministry of Agrarian Reform, Government of the Republic of Iraq.

NRDC (2004). Global warming puts the Arctic on thin ice. Natural Resources Defense Council, New York, USA. http://www.nrdc.org/globalWarming/qthinice.asp on 8 March 2004.

NSIDC (n.d.) The National Snow and Ice Data Center: State of the Cryosphere: Global Sea Ice Extent and Concentration: What sensors on satellites are telling us about sea ice. http://nsidc.org/sotc/sea_ice.html on 23 September 2004.

NOAA (n.d.) NOAA Coasts: http://www.noaa.gov/coasts.html on 8 November 2004.

Northoff, E. (2003). Fires are increasingly damaging the world's forests. Food and Agriculture Organization of the United Nations, Rome, Italy. http://www.fao.org/english/newsroom/news/2003/21962-en.html on 24 March 2004.

NSW EPA (2003). Water pollution and the freshwater crisis. New South Wales Environmental Protection Authority. http://www.epa.nsw.gov.au/stormwater/hsieteachguide/waterpoln.htm on 11 April 2004.

Paden, M. (2000). New vision for world water. Human Nature 5(1): 1-2. http://www.greencom.org/greencom/pdf/hnenjul00.pdf on 9 March 2004.

Pearce, F. (1999). Coral grief. Our Planet 10(3). http://www.ourplanet.com/ on 12 March 2004.

Phillips, S. (1995). The atmosphere. Liftoff. NASA. http://liftoff.msfc.nasa.gov/academy/space/atmosphere.html on 13 July 2004.

Polley, H.W., Johnson, H.B., Tischler, C.R. (2003). Woody invasion of grasslands: evidence that CO^2 enrichment indirectly promotes establishment of Prosopis glandulosa. Plant Ecology 164, 85-94. http://www.kluweronline.com/issn/1385-0237/current and http://www.nal.usda.gov/ttic/tektran/data/000010/88/0000108835.html on 4 March 2004.

Pomerance, R. (1999). Coral bleaching, coral mortality, and global climate change. Report. U.S. Department of State, Bureau of Oceans and International Environmental and Scientific Affairs, Washington, DC, USA. http://www.state.gov/www/global/global_issues/coral_reefs/990305_coralreef_rpt.html on 12 March 2004.

Postel, S.L., Daily, G.C., Ehrlich, P.R. (1996). Human appropriation of renewable fresh water. Science, 271:785-788.

Pullen, S. (1996). The Tundra Biome. Biomes Group of the Fall 96 Biology 1B Class, Section 115, University of California, Berkley, California, USA. http://www.ucmp.berkeley.edu/glossary/gloss5/biome/tundra.html

Ramsar Convention Bureau (1998). What are wetlands? Ramsar Information Paper No. 1. Ramsar Convention Secretariat, Gland, Switzerland. http://www.ramsar.org/about_infopack_1e.htm on 12 July 2004.

Remote Sensing Solutions GmbH. (1998). Fires – A global view. ESRIN/ESA (Contract No.: 13154/98/I/WE). http://www.rssgmbh.de/ESA%20Fire/Intro/global_1.html on 25 March 2004.

Revenga, C., Brunner, J., Henninger, N., Kassem, K., Payne, R. (2000). Pilot analysis of global ecosystems: freshwater systems. World Resources Institute Washington, DC, USA. http://www.wri.org/wr2000/freshwater_page.html on 11 April 2004.

Revenga, C. and Mock, G. (2000). Dirty water: pollution problems persist. Pilot Analysis of Global Ecosystems: Freshwater Systems. World Resources Institute, Washington, DC, USA. http://earthtrends.wri.org/pdf_library/features/wat_fea_dirty.pdf on 11 April 2004.

Rivero, Y. (2004). Water Lentils on the Offensive. TierraAmerica, 5/29: Acentos.

Rosegrant, M. W., Paisner, M. S., Meijer, S.T., Witcover, J. (2001). Global food projections to 2020: Emerging trends and alternative futures. International Food Policy Research Institute, Washington, DC, USA, 206. http://www.ifpri.org/pubs/books/globalfoodprojections2020.htm on 6 March 2004.

Schmidt, L.J. (2000). Disintegration of the Glacier Tongue: NASA Earth Science Enterprise Data Centers, 1 December 2000. http://earthobservatory.nasa.gov/Study/Ninnis/ on 26 February 2005.

Schowengerdt, R. A. (1997). Remote Sensing Models and Methods For Image Processing, Second Edition. Academic Press, San Diego, California, USA.

Schwela, D. (1995). Public health implications of urban air pollution in developing countries. Proceedings of the 10th World Clean Air Congress, Espoo, Finland, 28 May - 2 June 1995, 1, World Health Organization, Geneva, Switzerland.

SIDA (n.d.). Swedish International Development Cooperation Agency: Oceans and coastal zones ecosystems of vital importance to the human population of the world. http://www.sida.se/Sida/jsp/polopoly.jsp?d=3821 on 8 November 2004.

Shah, A. (2002). The ozone layer and climate change. Global issues. http://www.globalissues.org/EnvIssues/GlobalWarming/Ozone.asp on 13 July 2004.

Shanklin, Jonathan (n.d.). http://www.theozonehole.com/fact.htm on 4 April 2004.

Sherwood, K. and Idso, C. (2003). On Assessing Surface Air Temperature Trends. Editorial. CO2 Science Magazine 6(23) 4 June 2003. http://www.co2science.org/edit/v6_edit/v6n23edit.htm on 29 March 2004.

Sohngen, B., R. Mendelsohn, R. S. (1999). Forest management, conservation, and global timber markets, Abstract, American Journal of Agricultural Economics, 81(1): 1-10. http://www.acs.ohio-state.edu/units/research/archive/timber.htm on 24 February 2004.

Soon, W., Baliunas, S., Idso, C., Idso, S., Legates, D.R. (2003). Reconstructing climatic and environmental changes of the past 1000 years: A reappraisal. Energy & Environment 14: 233-296. http://www.kolumbus.fi/boris.winterhalter/EnEpreprintFeb03.pdf on 29 March 2004.

Soussa, A. (1944). Iraq Irrigation Handbook. Directorate General of Irrigation, Baghdad, Iraq.

SPRI (n.d.). An appeal. The University of Cambridge, Scott Polar Research Institute, UK, 6. http://www.spri.cam.ac.uk/about/appeal/brochure.pdf on 9 March 2004.

SRM (n.d.). Biodiversity of rangelands – An issue paper created by the Society for Range Management, Society for Range Management, Lakewood, Colorado, USA, 6. http://www.rangelands.org/pdf/Biodiversity_Issue_Paper_FA.pdf on 4 March 2004.

Stoddard J. L., Jeffries, D. S., Lükewille, A., Clair, T. A., Dillon, P. J., Driscoll, C. T., Forsius, M., Johannessen, M., Kahl, J. S., Kellogg, J. H., Kemp, A., Mannio, J., Monteith, D. T., Murdoch, P. S., Patrick, S., Rebsdor, A., B. Skjelkvåle, L., Stainton, M. P., Traaen, T., Van Dam, H., Webster, K. E., Wieting, J., Wilander, A. (1999). Regional trends in aquatic recovery from acidification in North America and Europe. Nature 401:575-578. http://www.nature.com/cgi-taf/DynaPage.taf?file=/nature/journal/v401/n6753/abs/401575a0_fs.html on 4 April 2004.

The Wilderness Society (2002). Summary of the Rodeo-Chediski Fire, Arizona http://www.wilderness.org/Library/Documents/WildfireSummary_RodeoChediski.cfm on 16 December 2004.

Thompson, O. E. (1996). Ozone in Earth's atmosphere. University of Maryland Department of Meteorology, College Park, Maryland, USA. http://www.atmos.umd.edu/~owen/CHPI/IMAGES/ozonefig1.html on 13 July 2004.

UCL (2003). Polar bears' habitat threatened by thinning of Arctic sea ice. Press Release. London, University College London, UK. http://www.ucl.ac.uk/media/archive/archive-release/?arctic on 5 March 2005.

UN (2002). United Nations Atlas of the Oceans. United Nations, Division for Ocean Affairs and the Law of the Sea, New York, USA. http://www.oceansatlas.org/index.jsp on 12 March 2004.

UN-HABITAT (2005). Brochure on Istanbul +5: Urban Environment Waste. United Nations Human Settlement Programme. http://www.unchs.org/Istanbul+5/70.pdf on 3 April 2005.

UNEP (1999a). Taking Action – An Environmental Guide for You and Y our Community. http://www.nyo.unep.org/action/09.htm on 4 April 2004.

UNEP (1999b). Global Environment Outlook 2000. United Nations Environment Programme, Earthscan, London, UK. http://www.grida.no/geo2000/index.htm on 5 March 2004.

UNEP (2002a). Climate change and waste: gas emission from waste disposal and E-waste: the great e-waste recycling debate. http://www.grid.unep.ch/waste/html_file/42-43_climate_change.html and http://www.grid.unep.ch/waste/html_file/36-37_ewaste.html on 6 June 2004.

UNEP (2002b). Global Environment Outlook 3 (GEO3) – Past, present and future perspectives, Earthscan, London, UK, 446. http://www.unep.org/geo/geo3/ on 4 March 2004.

UNEP-WCMC (2003). Global marine assessments - a survey of global and regional marine environmental assessments and related scientific activities. United Nations Environment Programme - World Conservation Monitoring Center, Cambridge, UK. http://valhalla.unep-wcmc.org/resources/publications/gma/ on 12 March 2004.

UNESCO (2000). The world's water, is there enough? United Nations Educational, Scientific and Cultural Organization, Paris, France. http://www.unesco.org/science/waterday2000/Brochure.htm#Up on 11 April 2004.

UNESCO (2003). Water for our future: what are the trends? United Nations Educational, Scientific and Cultural Organization, Paris, France. http://www.wateryear2003.org/en/ev.php@URL_ID=3697&URL_DO=DO_TOPIC&URL_SECTION=201.html on 9 March 2004.

UNFCCC (2000). National Communications from Parties included in Annex I to the Convention: Greenhouse Gas Inventory Data from 1990 to 1998. FCCC/SBI/2000/11, 5 September 2000. United Nations Framework Convention on Climate Change. http://unfccc.int/resource/docs/2000/sbi/11.pdf on 29 March 2004.

UNFPA (2001). The State of World Population 2001 - footprints and milestones: population and environmental change. United Nations Population Fund, New York, USA. http://www.unfpa.org/swp/2001/english/index.html on 9 March 2004.

UPI (2003). Sea ice changing differently at each pole. United Press International: Washington Time, Washington, DC, USA, 11 November 2003. http://washingtontimes.com/upi-breaking/20031111-031021-4917r.htm on 9 March 2004.

USCCSP (2003). Strategic plan for the climate change science program. Final Report. 202. U.S. Climate Change Science Program, Washington, DC, USA. http://www.climatescience.gov/Library/stratplan2003/final/ccsp-stratplan2003-all.pdf on 29 March 2004.

USDA Forest Service (2002). United States Geological Survey, EROS Data Center in collaboration with United States Forest Service. Hayman Fire & BAER Information. http://www.fs.fed.us/r2/psicc/fire/hayman/ on 31 December 2002.

USGS (2002). Rodeo-Chediski Fires – During, http://landsat.usgs.gov/gallery/detail/358/ on 8 July 2004.

USGS (1999). Coastal-Change and Glaciological Maps of Antarctica. http://pubs.usgs.gov/fs/fs50-98/ on 13 December 2004.

USGS (2003). Coral mortality and African dust. U.S. Department of Interior, Geological Survey, Washington, DC, USA. http://coastal.er.usgs.gov/african_dust/ on 12 March 2004.

UTEXAS (n.d.). University of Texas at Austin Library, Austin, Texas, USA. http://www.lib.utexas.edu/maps/islands_oceans_poles/arctic_region_pol02.jpg on 11 November 2004.

Wall, R. (2001). Current Issues Affecting World Water Supply. Know Your Environment Academy of Natural Sciences, Philadelphia, Pennsylvania, USA. http://www.acnatsci.org/research/kye/kye42001.html#h1 on 9 March 2004.

Warah, R. (2003). The Challenge of Slums: Global Report on Human Settlements 2003. United Nations Habitat Settlement Programme UN-HABITAT, 301. http://www.globalpolicy.org/socecon/develop/africa/2003/03slums.htm on 1 August 2004.

Weld, M. (1995). Losing the race - population outpacing food. Humanist in Canada 28(1):27. http://216.239.37.104/search?q=cache:E85BOqUINTcJ:www.populationinstitute.ca/archives/losing_the_race.htm+%22major+oceanic+fisheries+are+now+being+fished%22+&hl=en&ie=UTF-8 on 11 March 2004.

White, R., Murray, S., Rohweder, M. (2000). Pilot analysis of global ecosystems: Grassland ecosystems. World Resources Institute, Washington, DC, USA, 100. http://forests.wri.org/pubs_description.cfm?PubID=3057 on 4 March 2004.

Whittow, J. B. (2000). The Penguin Dictionary of Physical Geography, Second Edition. Penguin Books, London, UK.

Wiebe, K. (2003). Linking land quality, agricultural productivity, and food security. Agricultural Economic Report No. (AER823) U.S. Department of Agriculture: Economic Research Service, Washington, DC, USA, 63. http://www.ers.usda.gov/publications/aer823/ on 2 March 2004; http://www.ers.usda.gov/publications/aer823/aer823c.pdf on 2 March 2004.

Williams, M. (1990). Forests. Chapter 11 in The Earth As Transformed by Human Action (B.L. Turner II, ed.), Cambridge University Press, New York, USA, 179-201.

Williams, M. (1994). Forest and Tree Cover. Chapter 5 in Changes in Land Use and Land Cover: A Global Perspective (W.B. Meyer and B.L. Turner II, eds.) Cambridge University Press, New York, USA, 97-124.

WHO (2000). Fact Sheet No. 254: Vegetation Fires. http://www.who.int/mediacentre/factsheets/fs254/en/ on 1 November 2004.

WMO-UNEP (2002). Scientific Assessment of Ozone Depletion, World Meteorological Organization (WMO) and United Nations Environment Programme.

Wookey, P. A. (2002). Tundra. Volume 2, The Earth system: biological and ecological dimensions of global environmental change, Encyclopedia of Global Environmental Change. Chichester: John Wiley & Sons, Ltd. 593–602. http://www.eld.geo.uu.se/swe/hemsidor/phil/EGEC.pdf on 11 July 2004.

WRI (1998). World Resources 1998-99: Environmental change and human health. A joint publication by the World Resources Institute, the United Nations Environment Programme, the United Nations Development Programme, and The World Bank, Oxford University Press, 384. http://population.wri.org/pubs_description.cfm?PubID=2889 and http://population.wri.org/pubs_content_text.cfm?ContentID=1449 on 6 April 2004.

WRI (2000). World Resources 2000-2001: People and ecosystems: The fraying web of life. World Resources Institute in cooperation with United Nations Development Programme, United Nations Environment Programme, and World Bank, Washington DC, USA. http://pubs.wri.org/pubs_content_text.cfm?ContentID=218 on 2 March 2004.

Site References

Coastal

Gulf of Fonseca, Honduras

Valderrama, D. and Engle, C. (2001). A Risk Programming Model for Shrimp Farming in Honduras, Aquaculture/Fisheries Center University of Arkansas at Pine Bluff, USA. http://www.orst.edu/Dept/IIFET/2000/abstracts/valderrama.html on 31 December 2002.

World Rainforest Movement (2001). Honduras: Shrimp farming destruction. http://www.wrm.org.uy/bulletin/51/Honduras.html on 31 December 2002.

Gulf of Guayaquil, Ecuador

Earth Summit (2002). Watch National Reports. http://earthsummitwatch.org/shrimp/national_reports/ on 30 December 2002.

Ecuador Government Links (2002). http://www.consecuadorquebec.org/datoseen.htm on 30 December 2002.

Encyclopedia.com (2002). Guayaquil. http://www.encyclopedia.com/html/g/guayaqui.asp on 30 December 2002.

Greenpeace (n.d.). http://archive.greenpeace.org/~oceans/shrimpaquaculture/ifmangrovesdisappear.html on 7 January2003; http://archive.greenpeace.org/pressreleases/oceans/2000dec18.html on 7 January 2003.

Tradeport (1999). http://www.tradeport.org/ts/countries/ecuador/mrr/mark0083.html on 7 January 2003.

UN Commission on Sustainable Development (1997). http://www.earthsummitwatch.org/shrimp/national_reports/ecuaceli.html on 7 January 2003.

World Rainforest Movement (2002). Mangroves and shrimp farming companies, Ecuador. http://www.wrm.org.uy/bulletin/51/Ecuador.html on 30 December 2002.

WWF (2001). Gulf of Guayaquil-Tumbes mangroves. National Geographic Society. http://www.nationalgeographic.com/wildworld/profiles/terrestrial/nt/nt1413.html.on 30 December 2002.

Huang He Delta, China

Brismar, A. (1999). Environmental challenges and impacts of land use conversion in the Yellow River basin. IIASA Interim Report IR-99-016. http://www.iiasa.ac.at/Publications/Documents/IR-99-016.pdf on 31 December 2002.

Famouschinese.com (n.d.). http://www.famouschinese.com/virtual/Huang_He on 22 February 2005.

Ren, M. (1995). Anthropogenic effect on the flow and sediment of the Lower Yellow River and its bearing on the evolution of Yellow River delta, China. GeoJournal, Volume 37, No. 4, 473-478.

Ren, M. and Walker, H.J. (1998). Environmental Consequences of human activity on the Yellow River and its delta, China. Physical Geography, 19, 5, 421-432.

Strain, P., Engle, F. (1993). Looking At Earth. Turner Publishing, Inc. Atlanta Georgia, USA, 74-75.

Van Gelder, A., van den Berg, J. H., Cheng, G., and Xue, C. (1994). Overbank and channelfill deposits of the modern Yellow River delta. Sedimentary Geology, Volume 90, 293-305.

Xue, C. (1993). Historical changes in the Yellow River delta, China. Marine Geology, 113, 321-329.

IJsselmeer, Netherlands

Royal Netherlands Meteorological Institute (2002). EUMETSAT, http://www.eumetsat.de/en/index.html?area=left2.html&body=/en/area2/cgms/ap5-04.htm&a=284&b=2&c=280&d=200&e=0 on 31 December 2002.

USGS (2001). USGS Earthshots, Satellite Images of Environmental Change. Ijsselmeer, Netherlands. http://edc.usgs.gov/earthshots/slow/IJsselmeer/IJsselmeer on 31 December 2002.

Isahaya Bay, Japan

Asiaweek (2001). Public Works Time Bombs - An ecological disaster dents the ruling party, Asiaweek, Volume 27, No. 6, 16 February 2001. http://www.timeinc.net/asiaweek/magazine/nations/0,8782,98449,00.html on 22 February 2005.

Bremner, B. (2001). BusinessWeek. Big Spenders and the Great Seaweed Slaughter. http://www.businessweek.com/bwdaily/dnflash/feb2001/nf2001026_998.htm on 31 December 2002.

The Ramsar Convention on Wetlands (1998). Monitoring project for Isahaya Bay coordinated with World Wetlands Day (WWD). http://www.ramsar.org/wwd_isahaya_bay.htm on 31 December 2002.

Knife River Delta, Canada

Alisauska, R. T. (1992). Spring Habitat Use and Diets of Midcontinent, Adult Lesser Snow Geese: Journal of Wildlife Management, Volume 56, 1992, 43.

Audubon (n.d.). http://www.audubon.org/local/cn/97december/sgnth.html on 31 December 2002; http://www.audubon.org/news/release98/gill-hrs.html on 31 December 2002.

Batt, B. (1998a). A Perilous Abundance: Ducks Unlimited, January/February 1998, 56-58.

Batt, B. (1998b). U.S., Canada May Declare War on Geese. USA Today, 28 April 1998, Science Section, 59.

Bird Studies Canada (n.d.). Important Bird Areas of Canada. http://www.bsc-eoc.org/iba/site.cfm?siteID=MB033&lang=en on 31 December 2002.

Bihrle, C. (1998). Greater Goose Harvest. Higher Limits, Periodic All-Day Hunting Create Possibilities, North Dakota Outdoors, September/October 1998.

Drew, L and Madson, C. (1998) Out of Control! National Wildlife, April 1998, 28-31.

Hudson Bay Project (2002). The Effects of a Trophic Cascade. http://research.amnh.org/users/rfr/hbp/kenshow/ken26.html on 16 December 2004.

Johnson, M. (1996). The Snow Goose Population Problem, Part I. North Dakota Outdoors, Volume LIX, No. 2, August 1996, 14-176.

Johnson, M. (1996). The Snow Goose Population Problem, Part II: Working Toward a Solution. North Dakota Outdoors, Volume. LIX, No. 3, September/October 1996, 20-22.

Johnson, M. (1997). The Snow Goose Population Problem, Part III. Arctic Ecosystems in Peril, North Dakota Outdoors, Volume LIX, No. 8, March 1997, 3-4.

Madson, C. (1997). Snow Drifts. The Mid-Continent's Lesser Snow Geese: Wildfowl Magazine, April/May 1997, 48-49.

North Dakota Outdoors Magazine (n.d.). http://www.und.nodak.edu/org/ndwild/ndmag.html on 16 December 2004.

Texas Parks and Wildlife (n.d.). Snow Geese Damage Arctic Habitat. http://www.tpwd.state.tx.us/nature/research/snogeese/snogeese.htm on 16 December 2004.

These NDVI images were created directly from the digital number (DN) values of registered Landsat data (MSS bands 4 and 2, TM bands 4 and 3), without converting them to reflectance or correcting for atmospheric conditions.

Tompkins, S. (1999). Snow Geese Debate Heating Up, But There's No Solution in Sight. Houston Chronicle, 4 February 1999, Outdoors section, 11.

USGS (2001). Earthshots: Satellite Images of Environmental Change. http://edcwww.cr.usgs.gov/earthshots/slow/Knife/Knife on 16 December 2004.

World Lakes Database. Lake Balkhash, Kazakhstan. http://Www.Ilec.Or.Jp/Database/Asi/Asi-54.Html on 16 December 2004.

World Wildlife (2001). Southern Hudson Bay taiga. http://www.worldwildlife.org/wildworld/profiles/terrestrial/na/na0616_full.html on 16 December 2004.

Phnom Penh, Cambodia

O'Neill, T. (1993). The Mekong River: National Geographic, Volume 183, No. 2, February 1993, 10, reproduced in USGS Earthshots 2001. http://edcwww.cr.usgs.gov/earthshots/slow/PhnomPenh/PhnomPenh on 31 December 2002.

Phnom Penh Municipality (n.d.). http://www.phnompenh.gov.kh/ on 7 January 2003.

Russell R. (ed.) (1990). Cambodia; a country study. Headquarters, Department of the Army, Washington, DC, USA, xvi.

Sundarban, India/Bangladesh

Bangladesh Observor. (2000). Eco-systems at great risk' Newpaper article published on 31 August 2000, Dhaka, Bangladesh.

Blasco, F. (1977). Outlines of Ecology, Botany and Forestry of the Mangals of Indian Subcontinent. Chapman, Bangladesh Observer, 241-260.

Chakrabarti, K. (1986). Tiger (Panthera tigris tigris) in the mangrove forest of Sundarbans – an ecological study. Tigerpaper 13(2): 8-11.

IPCC (1998). The Regional Impacts of Climate Change: An Assessment of Vulnerability. Intergovernmental Panel on Climate Change, Cambridge University Press, Cambridge, UK.

Lahiri, P.K. (1973). Management plan of tiger reserve in Sundarbans, West Bengal, India. Department of Forests, West Bengal, India, 101.

Thon Buri, Thailand

FAO (1999). The state of world fisheries and aquaculture. http://www.fao.org/docrep/w9900e/w9900e02.htm on 31 December 2002.

Flaherty, M. (2002). University of Victoria, Department of Geography, Victoria, Canada. http://office.geog.uvic.ca/dept/faculty/flaherty/ on 31 December 2002.

Water

Kara-Bogaz-Gol, Turkmenistan

Leont'yev, O. K. (1988). Problems of the Level of the Caspian and the Stability of its Shoreline: Soviet Geography, V. H. Winston and Son, Inc., Silver Spring, Maryland, USA, Volume 29, No. 6, June 1998, 608-616.

Lulla, K. P., Dessinov, L. V., Evans, C. A., Dickerson, P. W., and Robinson, J. A. (2001). Dynamic Earth Environments, 17 May 2001. http://eol.jsc.nasa.gov/newsletter/DynamicEarth/ on 29 August 2002.

Dickerson, P. W. (2000). A Caspian chronicle: Sea-level fluctuations between 1982 and 1997, in Dynamic Earth Environments: Remote Sensing Observations from Shuttle-Mir Missions (K. P. Lulla and L. V. Dessinov, eds.), John Wiley & Sons, New York, USA, 145-148, 266, 280-281.

USGS (2001). Earthshots, 8th ed., 12 January 2001, from the EROS Data Center of the U.S. Geological Survey, a bureau of the U.S. Department of the Interior.

Varushchenko, A. N., Lukyanova, S. A., Solovieva, G. D., Kosarev, A.N. and Kurayev, A. V. (2000). Evolution of the Gulf of Kara-Bogaz-Gol in the past century, in Dynamic Earth Environments: Remote Sensing Observations from Shuttle-Mir Missions (K. P. Lulla and L. V. Dessinov, eds.), John Wiley & Sons, New York, USA, 201-210.

Aral Sea, Kazakhstan

Development and Environment Foundation (2001). Environmental changes. http://ntserver.cis.lead.org/aral/enviro.htm on 30 December 2002.

FAO (2001a). Food and Agriculture Organization of the United Nations. "Time to Dave the Aral sea?" Agriculture 21. http://www.fao.org/WAICENT/FAOINFO/AGRICULT/magazine/9809/spot2.htm on 30 December 2002.

FAO (2001b). Food and Agriculture Organization of the United Nations. Water Report 15. http://www.fao.org/docrep/W6240E/w6240e03.htm#P1164_42540 on 30 December 2002.

GEF (1998). Global Environment Facility. Aral Sea Basin Program, Kazakhstan, Kyrgyz Republic, Tajikistan, Turkmenistan and Uzbekistan. Water and Environmental Management Project. Project Document. The World Bank, Washington, DC, USA.

IFAS –UNEP (2001). International Fund for the Aral Sea and the UN Environment Programme. Environment State of the Aral Sea Basin - Regional Report of the Central Asian States 2000. http://www.grida.no/aral/aralsea/english/arsea/arsea.htm on 30 December 2002.

UN (2001). United Nations Development Programme and the Government of the Republic of Uzbekistan. National Human Development Report 1996. http://www.cer.uz/NHDR/1996/index-e.htm on 30 December 2002.

USGS (2001). EarthShots USGS: Satellite Images of Environmental Change. http://edc.usgs.gov/earthshots/slow/Aral/Aral on 30 December 2002.

Atatürk Dam, Turkey

NASA (2004.). Ataturk Dam. http://earthobservatory.nasa.gov/Newsroom/NewImages/images.php3?img_id=16302 on 16 December 2004.

Southeastern Anatolia Project Regional Development Administration (2002). Southeastern Anatolia Project. Republic of Turkey Prime Ministry http://www.gap.gov.tr/gapeng.html on 31 December 2002.

Vesilind, P.J. (1993). The Middle East's Water Critical Resource. National Geographic, May 1993, Volume 183, No. 5.

Challawa Gorge Dam, Nigeria

Thompson, J. (2001). Projects: Hadejia-Jama'are River Basin. UCL Department of Geography. http://www.geog.ucl.ac.uk/~jthompso/hadejjia-jam.stm on 31 December 2002; http://www.geog.ucl.ac.uk/~jthompso/hadejjia-jam.stm on 6 January 2003.

UNEP (2003). http://grid2.cr.usgs.gov/publications/selected/Africa.pdf on 16 December 2004.

Dead Sea, Jordan

Libiszewski, S. (1995). Water disputes in the Jordan Basin Region, Environment and Conflicts Project ENCOP, ETHZ, Switzerland.

FAO (2000). Internal Renewable Water Resources, AQUASTAT Information System on Water in Agriculture.

Isreal Yearbook & Almanac (n.d.). Jerusalem Post: The America-Israeli Cooperative Enterprise. http://www.us-isreal.org/jsource/Peace/settlepop.html on 18 January 2004.

UN (2000). UN Department of Economic and Social Affairs. Population Growth Rate: World Population Years 1995-2000. http://www.fao.org/ag/agl/aghr/aquastat/main/index.htm on 18 January 2004.

Everglades, United States

Audubon Organization (2002). Audubon Campaigns. National Audubon Society. http://www.audubon.org/nas/ar97/campaign.html on 30 December 2002.

Conway Data (2002). Building Blocks: Schools, Water Top Florida's Smart-Growth Agenda. http://www.siteselection.com/features/2001/sep/fl/index.htm on 30 December 2002.

Everglades Plan (2002). Rescuing an endangered ecosystem-the journey to restore America's. http://www.evergladesplan.org/ on 30 December 2002.

Florida Population Components of Change (2002). Online article. http://www.state.fl.us/edr/Population/FLPopChange.pdf on 30 December 2002.

Starck, H. (2002). Lower East Coast: Everglades Restoration Plan. Nova Southeastern University. http://www.nova.edu/ocean/eglades/sum00/lowereast.html on 30 December 2002.

State Facts Florida (2002). Negative Population Growth. http://www.npg.org/states/fl.htm on 30 December 2002.

UNEP WCMC (2002). Everglades National Park. UNEP World Conservation Monitoring Centre. http://www.wcmc.org.uk/protected_areas/data/wh/everglad.html on 30 December 2002.

Gabcikova, Slovakia

Balon, E.K. and Holcik, J. (1999). Gabcikovo river barrage system: the ecological disaster and economic calamity for the inland delta of the middle Danube. Environment Biology of Fishes 54: 1-17.

Linnerooth-Bayer, J. and Murcott, S. (1996). The Danube River Basin: International Cooperation of sustainable Development. Natural Resources Journal vol. 36521-547.

Vranovsky, M (1997). Impact of the Gabcikovo hydropower plant operation on planktonic copepods assemblages in the River Danube and its floodplain downstream of Bratislava. Hydrobiologia 347:41-49.

WWF (1997). How to Save the Danube Floodplains: The Impact of the Gabcikovo Hydrodam System over Five Years. WWF Statement. Vienna, Austria.

Lesotho Highlands Water Project

Hoover, R. (2001). Promises for Power Go Unfulfilled. International Rivers Network. http://www.irn.org/programs/lesotho/index.asp?id=/programs/lesotho/001031.darkness.html on 31 December 2002; http://www.irn.org/wcd/index.asp?id=/wcd/lhwp.shtml on 16 December 2004.

McKenzie, R. et al. (2000). South African Department of Water Affairs and Forestry. http://www.dwaf.gov.za/orange/ on 31 December 2002.

Mochebelele, R. T. (2001). South Africa - Vanderkloof/Gariep Dams, Orange River http://www.dams.org/kbase/studies/za/ on 31 December 2002.

Lake Balkash, Kazakhstan

UNEP (2004). Activities: Sustainable Resource Use – Lake Balkash, Kazakhstan, Guillaume Le Sourd, Diana Rizzolio. UNEP GRID Geneva, Geneva, Switzerland. http://www.grid.unep.ch/activities/sustainable/balkhash/index.php on 6 March 2005.

Lake Chad, Africa

FEWS (1997). Famine Early Warning System. Special Report 97-4. http://www.fews.org/fb970527/fb97sr4.html on 31 December 2002.

IMF (2002). International Monetary Fund: Lake Chad Basin Commission (LCBC). http://www.imf.org/external/np/sec/decdo/lcbc.htm on 31 December 2002.

International Lakes environment Committee Foundation (1999). World Lakes Database. http://www.ilec.or.jp/database/afr/afr-02.html on 31 December 2002.

JSC (2000). Johnson Space Center Digital Image Collection. http://nsimages.jsc.nasa.gov/images/pao/STS36/10063865.htm on 31 December 2002.

USGS (2001). USGS EarthShots: Satellite Images of Environmental Change. USGS. http://edcwww.cr.usgs.gov/earthshots/slow/LakeChad/LakeChad on 31 December 2002.

Lake Chapala, Mexico

Campose, Jesus. Personal Communication. CentroGeo, Mexico City, Mexico.

De Anda, Jose, et al. (1998). Hydrologic Balance of Lake Chapala (Mexico). Journal of the American Water Resources Association, Vol. 34, no. 6, December 1998.

Lake Hamoun, Iran

Agrawala, S., Barlow, M., Cullen, H., and Lyon, B. (2001). The Drought and Humanitarian Crisis in Central and Southwest Asia: A Climate Perspective. International Research Institute for Climate Prediction, New York, USA.

Christensen, P. (1993). The Decline of Iranshahr: Irrigation and Environments in the History of the Middle East 500 B.C. to A.D. 1500. Museum Tusculanum Press, University of Copenhagen, Copenhagen, Denmark.

International Federation of Red Cross and Red Crescent Societies (2001). Despair on Iran's dusty plains, 19 July 2001.

New York Times (2001). Drought chokes off Iran's water and its economy by Neil MacFarquhar, 18 September 2001.

UN (2001). United Nations Inter-agency Assessment Report on the Extreme Drought in the Islamic Republic of Iran, by UN Country Team in Iran, July 2001.

Lake Nakuru, Kenya

Africa Environmental Outlook (2001). Water Quality in Eastern Africa. United Nations Environment Programme. http://www.unep.org/aeo/158.htm on 31 December 2002.

Forest Department (n.d.). Ministry of Environment and Natural Resources, KIFCON project.

Jones, T. (1993). Lake Nakuru Ramsar Convention Bureau. http://www.ramsar.org/lib_dir_1_3.htm on 31 December 2002.

Lambrechts, C. (2001). UNF, UNEP, KWS, University of Bayreuth, WCST. Personal Communication.

Thampy, R. J. Case Study: Kenya for the World Wide Fund for Nature (WWF). http://www.aaas.org/international/ehn/biod/thampy.htm on 12 March 2005.

Lake Victoria, Uganda

Albright, T., Moorhouse, T., McNabb, T. (2001). The Abundance and Distribution of Water Hyacinth in Lake Victoria and the Kagera River Basin, 1989-2001, U.S. Geological Survey, EROS Data Center, Sioux Falls, South Dakota, USA.

Neuville, G., Baraza, J., Bailly, J., Wehrstedt, Y., Hill, G., Balirwa, J. and Twongo, T. (1995). Mapping of the Distribution of Water Hyacinth Using Satellite Imagery, Technical Report, RCSSMRS/French Technical Assistance, April 1995.

Schouten, L., van Leuwen, H., Bakker, J., Twongo, T. (1999). Water Hyacinth Detection in Lake Victoria by Means of Satellite SAR, USP-2 report 98-28, Netherlands Remote Sensing Board, Delft, The Netherlands.

Mesopotamia Marshlands, Iraq

New York Times (2005). http://www.nytimes.com/imagepages/2005/03/07/science/20050308_MARS_GRAPHIC2.html on 28 March 2005.

Partow, H. (2001). The Mesopotamian Marshlands: Demise of an Ecosystem. Division of Early Warning and Assessment. United Nations Environment Programme. Nairobi, Kenya.

USGS (2001). USGS EarthShots: Satellite Images of Environmental Change. http://edcwww.cr.usgs.gov/earthshots/slow/Iraq/Iraq on 31 December 2002.

Three Gorges Dam, China

CNEMC (n.d.). Personal Communication and imagery courtesy of Wang Wenjie, China National Environmental Monitoring Center.

Jones, W. C., Freeman, M. (2001). Three Gorges Dam: The TVA on The Yangtze River Schiller Institute. http://www.schillerinstitute.org/economy/phys_econ/phys_econ_3_gorges.html on 31 December 2002.

NASA (2001). Three Gorges Dam, China Visible Earth. http://visibleearth.nasa.gov/cgi-bin/viewrecord?17955 on 31December 2002.

Tillou, S. L. and Honda, Y. (n.d). Trade and Environment Database. http://www.american.edu/ted/THREEDAM.htm on 7 January 2003.

Kennedy, B. (2001). China's Three Gorges Dam. CNN Interactive. http://www.cnn.com/SPECIALS/1999/china.50/asian.superpower/three.gorges/ on 7 January 2003.

PBS (n.d). Great Wall across the Yangtze, Public Broadcasting System. http://www.pbs.org/itvs/greatwall/index.html on 7 January 2003.

Forest

DMZ, Korea (Case Study)

Biodiversity Planning and Support Programme (2001). The Korean Demilitarized Zone: Eden of Wildlife. Northeast and East Central Asia - National Biodiversity Strategies Action Plans, Newsletter, Issue 3/4, 29 March 2001, UNDP – UNEP. http://www.bpsp-neca.brim.ac.cn/newsletter/issue3-4/4.html on 11 March 2005.

Ministry of Environment (n.d.). Republic of Korea. Systematic Conservation for an Eco-Network, Ministry of Environment. http://eng.me.go.kr/user/policies/policies_view.html?msel=b6&seq=7&filename=6_nature_08.html&table_name=me_new_nature on 11 March 2005.

Neufeld, A. N. (1997). Korean Demilitarized Zone as a Bioreserve. ICE Case Studies. http://www.american.edu/projects/mandala/TED/ice/DMZ.HTM on 6 March 2005.

The DMZ Forum (n.d.). For Peace and Nature Conservation. http://www.dmzforum.org/main.html on 11 March 2005.

USGS (2004). IUCN: The cranes-status survey and conservation action plan. Gland, Switzerland. http://www.npwrc.usgs.gov/resource/distr/birds/cranes/cranes.htm on 11 March 2005.

Angangueo, Mexico

Brower, et al. (2002). Quantitative Changes in Forest Quality in a Principal Overwintering Area of the Monarch Butterfly in Mexico, 1971 - 1999. Conservation Biology, Volume 16 Issue 2, 346, April 2002.

USGS (2001). USGS EarthShots: Satellite Images of Environmental Change. http://edcwww.cr.usgs.gov/earthshots/slow/Angangueo/Angangueo on 28 March 2005.

Arkhangelsk, Russia

WWF (2001). East Siberian Taiga. National Geographic Society. http://www.nationalgeographic.com/wildworld/profiles/g200/g084.html on 31 December 2002.

WWF (2001). Kamchatka Taiga and Grasslands. National Geographic Society. http://www.nationalgeographic.com/wildworld/profiles/g200/g085.html on 31 December 2002.

BorealForests.org. (2002). Forest Management in Russia. http://www.boreal-forest.org/world/rus_mgmt.htm on 31 December 2002.

Global Forest Watch (2002). Russia: Overview. World Resource Institute. http://www.globalforestwatch.org/english/russia/index.htm on 31 December 2002.

Taiga Rescue Network (2002). Russia's "Frontier Forests" Endangered by Multinational Corporations, Illegal Trade. http://www.taigarescue.org/publications/siberia on 31 December 2002.

UNEP-WCMC (2001). Russian Federation – Information Provided on a Regional Basis http://www.unep-wcmc.org/forest/rus_region.htm on 31 December 2002.

Finnish Nature League (2000). Old-growth forests in Russia. http://www.luontoliitto.fi/forest/russia/ on 31 December 2002.

British Columbia, Canada

Northwest Environment Watch (2004). Cascadia Scorecard, Seven Key Trends Shaping the Northwest. Transcontinental Printing, Canada.

Country Border Guatemala/Mexico, South America

Global Policy Forum (2002). Growing Poverty Is Shrinking Mexico's Rain Forest http://www.globalpolicy.org/socecon/develop/2002/1208chiapas.htm on 28 March 2005.

Tropical Forestry Projects information System (2000). http://www.odi.org.uk/tropics/projects/3228.htm on 18 June 2004.

Peters, G. (2002). The Christian Science Monitor, No quick solution to deforestation in lush Chiapas, 14 January 2002. http://www.csmonitor.com/2002/0114/p7s1-woam.html on 31 December 2002.

Iguazú National Park, Argentina

UNEP WCMC (1995). Protected Areas Programme. http://www.wcmc.org.uk:80/protected_areas/data/wh/iguazu.html on 31 December 2002.

Strain, P., Engle, F. (1999). Looking At Earth. Turner Publishing, Inc. Atlanta Georgia, USA, 231-232.

The Greatest Places.org (1999). http://www.greatestplaces.org/book_pages/iguazu2.htm on 7 January 2003.

WCMC (1995). Protected Areas Programme, Natural Heritage Sites. Iguazú National Park. http://www.wcmc.org.uk:80/protected_areas/data/wh/iguazu.html on 7 January 2003.

WCMC (1997). Protected Areas Programme, Natural Heritage Sites. Iguaçu National Park. http://www.wcmc.org.uk:80/protected_areas/data/wh/iguacu.html on 7 January 2003.

USF (n.d.). Itaipú Dam. http://ce.eng.usf.edu/pharos/wonders/modern/itaipu.html on 7 January 2004.

Itampolo, Madagascar

American University (1996). Deforestation in Madagascar. http://gurukul.ucc.american.edu/TED/MADAGAS.HTM on 31 December 2002.

Green, G.M. and R.W. Sussman (1990). Deforestation history of the eastern rainforests of Madagascar from satellite image. SCIENCE, Volume 248: 212-215.

Kisangani, DRC

Ivey, P.K. (2000). Cooperative reproduction in Ituri Forest hunter-gatherers: Who cares for Efe infants? Current Anthropology, Chicago, December 2000, Volume 41, Issue 5, 856-866.

NRC (1993). Sustainable Agriculture and the Environment in the Humid Tropics, National Research Council, 625-658.

P.S. (2002). Crisis of government, ethnic schisms, Civil War, and regional destabilization of the Democratic Republic of the Congo, World Affairs; Washington; Summer 2002.

Vogel, G (2000). Conflict in Congo threatens bonobos and rare gorillas, Science, 31 March 2000, Washington, DC, USA, Volume 287, Issue 5462, 2386-2387.

Weiss, H. (2001). Civil War in the Congo Society; New Brunswick; March/April 2001; Volume 38, Issue 3, 67-71.

Lappi, Finland

GreenPeace.org (n.d.). http://www.greenpeace.org.uk/forests/forests.cfm?ucidparam=20040815152527&CFID=2025372&CFTOKEN=22657474 on 18 March 2005.

Lappi, Finland (n.d.). http://smyhtml.tjhosting.com/yr2004/eng/4_1.html#sertivaatimuksia on 18 March 2004.

Olympic Peninsula, USA

ASA (1999). Zoom into Olympic National Forest Time Lapse: 1984 to 1995, 9 April 1999. http://svs.gsfc.nasa.gov/vis/a000000/a000900/a000900/ on 31 December 2002.

Dark, A. (1997). Landscape and Politics on the Olympic Peninsula: Social Agendas and Contested Practices in Scientific Forestry. Journal of Political Ecology, Volume 4, 1997. http://www.library.arizona.edu/ej/jpe/volume_4/5DARK.PDF on 7 January 2003.

USDA (n.d.). Olympic National Forest Geographic Information Systems. http://www.fs.fed.us/r6/data-library/gis/olympic/ on 23 June 2003.

Oudomxay, Laos

Giri, C., D. Pradhan, R.S. Ofren and Kratzschmar, E. (1998). Land Use/Land Cover Change in South and South-East Asia. UNEP Environment Assessment Programme for Asia and the Pacific, Bangkok, Thailand.

Papua (Irian Jaya), Indonesia

Australia West Papua Association (2002). West Papua Information Kit. University of Texas, Texas, USA. http://www.cs.utexas.edu/users/cline/papua/core.htm on 31 December 2002.

Cargill, Inc. (2002). http://www.cargill.com/today/releases/00_6_2palm.htm on 31 December 2002.

Library of Congress (2002). Indonesia – A Country Study. http://lcweb2.loc.gov/frd/cs/idtoc.html on 31 December 2002.

Myllyntaus, T., Hares, M, Kunnas, J. (2002). Sustainability in danger? Slash-and-burn cultivation in nineteenth-century Finland and twentieth-century southeast Asia Environmental History, Durham, April 2002.

Octovianus Mote (2001). From Irian Jaya to Papua: The limits of primordialism in Indonesia's troubled east; Octovianus Mote, Indonesia, Ithaca, New York, USA, October 2001, Issue 72, 115, 27.

OilPalm.net (2002). OilPalm.net. http://www.oilpalm.net on 31 December 2002.

Sharp, T. (2002). Irian Jaya dot org. http://www.irja.org/politics/tim.htm on 31 December 2002.

The Nation (1994). Suharto's cronies profit: The deforesting of Irian Jaya, New York, USA, 7 February 1994.

World rainforest movement (2001). http://wrm.org.uy/bulletin/47/viewpoint.html on 31 December 2002.

Rondônia, Brazil

Ellis, W. S., Allard, W.A. and McIntyre, L. (1988). Rondonia: Brazil's imperiled rain forest: National Geographic Magazine, Volume 174, No. 6, December 1998, 772-799.

Skole, D., and Tucker, C. (1993). Tropical Deforestation and Habitat Fragmentation in the Amazon: Satellite Data from 1978 to 1988: Science, American Association for the Advancement of Science, Washington, DC, USA, Volume 260, No. 5116, June 1993, 1905-1910.

USGS (2001). USGS EarthShots: Satellite Images of Environmental Change. Rondonia, Brazil. http://edcwww.cr.usgs.gov/earthshots/slow/Rondonia/Rondonia on 31 December 2002.

White, P. T., and Blair, J.P. (1983). Tropical rain forests: Nature's dwindling treasures: National Geographic Magazine, Volume 163, No. 1, January 1983, 2-46.

Sakhalin Island, Russia

Ardö, J., Lambert, N. J., Henzlik, V. and Rock, B. N. (1997). Satellite Based Estimations of Coniferous Forest Cover Changes.

CEZ (n.d.). Czech power Company CEZ, a.s. Personal Communication with Michaela Sabolovicova Chloupkova, Sefredaktorka, web stranek Prague, Czech Republik. http://www.cez.cz/eng/ on 7 January 2003.

Common Report on Air Quality in the Black Triangle Region (2000). Report prepared by the JAMS Working Group and published with the assistance of the PHARE Black Triangle Project. http://www.env.cz/envdn.nsf/0/1f9cf50ae801b07fc1256b5a00309f85/$FILE/Trojuhel.pdf on 27 July 2004.

Hory, K. (1997). Czech Republic, 1972-1989. Ambio, Volume 26, No. 3, May 1997, 158-166.

NASA (n.d.). NASA-Earth Observating System, Changes in Biochemical Cycles, www.eos-ids.sr.unh.edu/ids-cycles.html on 8 March 2004.

Photographic credits: Jonas Ardö, Ph.D. Department of Physical Geography Lund University, Sweden

Strub-Aeschbacher, N. (2002). Atlas of Global Change Project. UNEP/DEWA/GRID-Geneva, Switzerland.

Tai National Park, Coté D'Ivoire

Chatelain, C., Gautier, L. and Spichiger, R. (1996). "A Recent History of Forest Fragmentation in Southwestern Ivory Coast". Biodiversity and Conservation 5:1 Fall 1994, 37-53. Chapman & Hall, UK.

Chatelain,C., Gautier, L. and Spichiger, R. (2002). Forest Fragmentation in the Southwester Cote d'Ivoire. Conservatoire du Jardin Botanique, Geneva, Switzerland.

Valdivian Forest, Chile

Echeverria, C. and Antonio, L. (n.d.). Personal Communication. Forest Cover Change in South-Central Chile. Instituto de Silvicultura, Universidad Austral de Chile. Casilla 567, Valdivia. Chile.

Cropland

Al Isawiyah, Saudi Arabia

Saudi Cities (1999). http://saudicities.com/country2.htm on 31 December 2002; http://saudicities.com/country2.htm on 6 January 2003.

USGS (1998). Overview of Middle East Water Resources. http://exact-me.org/overview/p15.pdf on 7 January 2003.

Saudi Water Net (n.d.). http://www.kfupm.edu.sa/saudiwaternet/on 7 January 2003.

Almeria, Spain

WWF (n.d.). www.panda.org/new_facts/newsroom/news.cfm?uNewsld=12582&uLangld=1 on 16 February 2005.

Novovolyns'k, Ukraine

The World Bank Group. (n.d.) Ukraine Homepage. http://www.worldbank.org/tenthings/ECA/7-ukra.html on 9 July 2004.

FAO (n.d.). FAO Corporate Document Repository. Central and Eastern Europe and the Commonwealth of Independent States. http://www.fao.org/DOCREP/004/y6000e/y6000e11.htm on 3 July 2004.

Max Planck Institute for Social Anthropology (n.d.). http://www.eth.mpg.de/dynamic-index.html? On 28 June 2004; Land reform in two former republics of the Soviet Union: Georgia and Ukraine. http://www.eth.mpg.de/people/kaneff/comparisons.html. On 28 June 2004.

Paektu San, North Korea

Korea Herald (1996). A Handbook of North Korea. Studies. Mt. Peaktu and tourism, TED Case. http://www.american.edu/projects/mandala/TED/paektu.htm on 31 December 2002.

Korean News (2002). Mt. Peaktu. http://www.kcna.co.jp/item/1998/9801/news01/07.htm on 31 December 2002.

Santa Cruz, Bolivia

CountryWatch.com (2002). Bolivia. http://embassy.countrywatch.com/cw_country.asp?vCOUNTRY=021 on 31 December 2002.

Library of Congress (n.d.). Bolivia – A Country Study. http://lcweb2.loc.gov/frd/cs/botoc.html on 31 December 2002.

NASA (1986), Deforestation in Santa Cruz, Bolivia http://earthobservatory.nasa.gov/Newsroom/NewImages/images.php3?img_id=16274 on 18 February 2005.

NASA (2000). Deforestation in Bolivia. http://earthobservatory.nasa.gov/Newsroom/NewImages/images.php3?img_id=4544 on 6 January 2003.

Rainforest Action Network (2002). Rates of rainforest loss. http://www.ran.org/info_center/factsheets/04b.html on 31 December 2002.

Visible Earth (2001). Tierras Bajas Deforestation, Bolivia. http://www.visibleearth.nasa.gov/cgi-bin/viewrecord?8041 on 31 December 2002.

World Rainforest Movement (2001). Underlying Causes of Deforestation and Forest Degradation. http://www.wrm.org.uy/deforestation/LAmerica/Bolivia.html on 31 December 2002.

Tensas River Basin, USA

Louisiana Department of Environment Quality (n.d.). Nonpoint Source Pollution Program. http://nonpoint.deq.state.la.us/ws_tensas.html on 18 February 2005.

US Environmental Protection Agency (2002). An Ecological Assessment of the Louisiana Tensas River Basin http://www.epa.gov/nerlesd1/land-sci/tensas/tensas.html on 31 December 2002.

Torréon, Mexico

Coahuila State Government (2004). Geographic summary. http://en.coahuila.gob.mx/news20040704.htm on 28 February 2005.

Municipal Profile Torreon (n.d.). Secretariat of Planning and Development. http://servidor.seplade-coahuila.gob.mx/ohs_images/seplade/files/torreondic004.pdf on 9 July 2004.

Toshka Project, Egypt

Arabia.com. (2001). Final push on desert reclamation 'megaproject'. http://www.arabia.com/egypt/business/article/english/0,5127,20221,00.html on 31 December 2002.

Baker, M.M. (2002). Mubarak: Toshka Project Opens Way Towards New Civilization In Egypt. American Almanac. http://members.tripod.com/american_almanac/toshka.htm on 31 December 2002.

Knight-Ridder Tribune (2000). Lakes Bring New Hope For Life in Sahara. Earth Changes TV. http://www.earthchangestv.com/breaking/December2000/1206sahara.htm on 31 December 2002.

Market Access and Compliance (n.d.). Egypt FY2000 Country Commercial, Executive Summary. http://www.mac.doc.gov/tcc/data/commerce_html/countries/Countries/Egypt/CountryCommercial/2000/Egypt/body.htm on 18 February 2005.

UN (2001). Report on the United Nations/European Space Agency, Committee on Space Research Workshop on Data Analysis and Image-Processing Techniques Damascus, 25-29 March 2001. http://www.oosa.unvienna.org/Reports/AC105_765E.pdf on 31 December 2002.

Visible Earth (2001). Another New Lake in Egypt. http://visibleearth.nasa.gov/cgi-bin/viewrecord?7714 on 31 December 2002.

Grassland

Narok, Kenya

Le Monde diplomatique (2000). Clash Of Interests In Masai Country, Kenya's battle for biodiversity by By Alain Zecchini, November 2000. http://mondediplo.com/2000/11/21masai

NEMA (2004). National Environment Management Authority: Profile of Norak District. http://www.nema.go.ke/profile_of_narok_district.htm on 18 February 2005.

Kareithi, S (2003). Coping with Declining Tourism, Examples from Communities in Kenya. PhD Researcher, University of Luton, UK, 18 February 2003.

Peanut Basin, Senegal

Tappan, G. G. (2005). Personal Communication. SAIC, USGS National Center for Earth Resources Observation and Science. Sioux Falls, South Dakota, USA, 16 March 2005.

Revane, Senegal

Tappan, G. G. (2005). Personal Communication. SAIC, USGS National Center for Earth Resources Observation and Science. Sioux Falls, South Dakota, USA, 16 March 2005.

Upper Green River Basin, USA

Amos, J. F. (2003). Witness Statement. Environmental Aspects of Modern Onshore Oil and Gas Development. Testimony to the Committee on Resources of the United States House of Representatives, Subcommittee on Energy and Mineral Resources, 17 September 2003.

Upper Green River Valley Coalition (2003). Wyoming's Upper Green River Basin. Poster. 4 March 2003.

Wyperfeld, Australia

Parks Victoria (2002). Wyperfeld National Park. http://www.parkweb.vic.gov.au/1park_display.cfm on 31 December 2002.

USGS (2001). EarthShots: Satellite Images of Environmental Change. USGS. http://edcwww.cr.usgs.gov/earthshots/slow/Wyperfeld/Wyperfeld on 31 December 2002.

Urban

Banjul, The Gambia

CBD (1998). First National Report on The Implementation of the Convention on Biological Diversity. http://www.biodiv.org/doc/world/gm/gm-nr-01-en.pdf on 30 December 2002.

Consular Information Sheets and Travel Warnings (2001). The Gambia. United States Department of State. http://travel.state.gov/gambia.html on 30 December 2002.

ODCI (2002). The World Fact Book. Office of Director of Central Intelligence (ODCI), The Gambia. http://www.odci.gov/cia/publications/factbook/geos/ga.html on 30 December 2002.

The official website of the government of the Gambia (2002). The Gambia. The Republic of the Gambia. http://www.gambia.com/ on 30 December 2002.

UNEP (1999). Overview of Land-based Sources and Activities Affecting the Marine, Coastal and Associated Freshwater Environment in the West and Central African Region: http://www.gpa.unep.org/documents/technical/rseas_reports/171-eng.pdf on 30 December 2002.

Beijing, China

Gaubatz, P. (1995). Changing Beijing: The Geographical Review, Volume 85, No. 1, 79-96.

Huus, Kari (1994). No place like home; Beijing is remaking itself, but for whose benefit? Far Eastern Economic Review, Volume 157, No. 30, 72-73.

Laquian, A. A. (2000). The planning and Governance of Mega-Urban Regions: What Can We Learn from Asia and the Pacific Rim? Plenary Address delivered at the Conference of the Association of Collegiate Schools of Planning (ACSP), 1-5 November 2000, University of British Columbia, British Columbia, Canada.

UN (1996). Population Division, Urban Agglomerations, 1950-2015 (The 1996 Revision), on diskette. United Nations Publication, New York, USA.

UNFPA (2000). World urbanization prospects: The 1999 revision. United Nations Population Division, New York, USA, 128. http://www.igc.org/wri/wr-98-99/citygrow.htm on 30 December 2002.

USGS (2001). USGS EarthShots: Satellite Images of Environmental Change. http://edc.usgs.gov/earthshots/slow/Beijing/Beijing on 30 December 2002.

Brasilia, Brazil

Augusto Cesar Baptista Areal (2002). The city of Brasilia. http://www.geocities.com/TheTropics/3416/links_i.htm on 30 December 2002.

Augusto Cesar Baptista Areal (2002). Report of a Pilot Plan for Brasilia. http://www.infobrasilia.com.br/pilot_plan.htm on 30 December 2002.

Brazilian Embassy (1986). Capital of Dreams by Paul Forster. Geographical Magazine, Volume 58, No. 9, Washington, DC, USA, 462-467. http://www.civila.com/brasilia/index.html on 31 December 2002.

Dhaka, Bangladesh

Bangladesh Online (2002). Dhaka, a Tourist Spot. http://www.bangladeshonline.com/tourism/spots/dhaka.htm on 31 December 2002.

Bhuiyanm, S.H. Md. (1999). Asian City Development Strategy Tokyo conference. http://www.worldbank.org/html/fpd/urban/city_str/tokyo/vol1/chap3-1.pdf on 31 December 2002.

Dhaka (2002). Capital Development Authority. Facts about Dhaka. http://www.rajukdhaka.org/planning.htm on 31 December 2002.

Human settlements (1996). Statement by H.E. Mr. Anwarul Karim Chowdhury, Ambassador and Permanent Representative of Bangladesh to the United Nation at the Second Committee of the 51st session of the UNGA on Agenda Item 96(e) of the 51st UNGA: http://www.un.int/bangladesh/ga/st/51ga/51-96e.htm on 31 December 2002.

Rashid, H. Er., Babar, K. (2002). Case Study: Bangladesh Water Resources and Population Pressures in the Ganges River Basin. American Association for the Advancement of Science. http://www.aaas.org/international/ehn/waterpop/bang.htm on 31 December 2002.

UNFPA (2000). World urbanization prospects: The 1999 revision. United Nations Population Division, New York, USA, 128.

Delhi, India

Earth Satellite Corporation (2002). http://www.earthsat.com/env/sec_pol/land_use.html on 31 December 2002.

Ministry of Environment and Forests (2002). White Paper on Pollution in Delhi. Government of India. http://envfor.nic.in/divisions/cpoll/delpo-lln.html on 31 December 2002.

Rajagopal, K. (2001). State of Environment Report for Delhi 2001 (TERI). http://www.teriin.org/reports/rep09/rep09.htm on 31 December 2002.

Tata Energy Research Institute (2002). Transport in Delhi: Some Perspectives. http://www.teriin.org/urban/delhi.htm on 31 December 2002.

Z-news (2002). Delhi pollution: Accelerating life towards death. http://www.zeenews.com/links/articles.asp?aid=7359&sid=ENV on 31 December 2002.

Las Vegas, USA

USGS (2001). USGS EarthShots: Satellite Images of Environmental Change. http://edcwww.cr.usgs.gov/earthshots/slow/LasVegas/LasVegas on 31 December 2002.

William Acevedo et al. (1997). Urban Land Use Change in the Las Vegas Valley. http://geochange.er.usgs.gov/sw/changes/anthropogenic/population/las_vegas/ on 31 December 2002.

Mexico City, Mexico

National Research Council (1995). Mexico City's Water Supply: Improving the Outlook for Sustainability, National Academy Press, Washington, DC, USA. http://lanic.utexas.edu/la/Mexico/water/ch1.html on 31 December 2002.

Sustainable Development Information Service (2002). World Resources Institute. http://www.wri.org/wri/trends/citygrow.html on 31 December 2002.

Midrand, South Africa

Directorate Environmental Information and Reporting (2000). Midrand State of the Environment report. Department of Environmental Affairs and Tourism http://www.environment.gov.za/soer/reports/midrand/index.htm on 31 December 2002.

Fakir, S. and Broomhall, L. (1999). Midrand State of the Environment Report. http://www.environment.gov.za/soer/reports/midrand/index.htm on 7 January 2003.

Greater Johannesburg Metropolitan Council (2000). http://ceroi.net/reports/johannesburg/csoe/Default.htm on 31 December 2002.

Midrand Ecocity Project. (n.d.). http://www.midrandecocity.co.za/ on 7 January 2003.

Sugrue, A. (2002). Midrand Ecocity of the future. http://www.midrand-ecocity.co.za/overview.htm on 31 December 2002.

Moskva, Russia

UNEP (n.d.) Moscow Integrated Environmental Action Programme, State of the Environment in Moscow. http://www.md.mos.ru/eng/comp/c_pr.htm on 25 February 2005.

Bochin, L.A. (1998) Minister of the Moscow City Government for Nature Management and the protection of the Environment, 5 June 1998. http://www.ourplanet.com/imgversn/95/wed.html on 25 February 2005.

Paris, France

Demographia (2003). Paris Population Analysis and Data Product, 24 March 2001. http://www.demographia.com/db-paris-history.htm on 18 February 2005.

Infoplease (2004). Paris, City, France; History. http://www.infoplease.com/ce6/world/A0860241.html 2 March 2005.

MSN Encarta (2005). Paris City, France. http://encarta.msn.com/encyclopedia_761561798_6/Paris_(city_France).html on 4 March 2005.

Santiago, Chile

Architectural Resources Network (2002). Santiago 1863-1988. http://www.periferia.org/urban/santiago.html on 31 December 2002.

City Net Express (1994). Santiago, Computer Science Department, University of Chile. http://sunsite.dcc.uchile.cl/chile/turismo/santiago.html on 31 December 2002.

ICLEI (1999). Sustainable Santiago project report. http://www.cities21.com/iclei/finalrepeng.htm on 31 December 2002.

Luz Alicia Cárdenas Jirón (2001). Urban Form at the Fringe of Metropolitan Santiago. http://revistaurbanismo.uchile.cl/n1/13.html on 31 December 2002.

UN Cyber School Bus (2002). Santiago, Chile. http://www.un.org/cyberschoolbus/habitat/profiles/santiago.asp on 31 December 2002

UNEP (2000). GEO: Chapter Two: The State of the Environment - Latin America and the Caribbean. http://www.cger.nies.go.jp/geo2000/english/0091.htm on 31 December 2002.

USGS (2001) USGS Earthshots. USGS. Satellite Images of Environmental Change. http://edc.usgs.gov/earthshots/slow/Santiago/Santiago on 31 December 2002.

Sydney, Australia

Essex, Stephen, Chalkley, Brian. Olympic Games: Catalyst of Urban Change. Routledge, part of the Taylor & Francis Group. Volume 7, No. 3 September 1998, 187-206.

Australian Government (2001). Dept. of the Environment and Heritage. Human Settlement Theme Report. http://www.deh.gov.au/soe/2001/settlements/figures.html on 18 October 2004.

Tripoli, Libya

Country Studies. Population (n.d.). http://www.country-studies.com/libya/population.html on 21 March 2005.

UNEP (n.d.). Africa Environment Outlook, Past, Present and Future perspectives. http://www.unep.org/dewa/Africa/publications/AEO-1/207.htm on 9 December 2004.

Fedra, K. and Abdel-Rehim, A. (2003). Spatial Analysis for Coastal Zone Management: Beyond GIS. Sustainable Management of Scarce Resources in the Coastal Zone. http://www.ess.co.at/SMART/kfaafull.html on 9 December 2004.

Tundra and Polar Regions

Breidamerkurjökull, Iceland

Iceland (n.d.). http://www.ahojky.net/Pages/Iceland/Iceland%2020Vatnajokull%20Glacier%20Info.htm on 19 February 2005.

Filchner Ice Shelf, Antarica

Ferrigno, J. G. and Gould, W. G. (1987). Substantial changes in the coastline of Antarctica revealed by satellite imagery: Polar Record, Volume 23, No. 146, 577-583.

Korotkov, A. (n.d.). Personal communication. Department of Ice Regime and Forecasts, Arctic and Antarctic Research Institute.

Mark F. Meier (1993). Ice, climate, and sea level; do we know what is happening? In: W. R. Peltier, ed., Ice in the climate system: Berlin, Springer-Verlag, NATO ASI Series Volume I I2, 141-160.

Oerter, H. (1992). Evidence for basal marine ice in the Filchner-Ronne ice shelf. Nature, Volume 358, No. 6385, 30 July 1992, 3, 399.

USGS (2001). USGS EarthShots: Satellite Images of Environmental Change. http://edc.usgs.gov/earthshots/slow/Filchner/Filchner on 30 December 2002; Filchner Ice Shelf, Antarctica. http://edcwww.cr.usgs.gov/earthshots/slow/Filchner/Filchner on 19 February 2005.

Vaughan, D. (1993). Chasing the Rogue Icebergs. New Scientist, Volume 137, No. 1855, 9 January 1993, 26.

Williams, R.S. Jr. and Ferrigno, J.G. (1988). Satellite image atlas of glaciers of the world; Antarctica: U.S. Geological Survey professional paper, Washington, USA, 1386-B, 278, B103; Stewart 1990, 485.

Hubbard Glacier, USA

USGS (2001). USGS Earthshots. Satellite Images of Environmental Change, 8th ed., 12 January 2001, from the EROS Data Center of the U.S. Geological Survey, a bureau of the U.S. Department of the Interior. http://edcwww.cr.usgs.gov/earthshots/slow/Hubbard/Hubbard on 31 January 2002.

USGS (2002a). United States Geological Survey Water Resources of Alaska – Glacier & Snow Program. Hubbard Glacier, Alaska. http://ak.water.usgs.gov/glaciology/hubbard/index.htm on 18 March 2005.

USGS (2002b). U.S. Department of the Interior, U.S. Geological Survey. Advancing Glacier Coming Close to Blocking Fjord Near Yakutat, Alaska. http://www.usgs.gov/features/glaciers.html on 18 March 2005.

Mount Kilimanjaro, Tanzania

Grimshaw, J. M., Cordeiro, N. J., Foley, C. A. H. (1995). The mammals of Kilimanjaro. Journal of East African Natural History 84, 105-139.

Hemp, A. (2001). Ecology of the pteridophytes on the southern slopes of Mt. Kilimanjaro. Part II: Habitat selection. Plant Biology 3: 493-523.

Lambrechts, C. (2001). Personal Communication. UNF, UNEP, KWS, University of Bayreuth, WCST.

Prudhoe Bay, USA

Miller, P. A. (n.d.). The Impact of Oil Development on Prudhoe Bay, Arctic Connections. http://arcticcircle.uconn.edu/ANWR/arcticconnections.htm on 18 March 2005.

US Fish and Wildlife Service (2001). Potential Impacts of Proposed Oil and Gas Development on the Arctic Refuge's Coastal Plain: Historical Overview and Issues of Concern. http://arctic.fws.gov/issues1.htm on 23 February 2005.

Credit: Tim McCabe/UNEP/NRCS

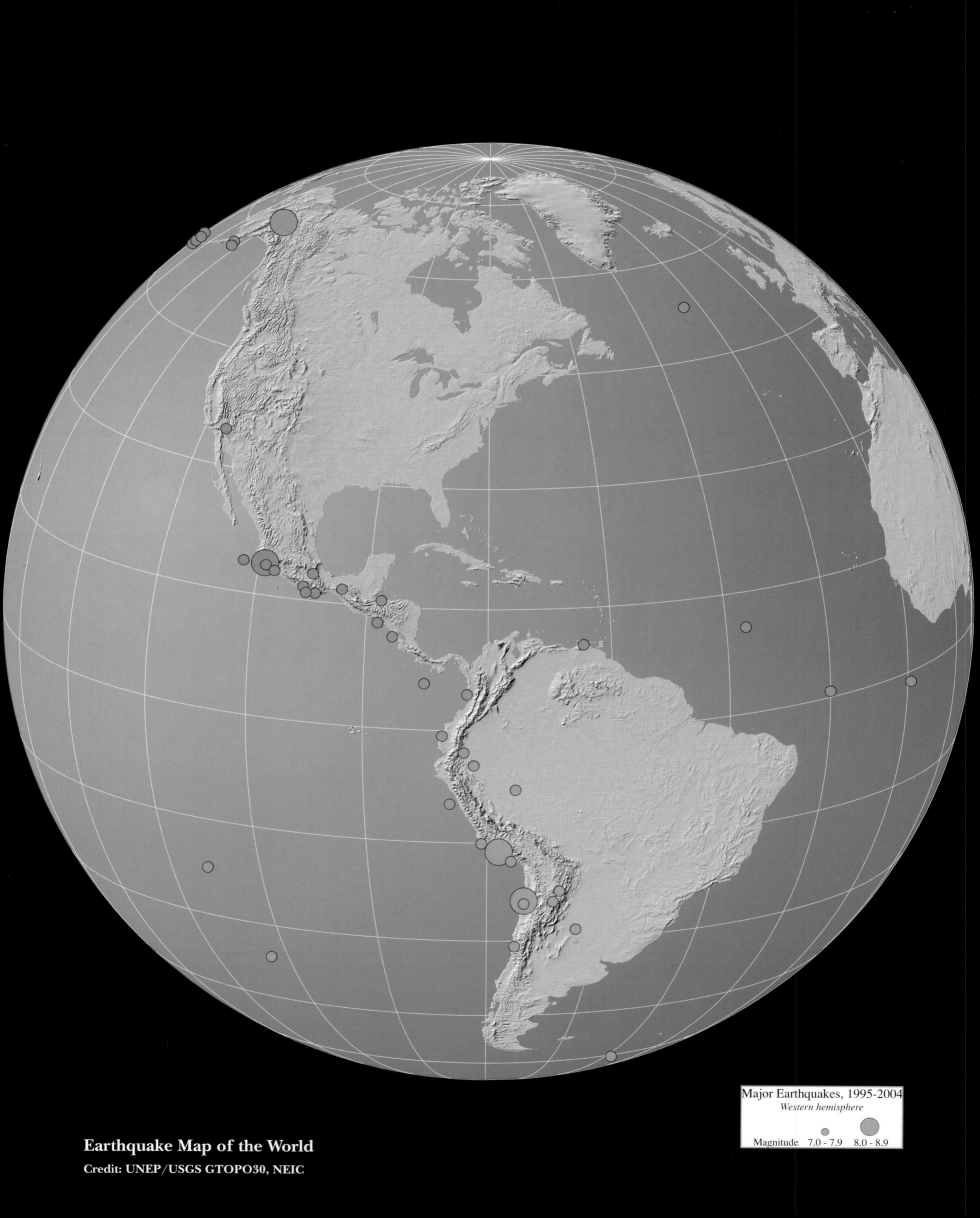

Earthquake Map of the World

Credit: UNEP/USGS GTOPO30, NEIC

Major Earthquakes, 1995-2004
Western hemisphere

Magnitude   7.0 - 7.9   8.0 - 8.9

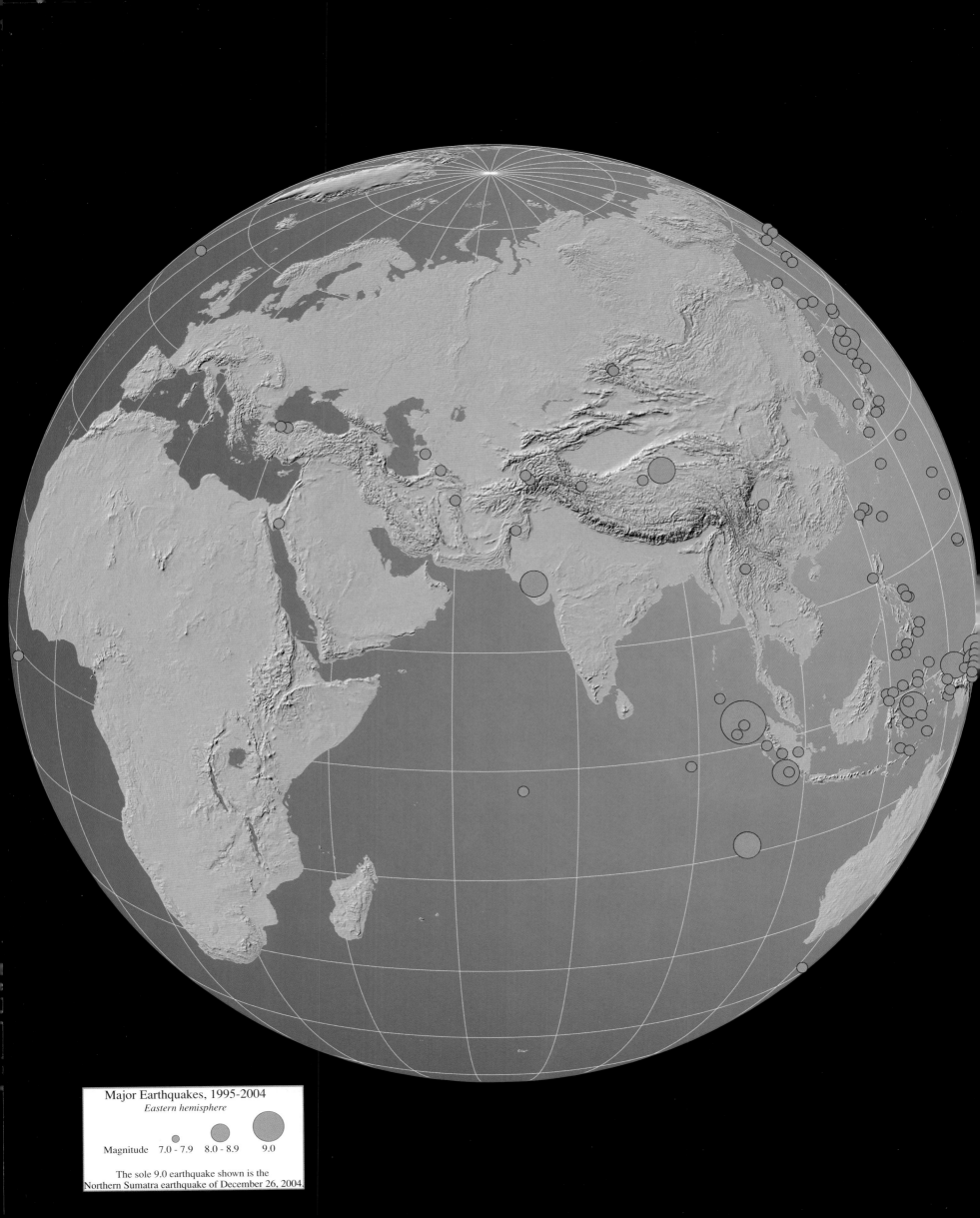

Major Earthquakes, 1995-2004

Eastern hemisphere

Magnitude   7.0 - 7.9   8.0 - 8.9   9.0

The sole 9.0 earthquake shown is the
Northern Sumatra earthquake of December 26, 2004.

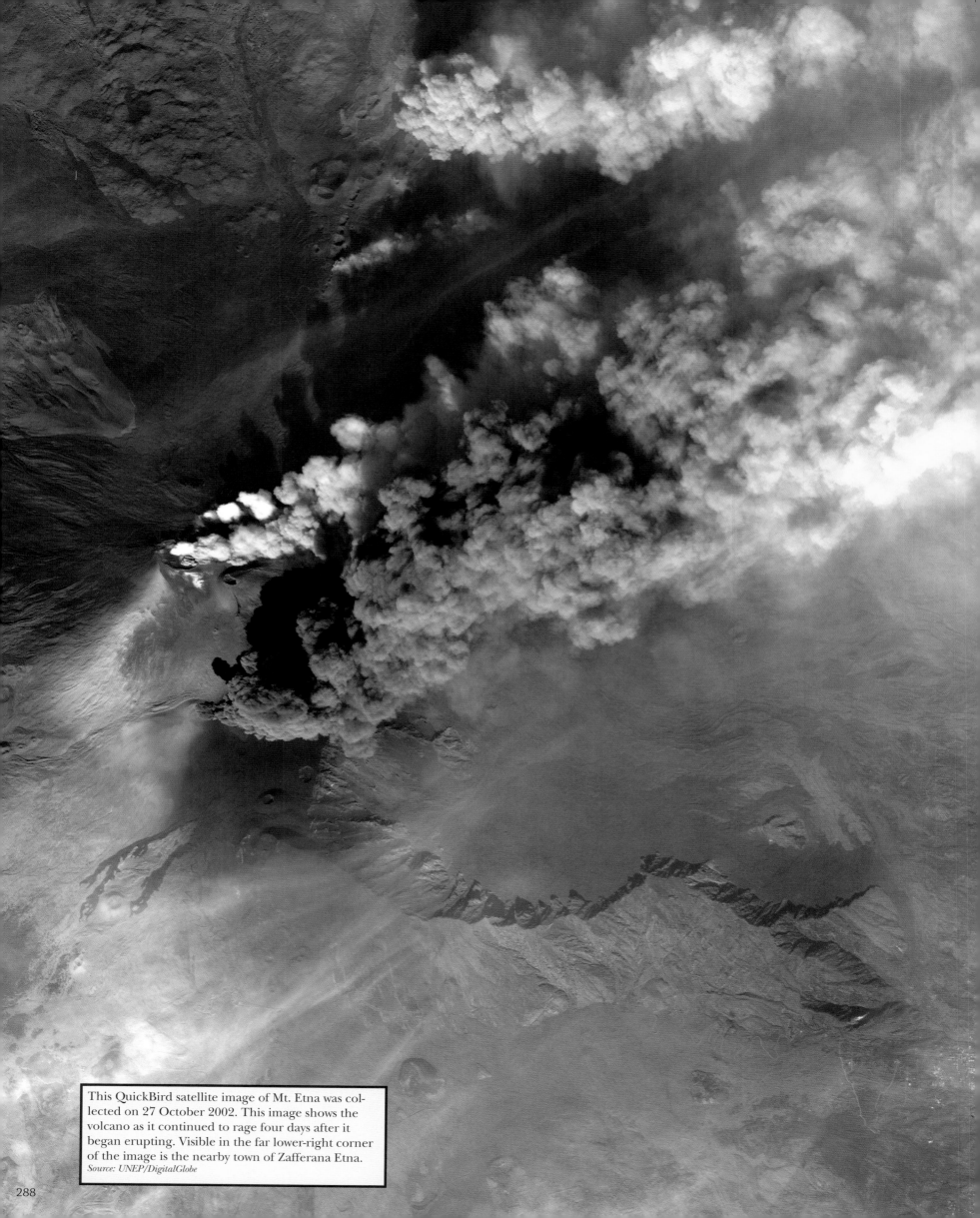

This QuickBird satellite image of Mt. Etna was collected on 27 October 2002. This image shows the volcano as it continued to rage four days after it began erupting. Visible in the far lower-right corner of the image is the nearby town of Zafferana Etna.
Source: UNEP/DigitalGlobe

Natural and Human–induced Extreme Events

4

Extreme events, whether natural or human-induced, can cause significant environmental change, not to mention their impacts on peoples' lives. But many types of natural hazards are also exacerbated by environmental degradation caused by humans. The examples illustrated by the pictures and stories in this chapter highlight the links between population growth and distribution, environmental damage, and natural disasters. They underscore the need to protect the natural environment of this, our only planet, and to strengthen its capacity to resist the impact of both increasing numbers of people and destructive natural events.

Extreme environmental events, or "natural" disasters, have generally been regarded as unpredictable and uncontrollable "acts of God." Increasingly, however, it is becoming clear that human activity can aggravate natural events. With a growing population, more people living in hazard-prone regions, and increased environmental degradation, the intensity, frequency, and impacts of natural hazards are also heightened. As forests are cut down, wetlands built over, and coral reefs disappear, for example, these ecosystems can no longer function as protective controls on the forces and impacts of hurricanes, floods, and tidal waves. Both the poor, sometimes pushed into vulnerable regions by economic and political forces, and the wealthy, who build expensive homes where they wish, move into disaster prone areas and are affected when natural events occur. With stricter building codes and generally less densely populated settlements, fewer deaths occur in developed nations than they do in developing countries, where these events cause very large numbers of

Tornado. *Credit: Unknown/UNEP/NOAA*

casualties and homelessness. Financial losses are easier to measure in developed countries; while the economic ramifications in developing nations may be extensive, they are more difficult to calculate. Finally, global climate change resulting from human activity is expected to increase the intensity, frequency, and impacts of weather-related "natural" hazards (UN/ISDR 2004).

Many countries are now engaged in disaster risk reduction and this activity is likely to increase in the wake of the Asian tsunami of 26 December 2004. The scientific community is making efforts to monitor numerous parametres related to hazardous events. By studying and understanding past events and monitoring on-going ones,

Number of people affected due to various disasters between 1995-2003

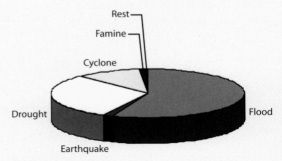

Figure 4.1: The number of people affected as a result of different types of natural hazards between 1995 – 2003. Floods were the most destructive hazard, followed by droughts, cyclone/hurricane/typhoons and earthquakes. *Source: UN/ISDR 2004*

we can glean information that will help to minimize the risk of disaster. While most natural hazards are inevitable, disasters are not (UN/ISDR 2004). Satellite imag-

ery, aerial photography and Geographic Information Systems (GIS) technology are important tools in monitoring and in providing early warning information about these natural hazards and the impacts they may have so that preventive measures can be taken against impending disasters (UN/ISDR 2004). This chapter includes case studies based on remote sensing data, providing visual examples of each type of extreme event discussed.

Extreme events are hazards that occur as consequences of the impacts of natural or a human-induced hazard. In this publication, extreme events are divided into three categories:

- Geo-hazards: volcanoes, earthquakes, tsunamis, landslides/mudslides;

- Climatic hazards: floods, drought, hurricanes, tropical cyclones, tornadoes, ice storms;

- Industrial hazards: oil spills, nuclear, and industrial accidents.

All of these events can expose people and ecosystems to danger. Proportionally, they tend to hurt the poor most of all. This is because the poor outnumber the rich and live in greater density in more poorly built housing on land most at risk. They also have fewer resources and capacity to prevent or cope with the impacts (UNEP 2002a). The number of disasters has increased more than four-fold since the 1960s, from an average of 44 disasters a year to an average of 181 disasters a year by the 1990s. Although some of this increase may be due to improved reporting of events, it is likely that the number, severity, and frequency of natural disasters

is increasing. In addition to improved reporting, the substantial growth in world population and the increasing vulnerability of marginal groups is a significant factor in the growth of natural disasters (Kasperson et al. 2001).

Since 1900, natural hazards have caused over 50 million deaths. Between 1995 and 2003, they affected 6 000 million people (some 2 500 million in Asia alone) and caused over 6 million deaths. Floods affect by far the most people (Figure 4.1). While the number of disasters appears to be increasing, the number of fatalities is declining. This fact may be attributed to improved forecasting, better preparedness, and quicker response to disasters. On the other hand, the number of persons affected has increased. This is not surprising, given the rapidly growing populations of most developing countries and the millions of people who depend directly on the natural resources in their immediate environments to sustain their livelihoods. When these resources disappear or are degraded in floods, earthquakes, tsunamis, and other disasters, the economies of families and whole communities are devastated.

Economic losses from natural disasters have also increased over the past 50 years. Part of this upward trend is linked to socioeconomic factors, such as population growth in, and migration to, large cities in vulnerable areas, and the increased wealth of some populations that choose to live in hazard-prone areas. Another factor is linked to climate change, such as changes in precipitation and flooding events that destroy property and businesses (OWF n.d.).

4.1 Geo-hazards

Volcanoes

A volcano is a vent in the surface of the Earth through which magma and associated gases and ash erupt; also, the word refers to the form or structure (usually conical) that is produced by the ejected material (UND n.d.). Volcanic eruptions are among the most impressive natural disasters, due to their unpredictable nature, which includes flying debris, streams of molten lava, emissions of toxic gases, seismic effects, and the many impacts these have on the people, animals, and plants in their way.

About 550 volcanoes have erupted in the Earth's recorded history and an equivalent number of dormant volcanoes have only erupted in the past 10 000 years. Both dormant and "active" volcanoes have the potential to erupt again. On any given day, about ten volcanoes are actively erupting (Camp 2000). Explosive eruptions give little warning, while effusive eruptions, which send out gently flowing lava, allow time for people to escape (Francis 1993).

Of all natural hazards, volcanic eruptions and earthquakes are the least exacerbated by human activity. These powerful events fit more neatly into the definition of truly "natural" disasters. Nevertheless, human activity can increase the risk of damage caused by such events. For example, volcanic regions are attractive sites because the soils are fertile and they provide valuable minerals, water reservoirs, geothermal resources, and scenic beauty. People become more vulnerable when they settle too close to active volcanoes. Poor people may move closer to these potentially dangerous sites as their populations grow and or when they have no access to other land. Out of necessity, they may cut trees on volcano slopes, increasing the danger from lava flows when they happen (Benson 2002).

A massive volcanic explosion can have important environmental consequences also, due to the blast of huge clouds of ash, dust, and gases into the atmosphere. Volcanic debris in the lower atmosphere falls out or is rained out within days. Volcanic gas can be directly harmful to humans, animals, plants, agricultural crops, and property. The most common consequence is the movement of large numbers of people fleeing the lava flow.

Environmentally, the hazards from volcanic gases are most severe in the areas immediately surrounding volcanoes, especially on volcano flanks downwind of active vents and fumaroles. These hazards can persist for long distances downwind, however, following large eruptions, or from volcanoes erupting gas-rich magma (McGee et al. 1997). The resulting veil of pollution in the upper atmosphere can have long-term and geographically extensive impacts on climate. Such pollution is in the stratosphere and may remain for several years, gradually spreading to cover much of the globe. The particles reflect energy from the sun back into space, preventing some of the sun's rays from heating the Earth, thus reducing global warming. The Mount Pinatubo eruption of 1991 was such a case. An individual eruption may generate global cooling amounting to two or three tenths of a degree Celsius with effects lasting for a year or two (Kelly 2000; Santer et al. 2001). Millions of tonnes of sulfur dioxide gas may reach the upper atmosphere where it transforms into tiny particles of sulfuric acid, known as aerosols, that can lead to acid rain (Kelly 2000; CSIRO 2002). Major eruptions have not been common this past century, occurring once every ten to twenty years, so the long-term influence has been slight (Kelly 2000).

Volcano *Credit: HVO/UNEP/USGS*

Case Study: Kilauea Volcano Eruption, Kalapana, Hawaii 1983-1991

An example of the kind of damage wrought by volcanoes is the eruption of Kilauea Volcano in Hawaii. Between 1983 and 1990, erupting lava repeatedly invaded communities along the southern coast of the Big Island of Hawaii, destroying more than 180 homes, a visitor center in Hawaii Volcanoes National Park, highways, and treasured historical and archaeological sites.

Source: USGS 2000; USGS 2002

Credit: J.P. Eaton/UNEP/USGS

Maps of lava-flow field from the Pu`u `O`o and Kupaianaha vents of Kilauea Volcano, Hawaii, January 1983 - January 1991

Orange colour shows areas covered by lava erupted from Pu`u `O`o between January 1983 and June 1986. Red colour shows areas covered by lava erupted from Kupaianaha between July and October 1986.

Photos by J.P. Eaton. Source: http://hvo.wr.usgs.gov/kilauea/history/1990Kalapana/#heart

These photos show the ecological effects of the eruption, including defoliated papaya plants in an orchard at the north edge of the Saefuji orchid farm. The leaves have been abraded and sheared off by falling pumice from the lava fountain in the background. The ridge between the orchard and the fountain in the photo on the right was formed by the advancing `a`a flow (an Hawaiian term for a type of lava flow that leaves rough-edged, porous lava). *Photos by J.P. Eaton/UNEP/USGS*
Source: http://hvo.wr.usgs.gov/kilauea/history/1960Jan13

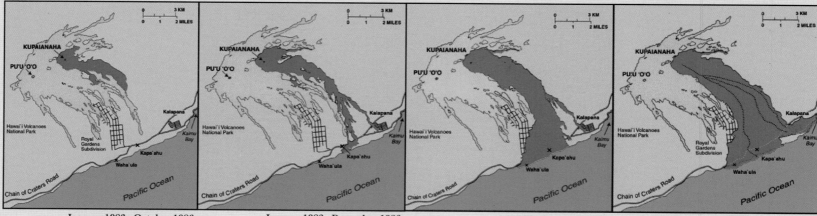

January 1983 - October 1986   **January 1983 - December 1986**   **January 1983 - December 1989**   **January 1983 - January 1991**

23 April 1990   **6 June 1990**   **13 June 1990**

Case Study: Eruption of Mount St. Helens 18 May 1980

Mount St. Helens is located in the state of Washington on the west coast of the United States. It is part of the Cascade Range, which is dominated by periodically active volcanic peaks. For most of the 20th century, the snow-covered mountain was known for its quiet beauty, until on 18 May 1980, the top 420 m (1 300 ft) disappeared within minutes. The blast leveled 400 km² (249 square miles) of forest, formed a deep horseshoe crater, and sent thousands of tonnes of ash into the upper atmosphere. A major debris flow filled a valley along 24 km (15 miles). Sixty-two people were dead or missing. This eruption of Mount St. Helens was the most destructive in the history of the United States, with total economic losses estimated at US$1.2 billion (NGDC 2004).

Growth of the new lava dome inside the crater of Mount St. Helens continues, accompanied by low rates of seismicity, low emissions of steam and volcanic gases, and minor production of ash (USGS 1999). Lessons learned from this and other volcanic activity in the Cascade Range will be invaluable to scientists for predicting such events and anticipating their ecological impacts (UNEP 2003).

Mount St. Helens and the devastated area is now within the Mount St. Helens National Volcanic Monument, under jurisdiction of the United States Forest Service. Visitor centers, interpretive areas, and trails are being established as thousands of tourists, students, and scientists visit the monument daily. Mount St. Helens is once again considered to be one of the most beautiful and interesting of the Cascade volcanic peaks.
Credit: Photograph taken on May 19, 1982. Lyn Topinka/UNEP/USGS Source: Poland 2002

15 Sep 1973

22 May 1983

25 Sep 2000

4 Oct 2004

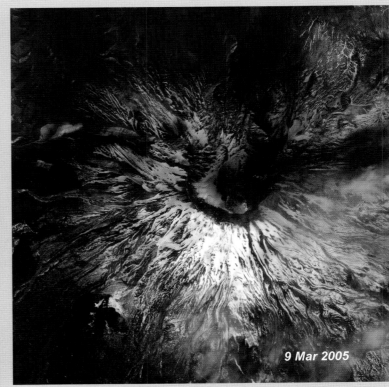

9 Mar 2005

Credit: Space Imaging

Credit: Space Imaging

293

Earthquake destruction *Credit: News Photo/UNEP/FEMA*

Earthquakes and Tsunamis

Earthquake refers to volcanic or magmatic activity or other sudden stress changes in the earth. The term is used to describe both a sudden slip on a fault and the resulting ground shaking and radiated seismic energy caused by the slip (USGS 2002).

Earthquake impacts are many and varied, ranging from minor structural damage in a few buildings to complete devastation over huge areas. The most powerful earthquakes are capable of annihilating major urban centers and severely disrupting the

Table 4.1 – Five largest earthquakes in the world since 1900

Year	Magnitude	Country
1960	9.5	Chile
1964	9.2	Prince William Sound, Alaska
1957	9.1	Andreanof Islands, Alaska
1952	9.0	Kamchatka
2004	9.0	Banda Aceh, Indonesia

Source: NEIC 2004, http://neic.usgs.gov/neis/eqlists/10maps_world.html December 30, 2004.

social and economic fabric of nations. For example, the Kobe Earthquake of 1995 resulted in over 6 000 deaths and estimates of repair costs in the range of US$95 billion to US$147 billion (EQE 1995).

A tsunami (soo-NAH-mee) is a series of extremely long traveling ocean waves generated primarily by underwater ground displacement due to an earthquake or volcanic eruption. In the deep ocean, tsunami waves propagate at speeds exceeding 800

km/h (500 mph). Here, the wave height is only a few tenths of metres(<1 foot) or less. Tsunamis differ from ordinary ocean waves because of the great distance and time between wave crests, which are often separated geographically by more than 100 km (60 miles) in the deep ocean and in time by 10 minutes to an hour. As they reach the shallow waters of the coast, the waves slow down and the water can pile up into a wall of destruction. The effect can be amplified where a bay, harbor, or lagoon funnels the wave as it moves inland. Large tsunamis have been known to rise over 30 m (100 ft). Even a tsunami 3–6 m (10–20 ft) high can be very destructive and cause many deaths and injuries.

Table 4.1 shows the five largest earthquakes since 1900 as measured on the Richter scale. Although it is the fifth largest, the Banda Aceh earthquake-tsunami that originated in Indonesia on 26 December 2004 affected two continents and led to the largest number of deaths.

According to long-term records (since about 1900), we can expect about 18 major earthquakes (7.0 - 7.9 on the Richter scale) and one great earthquake (8.0 or above) in any given year (NEIC 2003). The U.S. Geological Survey, however, estimates that several million earthquakes occur in the world each year. Many go undetected because they occur in remote areas or have very small magnitudes. The National Earthquake Information Center (NEIC) now locates about 50 earthquakes each day, or about 20 000 a year (NEIC 2004). An increase in the number of seismograph stations and the more timely receipt of data

has allowed scientists to locate earthquakes more rapidly and to detect ever-smaller seismic events (NEIC 2003).

The number of earthquakes and tsunamis resulting in fatalities has increased approximately in proportion to global populations, and although a decreasing fraction of the global population has been killed by earthquakes in the 20th century compared to past centuries, seismic risk in certain regions has increased substantially. The cause of the apparent paradox lies in the growth of urban agglomerations where most of the world's growing population will live, and the location of many of these cities near plate boundaries where earthquakes occur quasi-periodically (Bilham 1995).

The growth of giant urban cities near regions of known seismic hazard is a new experiment for life on the Earth. With few exceptions, recent large earthquakes (M>7.5) have spared the world's major urban centers. This will not persist indefinitely. The recurrence interval for damaging earthquakes varies from 30 years to 3 000 years; if population densities remain high in the 21st century, several megacities will be damaged by significant earthquakes (Bilham 1995).

Tsunamis are a threat to life and property for all coastal residents. There has been massive migration to coastal areas, and today, more than half the world's population lives close to the sea (Global Oceans 1999). This has caused the rapid degradation of these areas. As protective natural features, such as coral reefs and mangroves, are removed by human development for tourist hotels and shrimp

Case Study: Bhuj Earthquake, India 26 January 2001

R. P. Singh, S. Bhoi, A. K. Sahoo

The magnitude 7.6 Bhuj earthquake that shook the Indian Province of Gujarat on the morning of 26 January 2001 was one of the two most deadly earthquakes to strike India in its recorded history. One month after the earthquake the death toll had reached 19 727, and the number of injured reached 166 000 with at least 600 000 people left homeless. Government estimates placed direct economic losses at US$1.3 billion. Other estimates indicate losses were as high as US$5 billion. The earthquake brought significant changes to the land and surrounding ocean water bodies.

The images above show changes in chlorophyll concentration prior to and after the earthquake. High concentrations of chlorophyll, together with high ocean surface temperature, are favorable conditions for catching fish. The significant increase in the fish caught in February around the Gujarat coast after the earthquake was found to be double that of the normal February fish catch. *Source: Singh et al. 2002*

Chlorophyll Concentration

Pre (18 Jan 2001)　　　　　　　Post (26 Jan 2001)

1　　10　　100　　(mg m^{-3})

Mud volcano observed in Gujarat earthquake of 26 January 2001. *Credit: Ramesh P. Singh/UNEP/Indian Institute of Technology, Kanpur*

A big tentional crack (approximately 30 cm deep) in a nearby field on Bhuj, Khewda. Salt water had come up to the surface through the crack due to liquefaction. *Credit: Ramesh P. Singh/UNEP/Indian Institute of Technology, Kanpur*

acquaculture farms, for example, so the shoreline becomes increasingly vulnerable to the impacts of wave action and potential tsunamis. Coastal areas are also increasingly at risk due to the effects of burning of fossil fuels; climate change threatens to trigger more powerful storms and raise sea levels, exposing coasts to erosion (Doyle 2004). Global warming, poorly planned coastal development, and other threats over which humans have some control are weakening the coast's ecological defenses against natural disasters.

Of course, tsunamis can cause immeasurable damage to marine and terrestrial ecosystems, including coral reefs, mangroves, and forests. This in turn affects the livelihoods of coastal populations who depend directly on natural resources such as fish, food from household gardens, and forest products.

Case Study: Dust Storms Over China March and April 2002

Dust storms are increasing globally with far-reaching consequences for the environment and human health. Severe dust storms can reduce visibility to zero, making travel impossible, and can blow away valuable topsoil, while depositing soil in places where it may not be wanted. Drought and, of course, wind contribute to the emergence of dust storms, as do poor farming and grazing practices. The dust picked up in such a storm can be carried thousands of kilometres.

This pair of images, acquired 16 days apart, covers the Liaoning region of China and parts of northern and western Korea. They contrast a relatively clear day (23 March 2002) with one in which the skies were extremely dusty (8 April 2002). In the later view (right image), the dust obscures most of the surface, although the Liaodong peninsula extending between the Bo Hai Sea and Korea Bay is faintly visible at the lower left. Wave features are apparent within the dust layer.

Storms such as this transport mineral dust from the deserts of China and Mongolia over great distances, and pollution from agriculture, industry and power generation is also carried aloft. Thick clouds of dust block substantial amounts of incoming sunlight, which in turn can influence marine phytoplankton production and have a cooling effect on regional climates. *Source: NASA n.d.; Planetary Photo Journal; Wikipedia n.d.; Vince 2004.*

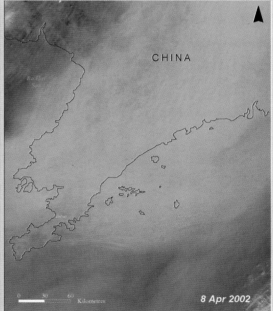

23 Mar 2002　　　　　　8 Apr 2002

Satellite images courtesy NASA/GSFC/LaRC/JPI, MISR Team

Credit: Unknown/UNEP/FAO

Case Study: Indian Ocean Tsunami
26 December 2004

On 26 December 2004, an undersea earthquake measuring 9.0 on the Richter scale took place in the Indian Ocean, off the west coast of northern Sumatra, Indonesia. It caused one of the deadliest disasters in recent times. Resulting tsunami waves crashed into the coastlines of twelve countries bordering the Indian Ocean, causing massive losses in human life and infrastructure, and damage to marine and terrestrial ecosystems. It is estimated that the tsunami killed more than 200 000 people, left up to 5 million in need of basic services, and caused billions of dollars of damage.

The effects of the disaster include massive changes in the physical environment. For example, it is possible that the ocean depth in parts of the Straits of Malacca, one of the world's busiest shipping channels off the coast of Sumatra, was reduced from about 1 200 m (4 000 feet) to perhaps only 30 m (98 ft), a depth that is too shallow for shipping (AP 2005).

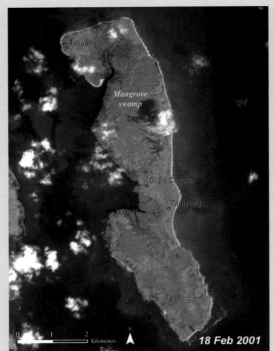

18 Feb 2001

28 Dec 2004

The island of Trinkat, part of the Nicobar Islands, India, appears to have been cut in half by the tsunami with a new channel of water approximately 5 km (3 miles) long stretching from the settlement of Tapiyang to a point on the opposite coast just west of Ol Ok Chuaka. Another channel has possibly been opened up to the southeast of Takasem separating the large mangrove area from the inhabited northern end of the island. The mangrove appears to be relatively intact though several inlets have been created in the east. The extensive coral reefs visible along the west and east coasts of Trinkat before the tsunami are largely obscured by large plumes of sediments presumably washed from the land. The coastline has retreated along the east coast enlarging the lagoon. This scouring of terrestrial matter into the lagoon and onto the reefs could have serious consequences for shallow water habitats if sediments settle for longer periods.
Source: UNOSAT

Photo taken in Kulmunai Kuddi on Sri Lanka's east coast.
Credit: Unknown/UNEP/USGS

The city of Banda Aceh, Indonesia, suffered catastrophic damage as a result of the tsunami that struck on 26 December 2004. These QuickBird Natural Colour images on 23 June 2000 and on 28 December 2004 (below) clearly show the city before the devastation and the extent of the damage after the tsunami. *Source: Digital Globe: http://www.digitalglobe.com/images/tsunami/Banda_Aceh_Tsunami_Damage.pdf*

23 June 2000

28 Dec 2004

QuickBird Natural Color Image
December 26, 2004

QuickBird High Off Nadir Image
December 31, 2004

These images are of the southwestern coast of Sri Lanka, taken shortly after the tsunami struck the coastline. The image dated 26 December 2004 was taken shortly after the shoreline was struck by the tsunami, while the second image was taken after the ocean returned to normal.
Source: http://www.digitalglobe.com/images/tsunami/Sri_Lanka_Tsunami_Damage.pdf

Case Study: Bam Earthquake
26 December 2003

Bam is located in the southeastern corner of Kerman province in Iran. Maintaining its position in the middle of the southern trade route, this small, fortified city on the outskirts of the vast Dasht-é-Lut Desert is just 350 km (217 miles) west of Pakistan and 450 km (280 miles) north of the Persian Gulf. Eighty-thousand people make their homes within Bam's boundaries.

A 6.6 magnitude earthquake struck southeastern Iran on 26 December 2003, killing over 40 000 people, injuring 16 000, leaving 70 000 homeless and destroying much of the city of Bam, the earthquake's epicenter. The quake destroyed the ancient citadel of Arg-e-Bam, located on the historic Silk Road and thought to be over 2 000 years old. This citadel was said to be the largest mud brick structural complex in the world. Apart from the toll on human lives, the loss of this ancient site represents an important cultural loss.

Although Iran is subject to frequent large quakes, it does not have strong building codes and buildings generally do not withstand the impact of these events. As a result, casualties and damage are much higher than might be expected from a similar quake elsewhere in the world (The Earthquake Museum 2003). *Source: NASA 2004e*

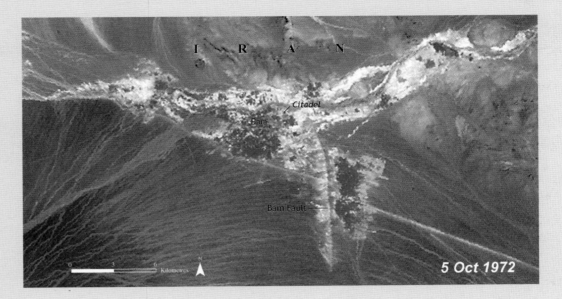

5 Oct 1972

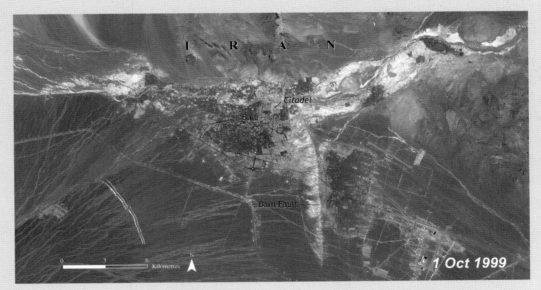

1 Oct 1999

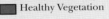 Healthy Vegetation

Credit: Unknown/UNEP/IIEES

In a region famous for the scarcity of its water, Bam thrived with extensive palm groves and citrus gardens (see images above). Benefiting from subterranean water reserves, surfacing through a number of several–km–long water canals, Bam was essentially an agricultural city famous for, and a major producer of, the very best date fruits in all of Iran. After the earthquake of 26 December 2003 that flattened the citadel and the mud-brick houses and destroyed 85 per cent of the city's buildings, just about the only things left standing tall above the ruins of Bam were the mainstays of the local economy: date palms. The date harvests that produced thousands of tonnes of dates each year were left undamaged in plantation fields and house gardens, offering hope for an agricultural-based recovery. Irrigation repairs have begun and agriculturists are optimistic that future date harvests could be as large as those before the earthquake. *Source: USGS n.d.*

Credit: Unknown/UNEP/IIEES

Mudslide *Credit: News Photo/UNEP/FEMA*

Landslides and Mudslides

Worldwide, thousands of people die every year from landslides and mudslides. In the United Sates alone, they cause an estimated US$1 billion in damage and kill 25 to 50 people every year. Earthquakes, volcanoes, and a number of types of weather events can trigger landslides, which are characterized by lethal mixtures of water, rocks, and mud. The two largest landslides in the world in the 20th century occurred at Mount St. Helens, Washington, in 1980 and at Usoy, Tajikistan, in 1911. Although Mount

St. Helens was the largest landslide recorded in historic time, fewer than 60 people were killed because most residents and visitors had been evacuated. The Usoy landslide, also triggered by an earthquake, moved 2.4 km^3 (1.5 cubic miles) of material and built a dam 573 m (1 880 feet) high (half again as high as the Empire State Building) on the Murgob River in Tajikistan; the dam still impounds a lake nearly 64 km (40 miles) long. This landslide took place in a sparsely populated area and thus caused few deaths (USGS 1999).

The deadliest landslide this century was also the result of an earthquake, which occured in western Iran on 20 June 1990. It caused 40 000–50 000 deaths. One of the world's other major landslides includes the rock and snow avalanche triggered by a magnitude 7.8 earthquake at Mount Huascaran, Peru, on 21 May 1970 that buried the towns of Yungay and Ranrahirca, killing perhaps as many as 20 000 people (NASA 1999).

Case Study: Mudslides in California 6-11 January 2005

Many days of storms across California in January 2005 led to flooding, mudslides, and huge snowfall totals. On 10 January, a landslide struck the town of La Conchita in Ventura County, destroying or seriously damaging 36 houses and killing ten people. It was not the first destructive landslide in the area and future landslides are likely to occur. The area is a narrow coastal strip of land between the shoreline and a high bluff above which rises a terrace covered by avocado and citrus orchards (Jibson 2005). Despite the landslide risk, a growing and generally wealthy population has expanded into fragile or risk-prone areas such as these in California, often building expensive homes like those destroyed in La Conchita. The population of Ventura County, for example, grew by five per cent in 2003, from 753 197 in 2000 to approximately 790 000 (US Census Bureau 2004).

In the image at right, Multi-satellite Precipitation Analysis (MPA) rainfall totals are shown for the period 6–11 January 2005. The red areas just off of the coast indicate the highest totals of more than 225 mm (about 9 inches) of rainfall. *Source: NASA 2005*

January 6 - 11, 2005

Rainfall Accumulation (mm)

25   75   125   175   225

NASA Earth Observatory: http://earthobservatory.nasa.gov/ NaturalHazards/natural_hazards_v2.php3?img_id=12669 7 February 2005

Case Study: Landslide Creates Lake in Tibet
2004

Tibet is the major source of Asia's great rivers. It also has the Earth's loftiest mountains, the world's most extensive and highest plateau, ancient forests, and many deep valleys untouched by human disturbance.

In early summer of 2004, a landslide in the Zaskar Mountains, a range of the Himalayas, created a natural dam blocking the Pareechu River in its course from the Tibet Autonomous Region of China to the Himachal Pradesh State of northern India. The dam is 35 km (22 miles) from India's border with China. The water is slowly building behind the dam, creating an artificial lake in the remote mountain region. By 13 August, the lake had spread over 188 hectares and had reached a depth of 35 m (115 feet), with water levels rising daily.

The new dam and lake pose a threat to communities downstream in northern India. Indian and Chinese officials fear that the unstable dam will burst, releasing a torrent of water on these populated regions. The remoteness of the area and the ruggedness of the terrain have precluded preventative measures that could control the potential catastrophic release of water, although people have been evacuated from villages in both the Chinese and Indian parts of the region (NASA 2004a).

These images show the area before the landslide (top) and the growing lake following the landslide (center and bottom) on 15 July 2004 and 1 September 2004. What previously had been a river valley around the meandering Pareechu River on has been entirely covered with dark blue water.

Satellite images courtesy NASA/GSFC/MITI/ERSDAC/JAROS, and U.S./Japan ASTER Science Team

Source: Zhu Pingyi/UNEP/ICIMOD and RRC-AP

4.2 Climatic Hazards

Climatic hazards include storms, floods, heat waves, droughts, and ice storms. The majority (two-thirds) of all natural disasters are climate or weather-related, principally through drought, flooding, and storms. Furthermore, of all natural hazards, human activity affects weather-related hazards the most. With a changing climate influenced by the burning of fossil fuels, extreme weather events are projected to increase in frequency and/or severity during the 21st century (IPCC 2001). Combined with population growth and increased settlement in risk-prone areas, the impacts of such events on humans and ecosystems will also increase.

A storm is a low pressure in the atmosphere marked by wind and usually by rain, snow, hail, sleet, or thunder and lightning. One of the most violent and destructive is the cyclone or hurricane.

A tropical cyclone is a large-scale closed circulation system in the atmosphere above the ocean with low barometric pressure and strong winds. The winds rotate clockwise in the southern hemisphere and counter-clockwise in the northern hemisphere. The system has wind speeds of 119 km/h (73 mph) or more (UN-DHA 1992).

Tropical cyclones are called "hurricanes" in the western Atlantic and "typhoons" in the western Pacific. These dangerous storms can be found in three of the Earth's four oceans and in both hemispheres. Even though Atlantic Ocean tropical cyclones (hurricanes) receive a lot of attention, only 12 per cent of tropical cyclones worldwide are located here. The northwestern Pacific Ocean averages more than 25 cyclones (typhoons) each year. Another location with great activity is the Indian Ocean. No other part of the world has so much activity in such a small area. The Southern Hemisphere also experiences tropical cyclones. However, they are confined to the Western Pacific and Indian Oceans (DAS n.d.).

A tropical cyclone's storm surge is the most destructive aspect of the storm. It kills the most people, destroys buildings, and erodes coastal shorelines. Hurricane Andrew, which landed in south Florida in 1992, was the most expensive cyclone to date, causing US$25 billion in property damage and killing 26 people. The cyclone that caused the highest mortality in the 20th century was an unnamed typhoon that struck Bangladesh in 1970, killing about 300 000 people.

Scientists predict that global warming will cause warmer ocean temperatures and associated increased moisture in the atmosphere—two variables that work to power hurricanes. As a result, more intense hurricanes that cause even more damage when they hit land are predicted (Henderson-Sellers et al. 1998).

Tropical cyclones also cause flooding. A flood is a significant rise of water level in a stream, lake, reservoir, or coastal region (UN-DHA 1992). Human actions can cause or contribute to flooding events through the impacts of dams, levees, the removal of wetlands (that store water), deforestation (resulting in erosion), and other means.

Flooding *Credit: Andrea Booher/UNEP/FEMA*

Case Study: Supercyclone hits eastern India 29 October 1999.

On 29 October 1999 a supercyclone with winds in excess of 257 km/h (160 mph) swept in from the Bay of Bengal to hit the eastern state of Orissa, India. An estimated 15 million people were left homeless by the storm, which had a death toll as high as 10 000.

Supercyclone approaching Orissa coast, India, October 1999

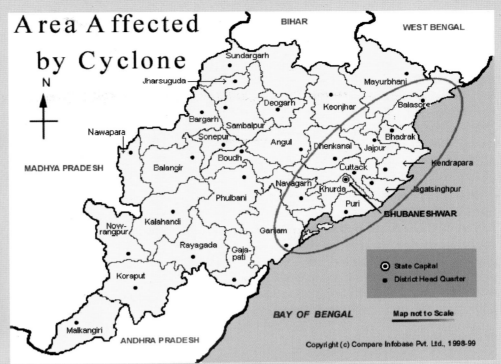

Coastal areas of Orissa were hard hit by the supercyclone. The oval outline encloses major areas affected by this supercyclone. *Source: Maps of India, 2004. http://www.mapsofindia.com/maps/mapin-news/22101999.htm July 18, 2004*

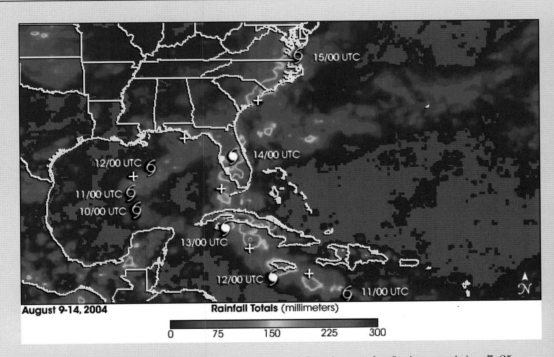

Case Study: Hurricane Charley August 2004

Hurricane Charley developed from a tropical wave that emerged off the African coast early in August 2004. It hit the western tip of Cuba and by the time it reached the greater Havana area, maximum sustained winds were nearly 165 km/hr (105 mph). Western Cuba suffered more than US$1 billion in property damage and three people died.

The map above shows Charley's path between the 9th and 14th of August 2004 as it traveled up from the Caribbean into Florida and the southeast United States. The map shows Multi-satellite Precipitation Analysis (MPA) rainfall totals for the period. A swath of 7–12 cm (3–5 in) rainfall (green area) extends from the central Gulf of Mexico into northern Florida as a result of Tropical Storm Bonnie, which landed in Florida on the 12th of August.

A heavier swath of rain containing 7–25 cm (3-10 in) amounts (darker red areas) extends from the north central Caribbean up through Cuba across Florida and merges with a heavy rain area along the Carolina coast.

In Florida, 25 of the state's 67 counties were declared federal disaster areas. Estimated insured losses from Charley were US$7 billion, while total economic loss was estimated at nearly US$15 billion. Charley was blamed for 22 deaths.

Despite its history of hurricanes, Florida's warm weather and beaches attract migrants, retirees, and tourists. Florida's population grew by 6.5 per cent between 2000 and 2003 (U.S. Census Bureau 2004). In some coastal areas, tourists and "snow birds" (northern Americans and Canadians who spend the winter in the south) swell populations by 10 to 100 fold. Large parts of densely populated coastal areas

Hurricane Charley blew ashore over Punta Gorda, Florida, on 13 August 2004, with winds topping 233 km/hr (145 mph). Two days later, the Ikonos satellite captured the top image above. The image shows the destruction the Category 4 hurricane wrought on the coastal city. Debris is scattered across roads, parking lots, and yards, giving the scene a "messy" appearance compared to the crisp, neat neighborhoods shown in the lower image, taken two years earlier on 28 July 2002. *Source: NASA 2004b, http://earthobservatory.nasa.gov/Newsroom/NewImages/images.php3?img_id=16639*

are subject to the inundation caused by hurricane storm surges and on numerous occasions have experienced heavy economic losses from these events (NOAA n.d.). *Source: NASA 2004b, http://earthobservatory.nasa.gov/NaturalHazards/natural_hazards_v2.php3?img_id=12339, NASA 2004b*

Floods

Worldwide, the number of major flood disasters has grown significantly, from 6 cases in the 1950s to 26 in the 1990s. With the changing climate, global precipitation has increased by about two per cent since 1900; during this time, rain patterns have changed, with some places becoming wetter and others, such as North Africa south of the Sahara, drier (Cosgrove 2003).

From 1971 to 1995, floods affected more than 1 500 million people worldwide, or 100 million people per year. In the most calamitous storm surge, the flood in Bangladesh in April 1991 killed thousands of people. The United Nations estimates that by 2025, half the world's population will be living in areas at risk from storms and other weather extremes (Cosgrove 2003).

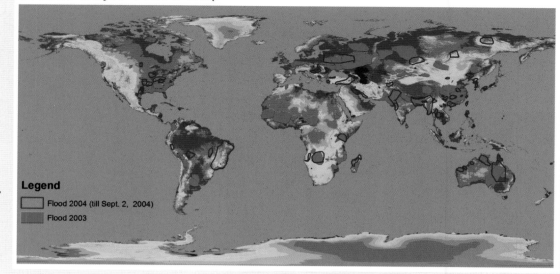

Global Major Flood Map

Legend

☐ Flood 2004 (till Sept. 2, 2004)
▨ Flood 2003

Major flood events around the world in 2003 and 2004 (updated through September 2, 2004)
Data Source: DFO 2004, http://www.dartmouth.edu/%7Efloods/Archives/index.html

Case Study: Flooding in Mozambique 2000 and 2001

The years 2000 and 2001 saw massive flooding in Mozambique, particularly along the Limpopo, Save and Zambezi valleys. In 2000 half a million people were made homeless and 700 lost their lives. The floods destroyed crops and overwhelmed water and sanitation infrastructure in many areas.

Southern Mozambique bore the full impact of the rains and rising waters. In the capital, Maputo, tens of thousands of people were forced to flee their homes. The worst hit were people living in makeshift homes in the slums around the capital. Maputo, the capital city of one million, was literally isolated as a result of the floods, and entry into the city was impossible.

Further north, hundreds of thousands of people were left homeless in Gaza province. Roads, homes, bridges and crops were destroyed. Electricity supplies were disrupted and towns left without clean water supplies after their pumping stations were swept away.

These two images show an area in Mozambique before the onset of flooding and during flooding. The 2000 image reveals a large area around the towns of Vila De Chibuto and Guija submerged under flood water from the Limpopo River. *Source: BBC 2000, Oxfam 2001, FEWS Net 2001. Satellite image: Kwabena Asante–SAIC-USGS National Center for EROS.*

Credit: Philip Wijmans/UNEP/ACT-LWF Trevo

Credit: Philip Wijmans/UNEP/ACT-LWF Trevo

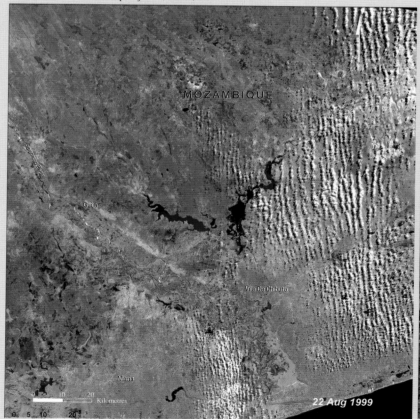

MOZAMBIQUE

Guija

Vila De Chibuto

Mana

22 Aug 1999

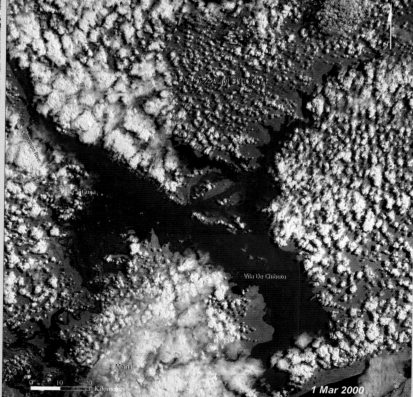

MOZAMBIQUE

Guija

Vila De Chibuto

Mana

1 Mar 2000

Credit: Cpl. Mike Escobar/UNEP

17 September 2000

22 September 2004

Case Study: Severe Flooding in Haiti and the Dominican Republic 23 – 25 May 2004

Several days of heavy rains in late May 2004 caused rivers to overflow in areas near the southern border between the Dominican Republic and Haiti. Heavy rains in this region of deforested hillsides generated rapid run-off and severe flooding. Floods and landslides devastated large areas of the island of Hispaniola, which the two countries share. The flooding demolished entire communities, caused massive loss of life, displaced tens of thousands of people on both sides of the border, and resulted in sizeable crop and livestock losses.

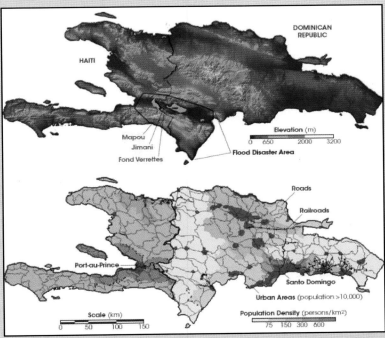

Tropical storm Jeanne struck the Island of Hispaniola on 18 September 2004; a wall of water and mud buried much of Gonaïves, Haiti as shown in this Ikonos imagery captured four days later, on 22 September 2004. Roads visible on 17 September 2000 image have disappeared, as have a number of buildings and adjacent farmlands submerged by water and mud. Note the damaged ship and changes in the water colour in the 22 September 2004 image. *Credit: Ikonos imagery provided on spaceimaging.com, courtesy of NASA's Earth Observatory*

Flooding as a result of this hurricane is blamed for over 3 000 lives lost, including 2 826 in the coastal city of Gonaïves, Haiti (USAID 2004).

Haiti, which is the poorest country in the Americas, has a population of about 8 million and is prone to deadly floods because 98 per cent of its forests have been chopped down, largely to make charcoal for cooking (Sustainable Institute 2004).

These maps (left) compare the topography of Hispaniola (top) with the island's population density (bottom). The flood disaster area around the Massif de la Salle is outlined in blue. *Source: Flood disaster hits Hispaniola, NASA 2004c*

These two Landsat images (below) contrast the two time periods during and after the floods. In the 12 May 2004 image, most of the region is covered by water (grayish colour) while the 26 September 2004 image water has receded, leaving behind green healthy vegetation especially in the area south and east of Gonaives. *Source: NASA Earth Observatory*

12 May 2004

26 Sep 2004

Drought

A drought is a period of dryness, especially when prolonged, that causes extensive damage to crops or prevents their successful growth. Droughts are often caused by heat waves. A heat wave is a period of unusually hot weather. High temperatures exacerbate the effects of drought, damage crops and their establishment, and reduce yields (FAO 1996). Climate change will potentially increase the likelihood of droughts in dry and semi-arid regions. There is already evidence that a number of such regions have experienced declines in rainfall. Droughts result in decreases in soil fertility and agricultural, livestock, forest, and rangeland production. They also exacerbate the process of desertification. (IPCC 2001).

Throughout history, various parts of the globe have suffered drought and subsequent famine, resulting in huge humanitarian and economic losses.

Case Study: Lake Mead–Drought in the Western United States 2003

The western half of the United States has suffered a sustained drought over the past several years, which has caused withering vegetation, more frequent and severe forest fires, and falling water levels in major reservoirs throughout the region.

This image of Lake Mead, Nevada, dramatically captures the result of decreased rainfall and snow in the western United States. As of 2003, water levels at Lake Mead dropped 18 m (60 ft). Lake Mead is formed by the Hoover Dam and is an important water source for the states of Arizona, Nevada, and California. About 25 million people live in the region and the lake supplies over 80 per cent of Las Vegas' drinking water. Population growth, the building of water-hungry golf courses, and the needs of irrigated agriculture in the region are taxing its water resources, however. Las Vegas is the country's fastest growing city and Nevada is its fastest growing state. Although temperatures in the Las Vegas Valley rise to 32°C (90°F) or more on more than 125 days of the year and it receives less than 1 000 mm (39 in) of rain a

Credit: Lynn Betts/UNEP/NRCS

year, Las Vegas has the highest per-capita consumption of water in the world (UNEP 2002b).

The combined effect of drought, population growth, unsustainable development, and climate change in this arid region of the United States could be a recipe for more disastrous droughts and potential conflict. Recently, conservation awareness campaigns and water–use restrictions have helped to lower water use, despite the addition of more than 60 000 new residents in 2003 (SNWA 2004). *Source: Images and text by NASA's Earth Observatory*

3 May 2000

The image to the right, acquired by the Landsat 7 satellite, shows the shoreline of Lake Mead in May 2000.

Water levels in the lake during the 3-year-span illustrated by the 2001, 2003, and 2004 images dropped 18 m (60 ft). In the Boulder Basin of Lake Mead, the lower water level has connected former islands like Saddle Island to the shoreline. *Source: UNEP/GRID - Sioux Falls*

Credit: Lynn Betts/UNEP/NRCS

Brilliant green golf course fairways contrast sharply against the drought-stricken landscape of the Boulder Basin. Despite the region's third-worst drought in recent history, new courses continue to be developed. *Source: UNEP/GRID - Sioux Falls*

28 May 2003

22 May 2001

28 Apr 2004

Case Study: Drought in Horn of Africa

What makes drought in the Horn of Africa an issue of global interest is its perennial recurrence and its extensive humanitarian impact. Poor agricultural practices and environmental degradation (catchments degradation) have greatly compounded the problem leading to serious food crises in the region. In 1984 and 1985, the Horn of Africa experienced one of the worst droughts of the twentieth century with a resultant famine that killed 750 000 people.

This Normalized Difference Vegetation Index (NDVI) image to the right shows the vegetation anomaly for August 1984. Dark red indicates the most severe drought, light yellow areas are normal, and green areas have denser than normal vegetation. *Source: NASA 2000, www.m-w.com, http://earthobservatory.nasa.gov/Library/DroughtFacts/*

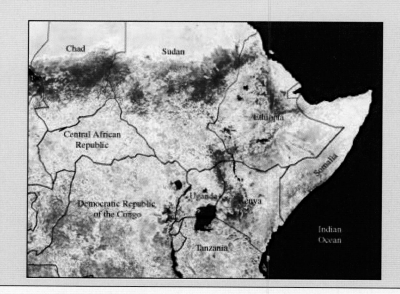

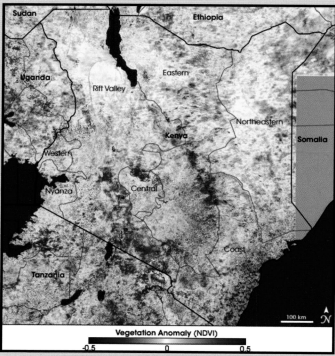

Vegetation Anomaly (NDVI)
-0.5   0   0.5

Case Study: Drought in Kenya January 2005

In Kenya, one of the countries in the Horn of Africa, drought has been looming for several years, leaving many regions of the country parched and hungry. As the 2004/2005 harvest drew to a close, the cereal deficit grew to 300 000 metric tonnes, which meant that up to 2.7 million people needed food aid that season—an unusually high number for Kenya. The second maize

Credit: Unknown/UNEP/African Wildlife Foundation

crop, scheduled to be harvested in March, was predicted to be 20 per cent below average because of a lack of rain. The 2005 shortages stem from a lack of rainfall during the short rainy season, which normally runs from November to January. Though some parts of Kenya received adequate rain, crop-growing regions in the Eastern, Central, and Coast Provinces received far-below-average rainfall. In Central Province alone, about 400 000 people face famine, according to government estimates. *Source: NASA 2004d, http://earthobservatory.nasa.gov/Newsroom/NewImages/images.php3?img_id=16816*

The impact of drought on the crops can be seen in this image, which shows the Normalized Difference Vegetation Index (NDVI) anomaly for Kenya as measured by the Moderate Resolution Imaging Spectroradiometer (MODIS) during the first two weeks of January 2005. NDVI is a measure of vegetation density and health. The anomaly image compares current conditions to average conditions in 2001, 2002, 2003, and 2004 during the first two weeks of January. Between 1-16 January 2005, brown clusters in the Coast and Eastern provinces show patterns of dryness where vegetation is less dense than it has been in the past. More pronounced drought areas surround Central Province. Grey pixels indicate regions where data were not available. An arch of green through the center of the country reveals where rainfall was plentiful and vegetation is thriving.

Case Study: Drought in Australia 2002-2004

After Australia's devastating drought in 2002, the 2003/2004 season saw record wheat and barley harvests, with the March crop up 119 per cent compared to the previous year's drought-stricken crop.

This pattern of large harvests after drought-stunted years is common. To recoup their losses, farmers increase the area they sow. In 2002, pasture land for livestock was so parched and the price of grain so high, that many farmers sold their livestock and converted their land to crops in 2003. In addition to the increase in cropland, well-timed rains in most parts of the country, particularly in Western Australia, combined to produce a bumper harvest that year. *Source: NASA 2003*

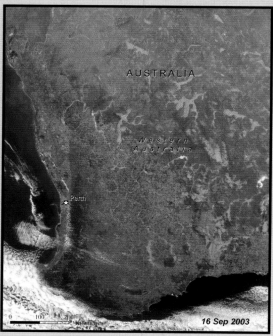

The difference between the two years is clearly visible in this image pair of the southwestern point of Australia, showing the expanded crop area. A larger portion of Western Australia is covered with greener vegetation in September 2003, right, compared to September 2002, a sign that all plants, including grain crops, were thriving in 2003. *Source: NASA 2003, http://earthobservatory.nasa.gov/NaturalHazards/natural_hazards_v2.php3?img_id=12010*

4.3 Industrial Hazards

Industrial hazards are threats to people and life-support systems that arise from the mass production of goods and services (Mitchell 1996). They can be intentional actions, such as the illegal discharge of oil into the environment, or accidental, such as toxic spills. Like natural hazards, they can expose people and ecosystems to danger, affecting lives, health, and socio-economic conditions (Draffan 2004).

One of the major industrial disasters occurred in Bhopal, killing at least 14 400 people and causing permanent disabilities to at least 50 000 others. In the early hours of 3 December 1984, gas leaked from a tank of methyl isocyanate (MIC), resulting in intense emission of toxic gases at a plant in Bhopal , India, owned and operated by Union Carbide India Limited (UCIL). This event is considered to be the worst chemical accident in history. In February 1989, the Supreme Court of India directed Union Carbide Corporation (UCC) and UCIL to pay a total of US$470 million in full settlement of all claims arising from the tragedy.

Oil spills

Fossil fuels (oil, natural gas, and coal) account for the vast bulk of global energy supplies. These fuels, formed over millions of years, are finite and non-renewable. Population growth and increased affluence and consumption increase the demand for fuel. In due course, these resources will become scarce and costly, requiring the introduction of replacement energy sources (MacKenzie 2000). In addition, disputes over their ownership already occur and there is significant potential for increased conflict.

Petroleum is an integral part of our lives. It provides 80 per cent of the world's transportation fuel, supplies nearly half the world's primary energy demand, and provides feedstock for the petrochemical industry. Petroleum products account for about a third of global oil use today.

The exploration for, development, transportation, and use of petroleum causes environmental problems worldwide. The most critical issue today is that fossil fuel burning emits gases that contribute to global climate change (Cohen 1990). Oil spills, the focus of this section, can harm life by poisoning, by direct contact, and by destroying habitats, especially in the marine environment.

As shown in Figure 4.2, about 37 per cent of oil in the world's oceans is

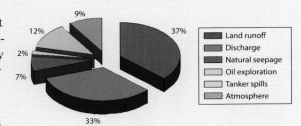

Figure 4.2: Sources of oil in the world's oceans. The highest contribution (about 37 per cent) results from urban run-off and the discharge from land-based industrial plants. These materials reach the sea via storm-water drains, sewage outfalls, creeks, and rivers. *Souce: APPEA n.d.*

the result of urban and industrial runoff. Another seven per cent is oil which seeps naturally out of fissures in the sea beds. About 14 per cent is caused directly by the

Oil spill *Credit: Khan Kuyucu/UNEP/Topfoto*

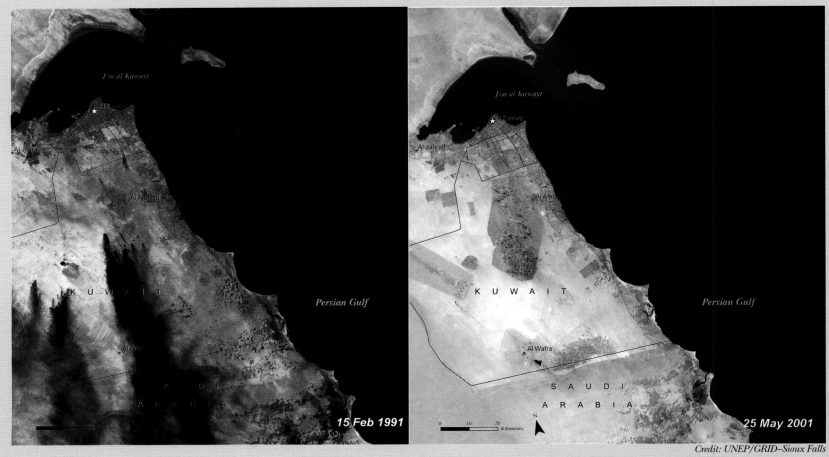

15 Feb 1991

25 May 2001

Credit: UNEP/GRID–Sioux Falls

Case Study: Gulf War
Kuwait and Persian Gulf
23–27 January 1991

During the Persian Gulf War, Iraq deliberately released 908–1 741 million litres (240–460 million gallons) of crude oil from tankers into the Persian Gulf 16 km (10 miles) off Kuwait. Oil spilled onto more than 1 287 km (800 miles) of Kuwait and Saudi Arabian beaches, devastating marine wildlife, especially birds (Krupa 1997).

The Persian Gulf war brought about some of the worst environmental pollution ever recorded as a result of oil spills and oil fires.

In the images, the blue shows water, green shows natural vegetation, light yellow shows desert areas and black shows pollution from oil spills and fire.

oil industry, of which 12 per cent is from accidents involving oil tankers. In the U.S., pipelines now spill considerably more than tankers. Another 33 per cent is the result of discharges to the environment and the remaining nine per cent of oil deposited in the oceans is absorbed from the atmosphere (APPEA n.d.).

As evident from Table 4.2, oil spills happen all around the world. Oil spills of at least 38 m³ (125 cubic feet) have occurred in the waters of 112 nations since 1960. The top four "hot spots" for oil spills from vessels include the Gulf of Mexico (267 spills); the northeastern U.S. (140 spills); the Mediterranean Sea (127 spills); and the Persian Gulf (108 spills) (Etkin 1997).

Despite overall increases in oil transport, the numbers of marine oil spills and the amount spilled have decreased significantly over the last two decades, particularly in the last few years. The average number of large spills per year during the 1990s was about a third of that witnessed during the 1970s (ITOPF 2003). This decrease can likely be attributed to reduced accident rates due to preventive measures and increased concerns over escalating financial liabilities (Etkin 2001).

Table 4.2 – Major oil-related industrial accidents between 1970-2004

Year	Location	Industry	Loss/description
1976	Massachusetts, USA	Oil spill	Argo Merchant runs aground on the Nantucket Shoals off Cape Cod (Massachusetts USA), spilling 29 million litres (7.6 million gallons) of No. 6 fuel oil.
1978	France	Oil tanker	Amoco Cadiz tanker runs aground off the coast of France, spilling 1.6 million barrels of crude oil.
1984	Cubatao, Brazil	Oil pipeline	Oil fire - 508 deaths
1988	Piper Alpha, North Sea	Oil rig	167 deaths from explosion of offshore oil platform
1989	Alaska, USA	Oil tanker	Exxon Valdez tanker spills 42 million litres (11 million gallons) of crude oil into Prince William Sound (Alaska USA)
1994	Seoul, S. Korea	Oil fire	500 deaths
1995	Taegu, S.Korea	Oil & gas explosion	100 deaths
1998	Warri, Nigeria	Oil pipeline	Pipeline at Jesse, Nigeria exploded, instantly killing more than 500 people and severely burning hundreds more. Up to 2 000 people had been lining up with buckets and bottles to scoop up oil. The fire spread and engulfed the nearby villages of Moosqar and Oghara, killing farmers and villagers sleeping in their homes.
2000	Adeje, Nigeria	Oil pipeline	250 deaths

Source: Compiled from Mitchell and Cutter 1997, Anon. 2004, Draffan 2004, and Uranium Information Centre Ltd. 2004

Nuclear Accidents

With increasing concern over potential energy shortages and the impacts of burning fossil fuels, the debate about nuclear power has been renewed. Like hydroelectricity generation, nuclear power has the merit of being a clean energy source in terms of emissions, however, there are risks associated with the release of dangerous radiation from potential nuclear meltdown and from nuclear waste.

Between 1940 and 2000, there were at least 120 notable accidents involving nuclear material. These ranged from a container of uranium hexafluoride exploding in Oak Ridge, Tennessee, in the United States in 1944 killing two people and injuring three others to the worst accident in the history of the nuclear power industry – Chernobyl, Ukraine in 1986 (Anon n.d.).

Credit: Warren Gretz/UNEP/NREL

Case Study: Chernobyl Nuclear Power Plant Accident, Ukraine 25-26 April 1986

The world's worst nuclear power accident occurred at Chernobyl in the former USSR (now Ukraine) on 25-26 April 1986. While testing a reactor, numerous safety procedures were disregarded and a chain reaction resulted in explosions and a fireball, which blew off the reactor's heavy steel and concrete lid. The explosion and fire released radioactive material that spread over parts of the Soviet Union, Eastern Europe, Scandinavia, and later, Western Europe. The Chernobyl accident killed more than 31 people immediately, and as a result of the high radiation levels in the surrounding 32–km (20–mile) radius, 135 000 people had to be evacuated. Some areas were rendered uninhabitable for years. As a result of the radiation released into the atmosphere, tens of thousands of excess cancer deaths (as well as increased rates of birth defects) were expected in succeeding decades (Anon. n.d.).

Credit: Unknown/UNEP/Ukrainianweb

The 31 May 1986 image was acquired about a month after the nuclear accident at Chernobyl's Reactor Number 4. The Chernobyl nuclear plant is located on the northwest shore of a cooling pond. Much of the farmland surrounding the plant was heavily contaminated with radio active nuclides and subsequently abandoned. The areas have changed from red and white patterns indicating planted agricultural fields and bare soil in the 1986 image to tan-gray tones indicating natural vegetation in the 1992 image. More than 120 000 people from 213 villages and cities were relocated outside the contamination zone. Pripyat, an abandoned city with a 1986 population of 45 000, is located 3 km (2 miles) northwest of the Chernobyl Nuclear Power Station. The wavy white line north of the Chernobyl plant in the 1992 image is a levee built to prevent the flow of contaminated water and soil into the Pripyat River. *Source: Earthshots 2001; Sadowski and Covington 1987; Stebelsky 1995; Mould 1988; Medvedev 1990; Williams 1995; Schmidt 1995; Park 1989; Marples 1996.*

31 May 1986

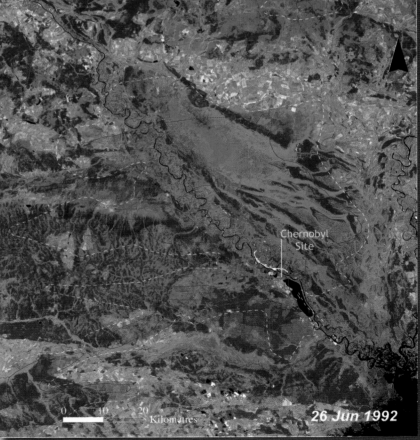

26 Jun 1992

Credit: Elena/UNEP

The lines overlaid on the 1992 image show the approximate extent of Cesium-137 radiation levels according to 1990 data. Locations within the solid red lines have radiation levels greater than 40 Curies per km^2, too high for life, and this area has been almost completely abandoned by people.

Caraz

P E R U

Huascarán

National

Park

Ancash

Mt. Huascarán N
△ 6655 m

Mt. Huascarán S
△ 6768 m

Yungay

Ranrahirca

Mancos

Images of the avalanche that covered Yungay City.
Credit: UNEP/Servicio Aereofotográfico Nacional, Lima, Perú

Legend
▲ Summits
• Towns
■ Source area of 1970 avalanche, W slope
■ 1970 avalanche, N slope
☐ 1962 avalanche
☐ 1970 avalanche, W slope

16 Oct 1973

0   3   6
Kilometres

N

Yungay

AVALANCHE
YUNGAY CITY, PERU

Andean glaciers have long been involved in numerous avalanches, which have caused considerable material losses and casualties by the thousands. The events of 1962 and 1970, originating from Mt. Huascarán's northern summit, were particularly deadly. On 31 May 1970,

Caraz

Huascarán

National

Park

P E R U

A n c a s h

Mt. Huascarán N
△
6655 m

Mt. Huascarán S
△
6768 m

Yungay

Ranrahirca

Mancos

Cemetery

School

Plaza

Watchtower

Almost nothing remains of Yungay City, literally erased by the 1970 avalanche.
Credit: Walter Silverio September, 1997/UNEP

Legend
▲ Summits
● Towns
■ Source area of 1970 avalanche, W slope
■ 1970 avalanche, N slope
□ 1962 avalanche
□ 1970 avalanche, W slope

0   3   6
Kilometres

N

26 May 2000

a 7.7 magnitude earthquake triggered a huge avalanche, 25 km (16 miles) long and moving at 280 km/h (174 mph), which wiped out the city of Yungay, claiming 18 000 lives. The scars are still visible today. Ice retreat has induced the formation of numerous peri-glacial lakes, dammed only by fragile moraine deposits. Subject to erosion, these walls may collapse, triggering flash floods—another threat for the local population.

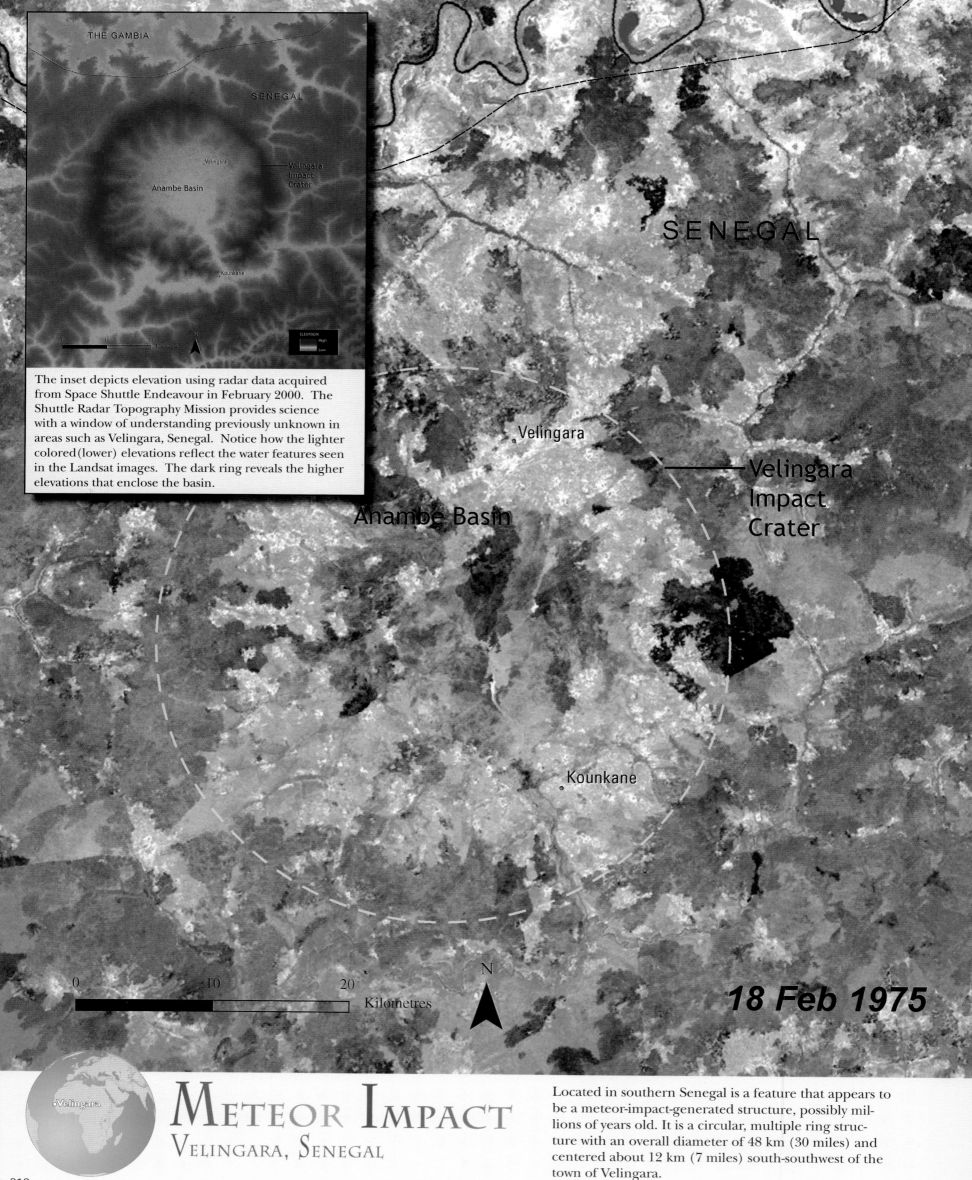

THE GAMBIA

SENEGAL

Velingara

Velingara
Impact
Crater

Anambe Basin

Kounkane

ELEVATION
High
Low

N

0   10   20
Kilometres

The inset depicts elevation using radar data acquired
from Space Shuttle Endeavour in February 2000. The
Shuttle Radar Topography Mission provides science
with a window of understanding previously unknown in
areas such as Velingara, Senegal. Notice how the lighter
colored (lower) elevations reflect the water features seen
in the Landsat images. The dark ring reveals the higher
elevations that enclose the basin.

SENEGAL

Velingara

Velingara
Impact
Crater

Anambe Basin

Kounkane

0   10   20
Kilometres

N

18 Feb 1975

Velingara

METEOR IMPACT
VELINGARA, SENEGAL

Located in southern Senegal is a feature that appears to
be a meteor-impact-generated structure, possibly millions of years old. It is a circular, multiple ring structure with an overall diameter of 48 km (30 miles) and
centered about 12 km (7 miles) south-southwest of the
town of Velingara.

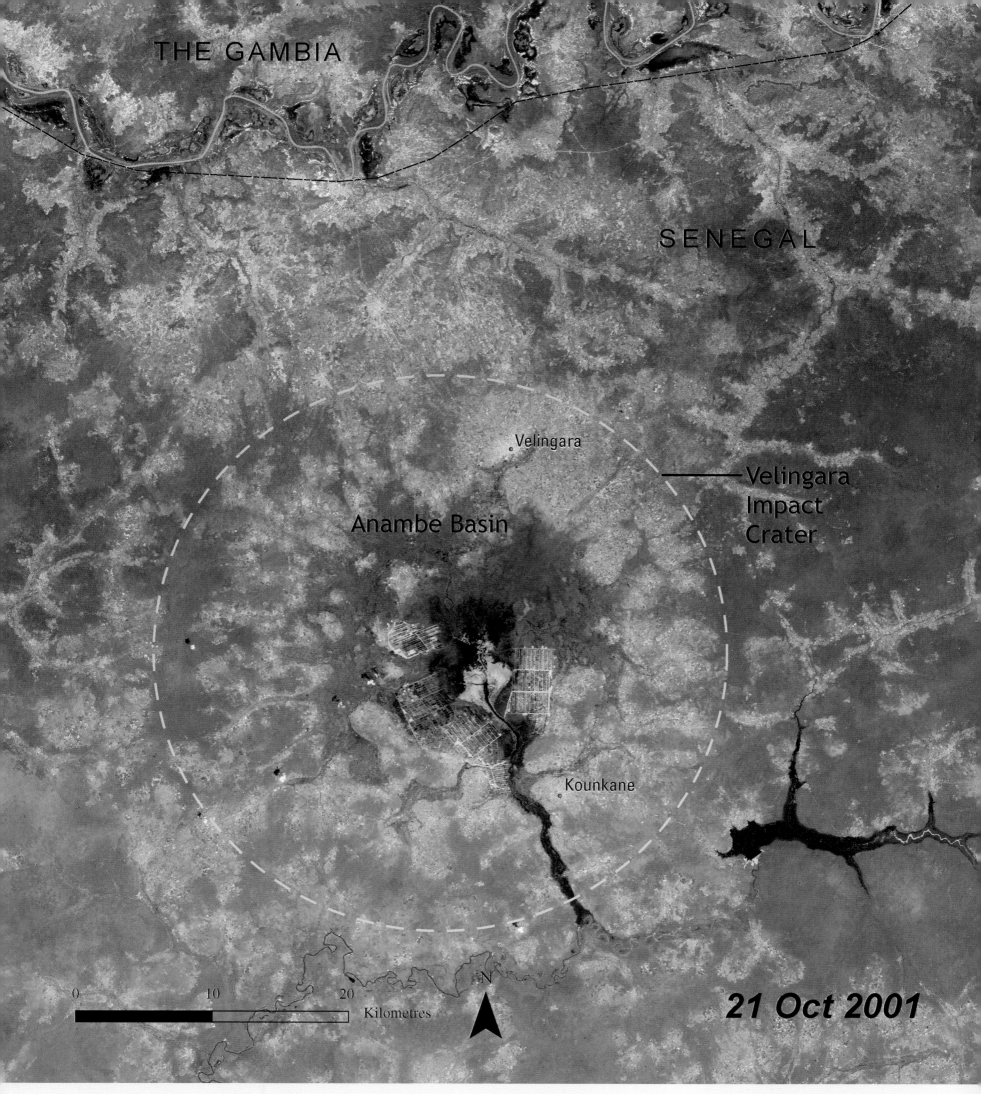

THE GAMBIA

SENEGAL

Velingara

——— Velingara
Impact
Crater

Anambe Basin

Kounkane

0   10   20
Kilometres

N

21 Oct 2001

The high rim structure of the Velingara Crater encloses the Anambe Basin. Water previously flowing out the south end of the basin was harnessed behind a dam in the mid-1970s as a source of irrigation for rice and other crops. The 1975 image predates the irrigation development.

By 2001 intense agricultural systems had appeared near the center of the crater (right image), contrasting sharply with the swampy areas (dark green) nearby. The Velingara Crater was first detected using Landsat data in the early 1970s.

References

Anon. (2004) Explosion disasters. http://www.factophile.com/Disasters/Explosions/ on 31 July 2004.

Anon. (n. d.). List of nuclear accidents, Wikipedia, the free encyclopedia. http://en.wikipedia.org/wiki/List_of_nuclear_accidents#1940s on 27 March 2004.

AP (2005). Tsunami redrew ship channels, ocean floor. The Associated Press, 5 January 2005. http://msnbc.msn.com/id/6791600/ on 10 January 2005.

APPEA (n.d.). Oceans & oil spills. Australian Petroleum Production and Exploration Association Ltd. http://www.appea.com.au/edusite/html/pt/oceans.html on 27 March 2004.

BBC (2000). Mozambique: How disaster unfolded, BBC, London, UK. http://news.bbc.co.uk/1/hi/world/africa/655227.stm on 29 March 2005.

Benson, C. (2002). Disaster Management. Department for International Development, Infrastructure and Urban Development Department and Overseas Development Institute, Pro-Poor Infrastructure Provision. http://www.odi.org.uk/keysheets/ppip/purple_2_disasters.pdf on 10 March 2005.

Bilham, R. (1995). Global fatalities from earthquakes in the past 2000 years: prognosis for the next 30. In Reduction and Predictability of Natural Disasters (eds.) Rundle, J, F. Klein and D. Turcotte. Santa Fe Institute Studies in the Sciences of Complexity, Vol. XXV, 19-31. Addison Wesley. http://cires.colorado.edu/~bilham/SantaFe.pdf on 12 April 2004.

Camp, V. (2000). Eruption variability. How volcanoes work. San Diego State University, Department of Geological Sciences, San Diego, California, USA. http://www.geology.sdsu.edu/how_volcanoes_work/ on 12 April 2004.

Cohen, B. L. (1990). The nuclear energy option. Plenum Press. http://www.phyast.pitt.edu/~blc/book/index.html on 27 March 2004.

Cosgrove. W. (2003). Number of killer storms and droughts increasing worldwide. World Water Council, Marseille, France. http://www.mindfully.org/Air/2003/Storms-Droughts-Increasing27feb03.htm on 14 April 2004.

CSIRO (2002). Volcanic eruptions and climate change. CSIRO Atmospheric Research Greenhouse Information Paper. Commonwealth Scientific and Industrial Research Organisation, Victoria, Australia. http://www.dar.csiro.au/publications/greenhouse_2000e.htm on 12 April 2004.

Draffan, G. (2004). Chronology of Industrial Disasters. http://www.endgame.org/industrial-disasters.html on 31 July 2004.

DAS (n.d.). Global activity - tropical cyclones around the world. University of Illinois at Urbana-Champaign, Department of Atmospheric Sciences, Weather World 2010 project, Champaign, Illinois, USA. http://ww2010.atmos.uiuc.edu/(Gl)/guides/mtr/hurr/glob.rxml on 24 July 2004.

DFO (2004). Dartmouth Flood Observatory inundation maps. http://www.dartmouth.edu/%7Efloods/Archives/index.htm on 18 December 2004.

Digital Globe (2005). Tsunami Aftermath: Sri Lanka, QuickBird Imagery, 26-31 December 2004. http://www.digitalglobe.com/images/tsunami/Sri_Lanka_Tsunami_Damage.pdf on 7 February 2005; Tsunami Aftermath: Banda Aceh, Indonesia, QuickBird Imagery, 28 December 2004. http://www.digitalglobe.com/images/tsunami/Banda_Aceh_Tsunami_Damage.pdf on 7 February 2005; Mt. Etna. http://rst.gsfc.nasa.gov/Sect13/Sect13_4d.html on 23 March 2005.

Doyle, A. (2004). Global Warming, Pollution Add to Coastal Threats. Reuters. December 28, 2004. http://enn.com/ch_clim.html?id=49 on 10 January 2005.

Elena (2004). http://www.kiddofspeed.com/chapter3.html on 10 February 2005.

Etkin, D.S. (1997). Oil spills from vessels (1960-1995): an international historical perspective. Cutter Information Corporation, Cambridge, Massachusetts, USA, 72.

Etkin, D. S. (2001). Analysis of oil spill trends in the United States and worldwide. International Oil Spill Conference, 1291-1300. http://www.environmental-research.com/publications/pdf/spill_statistics/paper4.pdf on 27 March 2004.

EQE (1995). The January 17, 1995 Kobe Earthquake - An EQE summary report. EQE International Limited, Warrington, UK. http://www.eqe.com/publications/kobe/kobe.htm on 12 April 2004.

FAO (1996). World Food Summit - Food for all. Food and Agriculture Organization of the United Nations. 13-17 November 1996, Rome, Italy. http://www.fao.org/DOCREP/x0262e/x0262e00.htm#TopOfPage on 9 July 2004.

FEWS Net (2001). Famine Early Warning Systems Network, Emergency declared following floods. http://fews.net/centers/innerSections.aspx?f=mz&m=1000116&pageID=monthliesDoc on 29 March 2005.

Francis, P. (1993). Volcanoes: a planetary perspective. Claredon Press, New York, USA, 443.

GlobalCoordinate.com (n.d.). http://redtailcanyon.com/default.aspx on 8 March 2005.

Global Oceans (1999). International Workshop on Coastal Mega-cities: Challenges of Growing Urbanisation of the World's Coastal Areas, Hangzhou, People's Republic of China. Global Forum on Oceans, Coasts, and Islands, 27-30 September 1999. http://www.globaloceans.org/globalinfo/hangzhou.html on 8 March 2005.

IIEES (n.d.). International Institute of Earthquake Engineering and Seismology, Tehran, Iran. http://www.iiees.ac.ir/English/eng_index.html on 16 May 2004.

IPCC (2001). Climate Change 2001: Impacts, Adaptation and Vulnerability. International Panel on Climate Change. http://www.grida.no/climate/ipcc_tar/vol4/english/ on 8 March 2005.

ITOPF (2003). Historical data – statistics. International Tanker Owners Pollution Federation Limited. http://www.itopf.com/stats.html on 27 March 2004.

Jibson, R. W. (2005). Landslide Hazards at La Conchita, California. U.S. Department of the Interior, U.S. Geological Survey. Open-File Report 2005-1067. http://pubs.usgs.gov/of/2005/1067/pdf/OF2005-1067.pdf on 8 March 2005.

Kasperson, E., Kasperson, J.X. and Dow, K. (2001). Vulnerability, equity, and global environmental change. In Global Environmental Risk, edited by J.X. Kasperson and R.E. Kasperson. Earthscan Publications, London, UK, 247-72.

Kelly, M. (2000). The causes of climatic change. Climatic Research Information Sheet 2. University of East Anglia, Norfolk, UK. http://www.cru.uea.ac.uk/cru/info/causecc/ on 12 April 2004.

Krupa, M. (1997). ICE Case Studies: Environmental and Economic Repercussions of the Persian Gulf War on Kuwait. American University, The School of International Service, The Inventory of Conflict & Environment (ICE), Washington DC, USA. http://www.american.edu/projects/mandala/TED/ice/kuwait.htm on 9 March 2005.

MacKenzie, J. J. (2000). Oil as a finite resource: When is global production likely to peak? World Resources Institute. http://www.wri.org/wri/climate/jm_oil_000.html on 27 March 2004.

Maps of India (2004). http://www.mapsofindia.com/maps/mapinnews/22101999.htm on 18 July 2004.

Marples, D. R. (1986). Chernobyl and nuclear power in the USSR: St. Martin's Press, New York, USA, 160.

McGee, K. A.; Doukas, M. P.; Kessler, R.; Gerlach, T. M. (1997). Impacts of volcanic gases on climate, the environment, and people. U.S. Geological Survey Open-File Report 97-262. U.S. Department of Interior: Geological Survey; Cascades Volcano Observatory, Washington, USA. http://pubs.usgs.gov/of/of97-262/of97-262.html on 12 April 2004.

Medvedev, Z. A. (1990). The legacy of Chernobyl: New York, W.W. Norton, 108.

Merriam-Webster Online (n.d.). www.m-w.com on 10 March 2005.

Mitchell, J. K. ed. (1996). The long road to recovery: Community responses to industrial disaster. United Nations University Press. http://www.unu.edu/unupress/unupbooks/uu21le/uu21le00.htm#Contents on 28 July 2004.

Mitchell, J. T. and Cutter. S. L. (1997) Global Change and Environmental Hazards: Is the World Becoming More Disastrous? Active learning modules on the human dimensions of global change. University of South Carolina, Department of Geography, Columbia, South Carolina, USA. http://www.aag.org/HDGC/www/hazards/toc.html on 31 July 2004.

Mould, R. F. (1988). Chernobyl– the real story: Oxford, England, Pergamon Press, 116.

NASA (n.d.). NASA Planetary Photo Journal: Dust Obscures Liaoning Province, China. http://photojournal.jpl.nasa.gov/catalog/PIA03705 on 29 March 2005.

NASA (1999). Earth Observatory Media Alerts Archive. http://earthobservatory.nasa.gov/Newsroom/MediaAlerts/1999/199912301102.html on 10 March 2005.

NASA (2000). Drought: The Creeping Disaster. http://earthobservatory.nasa.gov/Library/DroughtFacts/ on 27 February 2005.

NASA (2003). Record Harvest in Australia. http://earthobservatory.nasa.gov/NaturalHazards/natural_hazards_v2.php3?img_id=12010 on 7 February 2005.

NASA (2004a). Landslide Lake, Tibet. http://asterweb.jpl.nasa.gov/gallery/gallery.htm?name=tibetlake on 7 February 2005.

NASA (2004b). Hurricane Charley. http://earthobservatory.nasa.gov/NaturalHazards/natural_hazards_v2.php3?img_id=12339 ; http://earthobservatory.nasa.gov/NaturalHazards/natural_hazards_v2.php3?img_id=16639 on 7 February 2005.

NASA (2004c). Flood Disaster Hits Hispanolia, Special Feature by Gubbels, T., Brakenridge. http://earthobservatory.nasa.gov/Study/Haiti2004/ on 24 July 2004.

NASA (2004d). Continuing Drought in Kenya. http://earthobservatory.nasa.gov/Newsroom/NewImages/images.php3?img_id=16816 on 13 March 2005.

NASA (2004e). Natural Disaster: Destructive Earthquake near Bam, Iran. http://earthobservatory.nasa.gov/Newsroom/NewImages/images.php3?img_id=16408; http://earthobservatory.nasa.gov/NaturalHazards/natural_hazards_v2.php3?img_id=11892 on 6 March 2005.

NASA (2005). Persistent Rains Bring Floods, Mudslides to California. http://earthobservatory.nasa.gov/NaturalHazards/natural_hazards_v2.php3?img_id=12669 on 7 February 2005.

NEIC (2003). Are earthquakes really on the increase? U.S. Department of Interior, Geological Survey, National Earthquake Information Center, Colorado, USA. http://neic.usgs.gov/neis/general/increase_in_earthquakes.html on 30 December 2004. http://neic.usgs.gov/neis/eqlists/10maps_world.html on 30 December 2004.

NEIC (2004). Frequency of occurrence of earthquakes. Earthquake Facts and Statistics. U.S. Department of Interior, Geological Survey, National Earthquake Information Center, Colorado. http://neic.usgs.gov/neis/eqlists/eqstats.html on 30 December 2004; Earthquake Hazards Program. http://neic.usgs.gov/neis/poster/2003/20031226.html on 6 March 2005.

NGDC (2004). Volcanoes in Eruption - Set 2. National Oceanic and Atmospheric Administration, Satellite and Information Service, National Geophysical Data Center. http://www.ngdc.noaa.gov/seg/hazard/slideset/30/30_614_slide.shtml on 30 March 2005.

NOAA (n.d.). The United States Hurricane Problem. National Oceanic and Atmospheric Administration. http://www.aoml.noaa.gov/general/lib/usahp.html NOAA Miami Regional Library on 9 March 2005.

Oxfam (2001). Mozambique floods 2000 – 2001. http://www.oxfam.org.uk/what_we_do/where_we_work/mozambique/floods/ on 29 March 2005.

OWF (n.d.) Disasters. Our World Foundation, London, UK. http://www.ourworldfoundation.org.uk/dis-nof.htm on 24 September 2004.

Park, C. C. (1989). Chernobyl; the long shadow: Routledge, New York, USA, 121.

Poland, M. (2002). Personal Communication. USGS Cascades Volcano Observatory. USGS Earthshots. http://edcwww.cr.usgs.gov/earthshots/slow/MtStHelens/MtStHelens on 31 December 2002.

Sadowski, F.G., and Covington, S.J. (1987). Processing and analysis of commercial satellite image data of the nuclear accident near Chernobyl, U.S.S.R. Washington, USA, USGS Survey Bulletin 1785, 10,19.

Santer, B. D.; Doutriaux, C.; Boyle, J. S.; Taylor, K. E.; Wigley, T. M. L.; Meehl, G. A.; Hansen, J. E.; Jones, P. D.; Roeckner, E. ;Sengupta, S. (2001). Accounting for the Effects of Volcanoes and ENSO in Comparisons of Modeled and Observed Temperature Trends. Journal of Geophysical Research-Atmospheres, November 2001. http://www.cgd.ucar.edu/cas/abstracts/files/Wigley2001_3.html on 12 April 2004.

Schmidt, K.F. (1995). The truly wild life around Chernobyl: U.S. News and World Report, 17 July 1995, 51-53.

Credit: Tan Kok Lian/UNEP/Topfoto

Singh, R.P.; Bhoi, S.; Sahoo, A.K. (2002). Changes Observed on Land and Ocean after Gujarat Earthquake of January 26, 2001 using IRS Data, International Journal of Remote Sensing, Volume 23, No. 16, 3123 – 3128. http://home.iitk.ac.in/~ramesh on 16 August 2004.

SNWA (2004). Southern Nevada Water Authority, Drought Handbook. http://www.snwa.com/html/wr_drought_handbook.html on 8 March 2005.

Stebelsky, I. (1995). Radionuclide contamination and settlement abandonment around Chernobyl: Annals of the Association of American Geographers, v. 85, 1995, 291.

Sustainability Institute (2004). Within Limits: News of Overshoot. Sustainability Institute, Hartland, Vermont, USA. http://www.sustainabilityinstitute.org/limits/overshoot.html on 10 March 2005.

The Earthquake Museum (2003). Reports on Recent Major Earthquakes. http://www.olympus.net/personal/gofamily/quake/2003quakes.html on 8 March 2005.

Ukrainian Web (n.d.): http://www.ukrainianweb.com/images/chernobyl/chernobyl_reactor.jpg on 10 February 2005.

UN-DHA (1992). International agreed glossary of basic terms related to disaster management. United Nations, Department of Humanitarian Affairs, International Decade for Natural Disaster Reduction, Geneva, Switzerland, 83. http://www.cred.be/emdat/Guide/glossary.htm on 13 April 2004.

UN-ISDR (2004). Living with Risk - A global review of disaster reduction initiatives

2004 version. United Nations Inter-Agency Secretariat of the International Strategy for Disaster Reduction, Geneva, Switzerland. http://www.unisdr.org/eng/about_isdr/bd-lwr-2004-eng.htm on 1 August 2004; http://www.unisdr.org/eng/about_isdr/basic_docs/LwR2004/ch1%20Section%201.pdf on 1 August 2004.

UND (n.d.). The cost of volcanic eruptions. University of North Dakota, Grand Forks, North Dakota, USA. http://volcano.und.nodak.edu/vwdocs/vw_hyperexchange/CostVolc.html on 12 April 2004; Volcanic and geologic terms. http://volcano.und.nodak.edu/vwdocs/glossary.html on 12 April 2004.

UNL (n.d.). The Southwestern U.S Drought of 2003. Some Hydrological Impacts. University of Nebraska-Lincoln High Plains Regional Climate Center, Lincoln, Nebraska, USA. http://www.hprcc.unl.edu/nebraska/swdrought-2003.html on 9 March 2005.

UNEP (2002a). Global Environment Outlook 3 (GEO3) – Past, present and future perspectives. United Nations Environment Programme. London: Earthscan. 446.

UNEP (2002b). North America's Environment: A Thirty-Year State of the Environment and Policy Perspective. United Nations Environment Programme, Nairobi, Kenya.

UNEP (2003). Selected Satellite Images of Our Changing Environment. Sioux Falls, SD: United Nations Environment Programme, UNEP/GRID – Sioux Falls and USGS/EROS Data Center.

UNOSAT (2004). UN for Outer Space Affairs. Satellite imagery for all. http://unosat.web.cern.ch/unosat/asp/default.asp on 5 December 2004.

Uranium Information Centre Ltd. (2004). Some energy-related accidents since 1977. In: Appendix 1. The Hazards of Using Energy, Appendix to Nuclear Issues Briefing Paper 14. Safety of Nuclear Power Reactors. http://www.uic.com.au/nip14app.htm on 31 July 2004.

USAID (2004). Flooding in Haiti & the Dominican Republic. http://www.usaid.gov/haiti/floods.html on 9 March 2005.

US Census Bureau (2004). Cumulative Estimates of Population Change for Counties and County Rankings: April 1, 2000 to July 1, 2003. http://www.census.gov/popest/counties/CO-EST2003-02.html on 8 March 2005.

USGS (n.d.). Earthquake Hazards Program: Poster of Bam, Iran Earthquake. Map prepared by U.S. Geological Survey National Earthquake Information Center on 14 January 2004. http://neic.usgs.gov/neis/poster/2003/20031226.html on 18 March 2005.

USGS (1999). USGS News Release: Most recent natural disasters were not the Century's worst, USGS says. U.S. Department of the Interior, U.S. Geological Survey. http://geography.about.com/gi/dynamic/offsite.htm?site=http://www.usgs.gov on 8 March 2005.

USGS (2000). The 1960 Kapoho Eruption of Kilauea Volcano, Hawai`i. http://hvo.wr.usgs.gov/kilauea/history/1960Jan13/ on 12 April 2004.

USGS (2001). USGS Earthshots, 8th ed., 12 January 2001, EROS Data Center of the U.S. Geological Survey, a bureau of the U.S. Department of the Interior.

USGS (2002). Earthquake hazard program. U.S. Department of Interior: Geological Survey, Reston, Virginia, USA. http://earthquake.usgs.gov/4kids/eqterms.html#earthquake on 12 April 2004.

Vince, G. (2004). New Scientist: Dust storms on the rise globally. http://www.newscientist.com/article.ns?id=dn6306 on 29 March 2005.

Wikipedia (n.d.). The Free Encyclopedia. http://en.wikipedia.org/wiki/Dust_storm on 29 March 2005.

Williams, N. (1995). Chernobyl; life abounds without people: Science, v. 269, 21 July 1995, 304.

Site Reference:

Avalanche,
Yungay City, Peru

Ericksen, G. E. and Plafker, G. (1978). Nevados Huascarán avalanches; in Rockslides and Avalanches, vol. 1, B. Voight editor; Elsevier Scientific Publishing Co., Amsterdam, 277-314.

Francou, B. and Wagnon, P. (1998). Cordillères andines, sur les hauts sommets de Bolivie, du Pérou et d'Equateur, Glénat, Grenoble, France, 127.

Hidrandina, S. A. (1988). Unit of Glaciology and Hydrology Huaraz, Glacier Inventory of Peru, Consejo Nacional de Cience y Tecnología (CONCYTEC), Lima, 105.

Sidjak R. W. and Wheate R. D. (1999). Glacier mapping of the Illecillewaet Icefield, British Columbia, Canada, using Landsat TM and digital elevation model data, International Journal of Remote Sensing, Vol. 20, No 2, 273-284.

Silverio, W. (2001). Elaboration d'un SIG pour la gestion d'une zone protégée de haute montagne : application au Parc National Huascarán, Pérou, MSc thesis in geomatics, University of Geneva, 112.

Silverio, W. (1999). Essai d'évaluation des instabilités de pente par un système d'information géographique et leur interprétation dans la région de Huascarán, Département d'Ancash, Pérou, MSc thesis in Analysis and Management of Geological Risk, University of Geneva, 65.

Meteor Impact
Velingara, Senegal

Master, S., Diallo, D.P., Kande, S., and Wade, S. (1999). The Velingara Ring Structure in Haute Casamance, Senegal: A Possible Large Buried Meteorite Impact Crater. Dept. of Geology, Univ. of the Witwatersrand, P. Bag 3, WITS 2050, Johannesburg, South Africa. http://www.ulrich-terhalle.de/~image-contrails/senegal/senegal-met-velingara-txt.html on 5 March 2005.

NASA (2000). Exploration: Impact Craters. http://liftoff.msfc.nasa.gov/Academy/SPACE/SolarSystem/Meteors/Craters.html on 7 March 2005.

Gray, T.G. (2005). Personal Communication. SAIC, USGS National Center for Earth Resources Observation and Science. Sioux Falls, SD, USA March 2005.

Wood, E., Tappan, G., Lietzow, R., Albright, T. (1998). The Impact of Agricultural Production Systems on Land Use/Cover Change: Senegal, West Africa. Center for Global Change, Raytheon STX Corporation. Vol. 5 No. 2, Summer 1998.

Epilogue
One Planet Many People

The history of the human race is filled with stories of ingenuity regarding our ability to harness the bounty of nature. Wind powered the sailing ships of explorers, wood and coal fueled railroads that threaded across our continents, and now petroleum fires the engines of our cars and airplanes and allows us to spread to all corners of the planet.

The goods and services from nature have sustained us, moved us, and inspired us. Our cultural heritage was shaped by the vast bounty of the Earth. Our ever-increasing demand for more of nature's goods has left a series of huge footprints—footprints visible from distant points in space. These footprints represent the places we live and work, the places where we gain food, fiber, and minerals, and the ribbons of transportation needed by our highly mobile societies to conduct our businesses.

As this volume illustrates in colorful and graphical ways, our successes may also be our failure. We have advanced our civilizations by conquering nature. As a people, we should respect what we have accomplished. However, we must ultimately ask ourselves the question—"have our efforts to tame the Earth ensured our permanence?" The evidence in the atlas suggests that our victories over nature are incomplete because in the course of our development, we have depleted our resources and contaminated our environment to the point where our future may be one full of struggles

and challenges as we try to access ever more precious commodities from nature on which we depend.

To survive, we must put the era of nature conquest behind us and embark on a new era—the sustainability and stewardship era. In this era, we must cleanse our air and water so that it supports life in the future. We must serve and renew our natural resources so that we have the food, fiber, and energy we need, and we must protect and preserve our remaining natural areas so that they can soothe our spirits and inspire our minds.

In W.L. Thomas's seminal volume on sustainable development published in 1956, Kenneth Boulding closed the dialog by providing the following point-counterpoint. He suggested that the moral of human exploitation of the Earth's resources was "The evolutionary plan went astray by evolving man." Boulding then offered the perspective of developers by writing "man's a nuisance, man's a crackpot, but only man can hit the jackpot."

Which perspective is right? From the vantage of space, we can clearly see our footprints on the Earth and we can over time see the expanding size and number of footprints. Our species can take pride in the complex patterns of our cities and farms as these demonstrate our ingenuity and industriousness. Our numbers have grown dramatically yet we can argue

that the overall quality of life has improved. At least on the surface... For while it appears that we have conquered nature, a closer look at the consequences of our footprint reveals the rest of the story. The Earth's environmental systems are changing fast—and maybe too fast. The impacts of our industriousness are changing as fast or maybe even faster than the pace of our footprints. The frequency of extreme events, such as droughts, floods, severe storms, and wildfires is accelerating faster than ever recorded. Our climate is changing more rapidly than ever before, and the rate of species extinction is going up at an alarming rate. From the vantage of space, we can see the footprints of the human race. Unfortunately, by the time we see those footprints, it may already be too late because the undesirable impacts of our actions are already spreading through the Earth's environment.

Boulding's message was simple: Sustain the Earth, keep it healthy, and make it thrive so that it continues to provide for the many people that use it as home. The view from space suggests that we have a lot of work ahead to tailor our behavior so that the Earth provides bounty for eons. And there's no time like the present to get started on the path to sustainability.

The Conservationist's Lament

The world is finite
Resources are scarce
Things are bad
And will be worse
Coal is burnt
And gas exploded
Forests cut
And soils eroded
Wells are drying
Air's polluted
Dust is blowing
Trees uprooted
Oil is going
Ores depleted
Drains receive
What is excreted
Land is sinking
Seas are rising
Man is far
Too enterprising
Fire will rage
With man to fan it
Soon we'll have
A plundered planet
People breed
Like fertile rabbits
People have
Disgusting habits

MORAL...

The evolutionary plan
Went astray
By evolving Man

The Technologist's Reply

Man's potential
Is quite terrific
You can't go back
To the Neolithic
The cream is there
For us to skim it
Knowledge is power
And the sky's the limit
Every mouth
Has hands to feed it
Food is found
When people need it
All we need
Is found in granite
Once we have
The men to plan it
Yeast and algae
Give us meat
Soil is almost
Obsolete
Man can grow
To pastures greener
Till all the earth
Is Pasadena

MORAL...

Man's a nuisance
Man's a crackpot
But only man
Can hit the jackpot

Kenneth Boulding in:
Thomas, W.L. ed. 1956. Man's Role in Changing the Face of the Earth.
    Chicago: University of Chicago Press.

Credit: Thomas Lang/UNEP/Topfoto

Acronyms and Abbreviations

AAAS	American Association for the Advancement of Science
ACT	Action by Church Together
AER	Agriculture Economic Research Service, United States Department of Agriculture
AEZ	Agro-ecological Zones
AMS	American Meteorological Society
AP	Associated Press
APPEA	Australian Petroleum Production and Exploration Association Ltd.
Ar	Argon
ASTER	Advanced Spaceborne Thermal Emission and Reflection Radiometer
BBC	British Broadcasting Corporation
BP	British Petroleum
BRIDGE	BRinging Information to Decision-makers for Global Effectiveness
Btu	British thermal units
°C	degree Centigrade
CFCs	Chlorofluorocarbons
CH_3Cl	Methyl chloride
CH_4	Methane
CIDA	Canadian International Development Agency
CIESIN	Center for International Earth Science Information Network
CIS	Commonwealth of Independent States
CITEPA	Inter-professional Technical Centre for Research into Air Pollution
CLIRSEN	Center for Integral Surveys of Natural Resources using Remote Sensing (Ecuador)
cm	Centimetres
CNPPA	Commission on National Parks and Protected Areas
CO	Carbon monoxide
CO_2	Carbon dioxide
CPI	Center-pivot irrigation
CSIRO	Commonwealth Scientific and Industrial Research Organisation
CSR	Climatological Solar Radiation
DAS	Department of Atmospheric Sciences - University of Illinois at Urbana-Champaign
DETR	Department of Environment, Transport and Regions (United Kingdom)
DEWA	Division of Early Warning and Assessment
DFO	Dartmouth Flood Observatory
DMZ	Demilitarized Zone
DMS	Defense Meteorological Satellite Program
DPRK	Democratic People's Republic of Korea
EEA	European Environment Agency
EIA	Energy Information Administration, United States Department of Energy
ENSO	El Niño/Southern Oscillation
EPA	Environmental Protection Agency
EQE	European Quality & Environment
EROS	Earth Resources Observation and Science (National Center)
ERSDAC	Earth Remote Sensing Data Analysis Center
ESA	Department of Economic and Social Affairs of the United Nations
ETM	Enhanced Thematic Mapper (ETM+).
FAO	Food and Agriculture Organisation of the United Nations
FEMA	Federal Emergency Management Agency
FEWS	Famine Early Warning Systems
FOEE	Friends of the Earth Europe
ft	Foot/Feet
GEF	Global Environment Facility
GEO	Global Environment Outlook
GEO3	Global Environmental Outlook Report 3 (UNEP Publication)
GHG	Greenhouse Gas
GIS	Geographic Information System
GLC	Global Land Cover
GLCF	Global Land Cover Facility
GPS	Global positioning system
GPW	Gridded Population of the World
GRID	Global Resource Information Database
GSFC	Goddard Space Flight Center (NASA)
H_2O	Hydrogen dioxide
HEAVEN	Healthier Environment through the Abatement of Vehicle Emissions and Noise
HFCs	Hydrofluorocarbons
HNO_3	Nitric acid
hPa	Hecto pascals, a unit for atmospheric pressure
IIASA	International Institute for Applied Systems Analysis
IAEA	The International Atomic Energy Agency
ICE	Inventory of Conflict and Environment
IIEES	International Institute of Earthquake Engineering and Seismology
IITK	Indian Institute of Technology Kanpur
IPC	International Programs Center, United States Census Bureau, Population Division
IPCC	Intergovernmental Panel on Climate Change
ISDR	International Strategy for Disaster Reduction
ITOPF	International Tanker Owners Pollution Federation Limited
IUCN	International Union for Conservation of Nature and Natural Resources
JAMS	Japanese Association of Mathematical Sciences
JAROS	Japan Resources Observation System Organization
KBG	Kara-Bogaz-Gol, Turkmenistan
kcal	kilocalories
kg	kilogrammes
km	kilometres
km/h	kilometers/hour
km^2	square kilometres
kWh	Kilo-watt hours
KWS	Kenya Wildlife Service
lb	pounds
LDCs	Least Developed Countries
LHWP	Lesotho Highlands Water Projet
LLDCs	Landlocked Developing Countries
LP DAAC	Land Processes Distributed Active Archive Center
LPG	Liquefied petroleum gas
LUT	Land Utilization Types
LWF	Lutheran World Federation
M	Magnitude
m	metres
MDG	Millennium Development Goals
MEA	Multilateral Environment Agreement
METI	Ministry of Economy Trade and Industry (Japan)
MIC	Methyl Isocyanate
MISR	Multi-angle Imaging SpectroRadiometer
mm	millimetres
MODIS	Moderate Resolution Imaging Spectroradiometer
MOPITT	Measurements of pollution in the troposphere instrument aboard NASA's Terra satellite
MPA	Multi-satellite Precipitation Analysis

MRS	Metropolitan Region of Santiago	UNESCO	United Nations Educational, Scientific and Cultural Organization
MSS	Multispectral scanner	UNF	United Nations Foundation
Mt.	Mount	UNFCCC	United Nations Framework Convention on Climate Change
n.d.	Not dated		
N_2	Nitrogen	UNFPA	United Nations Population Fund
N_2O	Nitrogen dioxide	UNHCR	United Nations High Commissioner for Refugees
NASA	National Aeronautics and Space Administration	UN-ISDR	United Nations Inter-Agency Secretariat of the International Strategy for Disaster Reduction
NCAR	The National Center for Atmospheric Research		
NCPPR	National Center for Public Policy Research	UPI	United Press International
NCR&LB	National Contractor Referrals and License Bureau	USAID	United States Agency for International Development
NDVI	Normalized Difference Vegetation Index		
NEIC	National Earthquake Information Center	USCCSP	United States Climate Change Science Program
NOAA	National Oceanic and Atmospheric Administration	USDA/FAS	United States Department of Agriculture/Foreign Agricultural Service
NOx	Nitrogen oxides		
NREL	National Renewable Energy Laboratory	USF	University of San Francisco
NRCS	Natural Resources Conservation Service	USGS	United States Geological Survey
NRDC	Natural Resources Defense Council	USSR	Union of Soviet Socialist Republics
NSIDC	National Snow and Ice Data Center	UTC	Universal Time
NSW EPA	New South Wales Environmental Protection Authority	UV	Ultraviolet
		VOCNM	Volatile organic compound (non-methane)
NWT	Northwestern Territories	VOC	Volatile organic compound
O_2	Oxygen	WCMC	World Conservation Monitoring Centre
O_3	Ozone	WCST	Wildlife Conservation Society – Tanzania
OECD	Organisation for Economic Co-operation and Development	WHO	World Health Organiation
		WMO	World Meteorological Organization
OWF	Our World Foundation	WRI	World Resources Institute
PBS	Public Broadcasting System	WWF	World Wildlife Fund
PFCs	Perfluorocarbons	WWF/DCP	World Wildlife Fund/Danube-Carpathian Programme
RFD	Reasonably Foreseeable Development		
ROK	Republic of Korea		
RRC-AP	Regional Resource Centre for Asia and the Pacific		
SAIC	Science Applications International Corporation		
SARCS	Southeast Asian Regional Committee for START		
SCOPE	Scientific Committee on Problems on the Environment		
SBSTTA	Subsidiary Body on Scientific, Technical and Technological Advice		
SF_6	Sulphur hexafluoride		
SIDS	Small Island Developing States		
SIDA	Swedish International Development Cooperation Agency		
SIO	Scripps Institution of Oceanography		
SNHP	Spanish National Hydrological Plan		
SNWA	Southern Nevada Water Authority		
SO_2	Sulfur dioxide		
SPRI	Scott Polar Research Institute		
SRM	Society for Range Management		
SWERA	Solar and Wind Energy Resource Assessment		
TBR	Transboundary Biosphere Reserve		
TM	Thematic Mapper		
TOMS	Total Ozone Mapping Spectrometer		
TSSC	Technical Support Services Contractor		
UCC	Union Carbide Corporation		
UCIL	Union Carbide India Limited		
UCL	University College London		
UCS	Union of Concerned Scientists		
UGRB	Upper Green River Basin		
UN	United Nations		
UND	University of North Dakota		
UN-DHA	United Nations, Department of Humanitarian Affairs		
UNDP	United Nations Development Programme		
UNDRO	United Nations Disaster Relief Organization		
UNEP	United Nations Environment Programme		

ETM/LANDSAT Equipped with high resolution instruments, Landsat- 7 was successfully launched on 15 April 1999. This satellite carries the Enhanced Thermal Mapper Plus (ETM+), which is an eight-band, multispectral scanning radiometer. The ETM+ is capable of resolving distances of meters in the panchromatic band; 30m (98 feet) in the visible, near and short-wave infrared band; and 60m (197 feet) in the thermal infraredband.

LANDSAT On 23 July 1972, NASA launched the first in a series of satellites designed to provide repetitive global coverage of the Earth's land masses. It was designated initially as the 'Earth Resources Technology Satellite-A'. The second in this series of Earth resources satellites (designated 'ERTS-B') was launched on 22 January 1975. It was renamed 'Landsat 2' by NASA, which also renamed 'ERTS-1' as 'Landsat 1'. Four additional Landsats were launched in 1978, 1982, and 1999 (Landsat 3, 4, 5 and 7), respectively.

SCANSAR Scanning synthetic aperture radar (ScanSAR) data is acquired on board the Canadian satellite RADARSAT-1. The RADARSAT-1 satellite was launched on 4 November 1995 and has been providing imagery for operational monitoring services on a global basis ever since. The state-of-the-art Synthetic Aperture Radar (SAR) can be steered to collect data over a 1 175 km (730 miles) wide area using 7 beam modes. This provides users with superb flexibility in acquiring images with a range of resolutions, incidence angles, and coverage area.

IKONOS Since its launch in September 1999, Space Imaging's IKONOS earth imaging satellite has provided a reliable stream of image data. IKONOS produces 1-meter black-and-white (panchromatic) and 4-meter multispectral (red, blue, green, near infrared) imagery that can be combined in a variety of ways to accommodate a wide range of high-resolution imagery applications.

QUICKBIRD The QuickBird satellite, launched in October 2001on a Boeing Delta II rocket from Vandenberg Air Force Base, California, is the first in a constellation of spacecraft that DigitalGlobe® is developing. QuickBird offers sub-meter resolution imagery, geolocational accuracy, and large on-board data storage. QuickBird's global collection of panchromatic and multispectral imagery is designed to support applications ranging from map publishing to land and asset management to insurance risk assessment.

PHOTOS Africa Focus; African Wildlife Foundation; Beth Allen; Bigfoto (www.bigfoto.com); Canadian Auto Workers (CAW); Chandra Giri; Christian Lambrechts; Cpl. Mike Escobar; David McKee; David P. Shorthouse; Digital Globe; Dmitry Petrakov; Ed Simpson; Elena; FEMA; Freefoto (freefoto.com); FAO; Gray Tappan; H. Gyde Lund; Hassan Partow; International Centre for Integrated Mountain Development (ICIMOD); IIEES; Invasive.org; John Welch; John Townshend; José de Jesús Campos Enríquez; J.P. Eaton; Juan Schlatter; Claudio Donoso; Lorant Czaran; Lumbuenamo Raymond; Lyn Topinka; Lynn Betts;Morgue File (www.Morguefile.com), DT Creations, Kevin Connors; NASA; NOAA; NREL; NRCS; Nik Wheeler; Olga Tutubalina; Peter Aengst; Peter Bardos-Déak, Philip Wijmans; Prof. Dr.-Ing.habil. Volker Quaschning; Ramesh P. Singh; Randy Cyr; Regional Resource Centre for Asia and Pacific (RRC-AP); Saman Salari Sharif,; Sergey Chernomorets ; Servicio Aerofotográfico Nacional, Lima, Perú; Simon Tsuo; South Florida Water Management District; Stephan Volz; Teal H.F. Smith; Topfoto (http://www.topfoto.co.uk/); Topham Photos; Ukrainianweb; UNEP-GRID; USGS; USDA; United States National Park Service; V. Sahanatien; Walter Silverio.

Acknowledgements

UNEP would like to thank the following for their contributions to **One Planet Many People: Atlas of Our Changing Environment:**

Jonathan M. Adams, Department of Biology, Providence College, United States

Robert G. Bailey, United States Department of Agriculture Forest Service, United States

Elgene Box, University of Georgia, United States

Robert Campbell, South Dakota School of Mines and Technology, United States

Ellen Carnevale, Population Reference Bureau, United States

Glenn Carver, Centre for Atmosphere Science, University of Cambridge, United Kingdom

Cyrille Chatelain, Conservatoire du Jardin Botanique, Switzerland

Sergey Chernomorets, University Centre for Engineering Geodynamics and Monitoring, Russia

Lorant Czaran, United Nations, New York

Paul Davis, University of Maryland, United States

Timothy Foresman, International Center for Remote Sensing Education, United States

Johann G. Goldammer, Freiberg University, Germany

David Herring, Earth Observatory, United States

Jean-Michel Jaquet, University of Geneva and United Nations Environment Programme, Global Resources Information Database, Switzerland

Satya P. S. Kushwaha, Forestry & Ecology Division, Indian Institute of Remote Sensing, India

Rebecca Lindsey, Science Systems and Applications, Inc., National Aeronautics and Space Administration, United States

Luisa Maffi, Terralingua, Canada

Martha Maiden, National Aeronautics and Space Administration, United States

James W. Merchant, University of Nebraska – Lincoln, United States

Eleanore Meredith, Earth Satellite Corporation, United States

Roger Mitchell, Earth Satellite Corporation, United States

Erika Monnati, Italy

Sumith Pathirana, Southern Cross University, Australia

Dmitry Petrakov, Moscow State University, Russia

Volker Quaschning, University of Applied Sciences, Germany

Navin Ramankutty, Center for Sustainability and the Global Environment, Institute for Environmental Studies, University of Wisconsin-Madison, United States

Era Singh, United States

Ramesh Singh, Indian Institute of Technology Kanpur, India

Leena Srivastava, The Energy and Resources Institute (TERI), India

Woody Turner, National Aeronautics and Space Administration, United States

Olga Tutubalina, Moscow State University, Russia

Antoinette Wannebo, Center for International Earth Science Information Network, Columbia University, United States

Wang Wenjie, Intergraph Mapping and Geospatial Solutions, China

Wesley Wettengel, World Wildlife Federation, United States

Ben White, University of Maryland, United States

From the United States Geological Survey, National Center for Earth Resources Observation and Science, United States:

Ron Beck

John Faundeen

Tom Holm

Rachel Kurtz

Janice Nelson

From Science Applications International Corporation, contractor to the United States Geological Survey, National Center for Earth Resources Observation and Science, United States:

Kwabena Asante

Roger Auch

Jon Christopherson

Jeff Danielson

Chandra Giri

Nazmul Hossain

Rynn Lamb

Lee McManus

Sandra Prince

James Rowland

Pat Scaramuzza

G. Gray Tappan

From the United Nations Environment Programme, Division of Early Warning and Assessment:

Johannes Akiwumi, Kenya

Dan Claasen, Kenya

Jesper Koefed, Kenya

Christian Lambrechts, Kenya

Dominique del Pietro, GRID – Geneva

Hassan Partow, GRID – Geneva

Pascal Peduzzi, GRID – Geneva

Walter Silverio, GRID – Geneva

Nicole Strub, GRID – Geneva

Tin Aung Moe, Thailand

Visiting scientists or interns at the United Nations Environment Programme, Global Resources Information Database - Sioux Falls, United States:

Daniel Amamoo-Otchere, Ghana

Lily-Rose Maida Awori, Kenya

Abdullah Daud, Bangladesh

José de Jesús Campos Enríquez, Mexico

Ragna Godtland, United States

Shingo Ikeda, Japan

Alfa N. Isiaku, Nigeria

John Molefe, Botswana

Elitsa Peneva, Bulgaria

Anup Prasad, India

S.K. Puri, India

Anil Raghavan, India

Ryan Reker, United States

Hua Shi, China

Shalini Venkataraman, Singapore

Special thanks goes to the Global Land Cover Facility (GLCF) of the University of Maryland and the National Aeronautics and Space Administration (NASA) Earth Observatory for providing access to satellite data.

Map Credits:

Topographic Map of the World

This image of the world was generated with data from the Global 30-arc second elevation (GTOPO30) dataset. The image is in the Orthographic Projection (Eastern hemisphere centered on 20 north latitude, 65 east longitude; Western hemisphere centered on 15 north latitude, 75 west longitude) commonly used for maps of the world. Elevation data used in this image were acquired by the SRTM aboard the Space Shuttle Endeavour, launched on 11 February 2000. The mission is a cooperative project between NASA, the National Geospatial-Intelligence Agency (NGA) of the U.S. Department of Defense and the German and Italian space agencies. It is managed by NASA's Jet Propulsion Laboratory, Pasadena, California, for NASA's Earth Science Enterprise, Washington, DC, USA. http://www2.jpl.nasa.gov/srtm/world.htm on 28 December 2004.

Nightlight Map of the World

This image of Earth's city lights was created with data from the Defense Meteorological Satellite Program (DMSP) Operational Linescan System (OLS). Originally designed to view clouds by moonlight, the OLS is also used to map the locations of permanent lights on the Earth's surface. Data courtesy Marc Imhoff of NASA GSFC and Christopher Elvidge of NOAA NGDC. Image by Craig Mayhew and Robert Simmon, NASA GSFC. http://visibleearth.nasa.gov on 30 December 2002.

Daylight Map of the World

NASA Goddard Space Flight Center Image by Reto Stöckli (land surface, shallow water, and clouds). Enhancements by Robert Simmon (ocean color, compositing, 3D globes, animation). Data and technical support: MODIS Land Group; MODIS Science Data Support Team; MODIS Atmosphere Group; MODIS Ocean Group Additional data: USGS EROS Data Center (topography); USGS Terrestrial Remote Sensing Flagstaff Field Center (Antarctica). http://visibleearth.nasa.gov on 30 December 2004.

Earthquake Map of the World

The earthquake map was produced by overlaying earthquake data (major earthquakes, 1995-2004), shown as dots of varying sizes depending on magnitude on the Richter scale, over a global elevation map produced from the Global 30-arc second elevation (GTOPO30) dataset. The earthquake data are from the U.S Geological Survey National Earthquake Information Centre, http://neic.usgs.gov/ on 15 February 2005. The GTOPO30 data are from the National Center for Earth Resources Observation and Science. http://edcdaac.usgs.gov/gtopo30/gtopo30.html on 15 February 2005.

Index